Basic Mathematics

Springer
New York
Berlin
Heidelberg
Barcelona
Hong Kong
London
Milan
Paris
Singapore
Tokyo

Books by Serge Lang
of Interest for High Schools

Geometry (with Gene Murrow)

This high school text, inspired by the work and educational interests of a prominent research mathematician, and Gene Murrow's experience as a high school teacher, presents geometry in an exemplary and, to the student, accessible and attractive form. The book emphasizes both the intellectually stimulating parts of geometry and routine arguments or computations in physical and classical cases.

MATH! Encounters with High School Students

This book is a faithful record of dialogues between Lang and high school students, covering some of the topics in the *Geometry* book, and others at the same mathematical level. These encounters have been transcribed from tapes, and are thus true, authentic, and alive.

Basic Mathematics

This book provides the student with the basic mathematical background necessary for college students. It can be used as a high school text, or for a college precalculus course.

The Beauty of Doing Mathematics

Here, we have dialogues between Lang and audiences at a science museum in Paris. The audience consisted of many types of persons, including some high school students. The topics covered are treated at a level understandable by a lay public, but were selected to put people in contact with some more advanced research mathematics which could be expressed in broadly understandable terms.

First Course in Calculus

This is a standard text in calculus. There are many worked out examples and problems.

Introduction to Linear Algebra

Although this text is used often after a first course in calculus, it could also be used at an earlier level, to give an introduction to vectors and matrices, and their basic properties.

Math Talks for Undergraduates

Written in a conversational tone, Lang presents a collection of talks that he gave to undergraduates on selected topics in mathematics such as prime numbers, the abc conjecture, approximation theorems of analysis, the semi-parallelogram law and Bruhat-Tits spaces, harmonic and symmetric polynomials, and more.

Serge Lang

Basic Mathematics

With 223 Illustrations

 Springer

Serge Lang
Department of Mathematics
Yale University
New Haven, CT 06520
U.S.A.

Mathematics Subject Classifications (2000): 00-01, 00A05

Library of Congress Cataloging-in-Publication Data
Lang, Serge
  Basic mathematics.
  Reprint. Originally published: Reading, Mass. :
Addison-Wesley, c1971.
  Includes index.
  1. Mathematics—1961–   I. Title.
QA39.2.L33   1988   512   88-12334

Previously published by Addison-Wesley Publishing Company Inc. © 1971

© 1988 by Springer-Verlag New York, Inc.

Printed and bound by Sheridan Books, Inc., Ann Arbor, MI.
Printed in the United States of America.

9 8 7 6 5 4

ISBN 0-387-96787-7
ISBN 3-540-96787-7   SPIN 10841589

Springer-Verlag   New York Berlin Heidelberg
A member of BertelsmannSpringer Science+Business Media GmbH

To Jerry

Acknowledgments

I am grateful to Peter Lerch, Gene Murrow, Dick Pieters, and Gail Young for their careful reading of the manuscript and their useful suggestions.

I am also indebted to Howard Dolinsky, Bernard Duflos, and Arvin Levine for working out the answers to the exercises.

S.L.

Foreword

The present book is intended as a text in basic mathematics. As such, it can have multiple use: for a one-year course in the high schools during the third or fourth year (if possible the third, so that calculus can be taken during the fourth year); for a complementary reference in earlier high school grades (elementary algebra and geometry are covered); for a one-semester course at the college level, to review or to get a firm foundation in the basic mathematics necessary to go ahead in calculus, linear algebra, or other topics.

Years ago, the colleges used to give courses in "college algebra" and other subjects which should have been covered in high school. More recently, such courses have been thought unnecessary, but some experiences I have had show that they are just as necessary as ever. What is happening is that the colleges are getting a wide variety of students from high schools, ranging from exceedingly well-prepared ones who have had a good first course in calculus, down to very poorly prepared ones. This latter group includes both adults who return to college after several years' absence in order to improve their technical education, and students from the high schools who were not adequately taught. This is the reason why some material properly belonging to the high-school level must still be offered in the colleges.

The topics in this book are covered in such a way as to bring out clearly all the important points which are used afterwards in higher mathematics. I think it is important not to separate arbitrarily in different courses the various topics which involve both algebra and geometry. Analytic geometry and vector geometry should be considered simultaneously with algebra and plane geometry, as natural continuations of these. I think it is much more valuable to go into these topics, especially vector geometry, rather than to go endlessly into more and more refined results concerning triangles or trigonometry, involving more and more complicated technique. A minimum of basic techniques must of course be acquired, but it is better to extend these techniques by applying them to new situations in which they become

ix

motivated, especially when the possible topics are as attractive as vector geometry.

In fact, for many years college courses in physics and engineering have faced serious drawbacks in scheduling because they need simultaneously some calculus and also some vector geometry. It is very unfortunate that the most basic operations on vectors are introduced at present only in college. They should appear at least as early as the second year of high school. I cannot write here a text for elementary geometry (although to some extent the parts on intuitive geometry almost constitute such a text), but I hope that the present book will provide considerable impetus to lower considerably the level at which vectors are introduced. Within some foreseeable future, the topics covered in this book should in fact be the standard topics for the second year of high school, so that the third and fourth years can be devoted to calculus and linear algebra.

If only preparatory material for calculus is needed, many portions of this book can be omitted, and attention should be directed to the rules of arithmetic, linear equations (Chapter 2), quadratic equations (Chapter 4), coordinates (the first three sections of Chapter 8), trigonometry (Chapter 11), some analytic geometry (Chapter 12), a simple discussion of functions (Chapter 13), and induction (Chapter 16, §1). The other parts of the book can be omitted. Of course, the more preparation a student has, the more easily he will go through more advanced topics.

"More preparation", however, does not mean an accumulation of technical material in which the basic ideas of a subject are completely drowned. I am always disturbed at seeing endless chains of theorems, most of them of no interest, and without any stress on the main points. As a result, students do not remember the essential features of the subject. I am fully aware that because of the pruning I have done, many will accuse me of not going "deeply enough" into some subjects. I am quite ready to confront them on that. Besides, as I prune some technical and inessential parts of one topic, I am able to include the essential parts of another topic which would not otherwise be covered. For instance, what better practice is there with negative numbers than to introduce at once coordinates in the plane as a pair of numbers, and then deal with the addition and subtraction of such pairs, componentwise? This introduction could be made as early as the fourth grade, using maps as a motivation. One could do roughly what I have done here in Chapter 8, §1, Chapter 9, §1, and the beginning of Chapter 9, §2 (addition of pairs of numbers, and the geometric interpretation in terms of a parallelogram). At such a level, one can then leave it at that.

The same remark applies to the study of this book. The above-mentioned sections can be covered very early, at the same time that you study numbers and operations with numbers. They give a very nice geometric flavor to a slightly dry algebraic theory.

Generally speaking, I hope to induce teachers to leave well enough alone, and to avoid torturing a topic to death. It is easier to advance in one topic by going ahead with the more elementary parts of another topic, where the first one is applied. The brain much prefers to work that way, rather than to concentrate on ugly technical formulas which are obviously unrelated to anything except artificial drilling. Of course, some rote drilling is necessary. The problem is how to strike a balance. Do not regard some lists of exercises as too short. Rather, realize that practice for some notion may come again later in conjunction with another notion. Thus practice with square roots comes not only in the section where they are defined, but also later when the notion of distance between points is discussed, and then in a context where it is more interesting to deal with them. The same principle applies throughout the book.

The Interlude on logic and mathematical expression can be read also as an introduction to the book. Because of various examples I put there, and because we are already going through a Foreword, I have chosen to place it physically somewhat later. Take a look at it now, and go back to it whenever you feel the need for such general discussions. Mainly, I would like to make you feel more relaxed in your contact with mathematics than is usually the case. I want to stimulate thought, and do away with the general uptight feelings which people often have about math. If, for instance, you feel that any chapter gets too involved for you, then skip that part until you feel the need for it, and look at another part of the book. In many cases, you don't necessarily need an earlier part to understand a later one. In most cases, the important thing is to have understood the basic concepts and definitions, to be at ease with the simpler computational aspects of these concepts, and then to go ahead with a more advanced topic.

This advice also applies to the book as a whole. If you find that there is not enough material in this book to occupy you for a whole year, then start studying calculus or possibly linear algebra.

The book deals with mathematics on both the manipulative (or computational) level and the theoretical level. You must realize that a mastery of mathematics involves both levels, although your tastes may direct you more strongly to one or the other, or both. Here again, you may wish to vary the emphasis which you place on them, according to your needs or your taste. Be warned that deficiency at either level can ultimately hinder you in your work. Independently of need, however, it should be a source of pleasure to understand why a mathematical result is true, i.e. to understand its proof as well as to understand how to use the result in concrete circumstances.

Try to rely on yourself, and try to develop a trust in your own judgment. There is no "right" way to do things. Tastes differ, and this book is not meant to suppress yours. It is meant to propose some basic mathematical topics, according to my taste. If I am successful, you will agree with my taste, or you will have developed your own.

New York   S.L.
January 1971

Contents

Part One
ALGEBRA

In this part we develop systematically the rules for operations with numbers, relations among numbers, and properties of these operations and relations: addition, multiplication, inequalities, positivity, square roots, n-th roots. We find many of them, like commutativity and associativity, which recur frequently in mathematics and apply to other objects. They apply to complex numbers, but also to functions or mappings (in this case, commutativity does not hold in general and it is always an interesting problem to determine when it does hold).

Even when we study geometry afterwards, the rules of algebra are still used, say to compute areas, lengths, etc., which associate numbers with geometric objects. Thus does algebra mix with geometry.

The main point of this chapter is to condition you to have efficient reflexes in handling addition, multiplication, and division of numbers. There are many rules for these operations, and the extent to which we choose to assume some, and prove others from the assumed ones, is determined by several factors. We wish to assume those rules which are most basic, and assume enough of them so that the proofs of the others are simple. It also turns out that those which we do assume occur in many contexts in mathematics, so that whenever we meet a situation where they arise, then we already have the training to apply them and use them. Both historical experience and personal experience have gone into the selection of these rules and the order of the list in which they are given. To some extent, you must trust that it is valuable to have fast reflexes when dealing with associativity, commutativity, distributivity, cross-multiplication, and the like, if you do not have the intuition yourself which makes such trust unnecessary. Furthermore, the long list of the rules governing the above operations should be taken in the spirit of a description of how numbers behave.

It may be that you are already reasonably familiar with the operations between numbers. In that case, omit the first chapter entirely, and go right

3

ahead to Chapter 2, or start with the geometry or with the study of coordinates in Chapter 7. The whole first part on algebra is much more dry than the rest of the book, and it is good to motivate this algebra through geometry. On the other hand, your brain should also have quick reflexes when faced with a simple problem involving two linear equations or a quadratic equation. Hence it is a good idea to have isolated these topics in special sections in the book for easy reference.

In organizing the properties of numbers, I have found it best to look successively at the integers, rational numbers, and real numbers, at the cost of slight repetitions. There are several reasons for this. First, it is a good way of learning certain rules and their consequences in a special context (e.g. associativity and commutativity in the context of integers), and then observing that they hold in more general contexts. This sort of thing happens very frequently in mathematics. Second, the rational numbers provide a wide class of numbers which are used in computations, and the manipulation of fractions thus deserves special emphasis. Third, to follow the sequence integers–rational numbers–real numbers already plants in your mind a pattern which you will encounter again in mathematics. This pattern is related to the extension of one system of objects to a larger system, in which more equations can be solved than in the smaller system. For instance, the equation $2x = 3$ can be solved in the rational numbers, but not in the integers. The equations $x^2 = 2$ or $10^x = 2$ can be solved in the real numbers but not in the rational numbers. Similarly, the equations $x^2 = -1$, or $x^2 = -2$, or $10^x = -3$ can be solved in the complex numbers but not in the real numbers. It will be useful to you to have met the idea of extending mathematical systems at this very basic stage because it exhibits features in common with those in more advanced contexts.

1 Numbers

§1. THE INTEGERS

The most common numbers are those used for counting, namely the numbers

$$1, 2, 3, 4, \ldots,$$

which are called the **positive integers.** Even for counting, we need at least one other number, namely,

$$0 \ (\textbf{zero}).$$

For instance, we may wish to count the number of right answers you may get on a test for this course, out of a possible 100. If you get 100, then all your answers were correct. If you get 0, then no answer was correct.

The positive integers and zero can be represented geometrically on a line, in a manner similar to a ruler or a measuring stick:

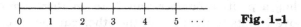

0	1	2	3	4	5 ···

Fig. 1-1

For this we first have to select a unit of distance, say the inch, and then on the line we mark off the inches to the right as in the picture.

For convenience, it is useful to have a name for the positive integers together with zero, and we shall call these the **natural numbers.** Thus 0 is a natural number, so is 2, and so is 124,521. The natural numbers can be used to measure distances, as with the ruler.

By definition, the point represented by 0 is called the **origin.**

The natural numbers can also be used to measure other things. For example, a thermometer is like a ruler which measures temperature. However, the thermometer shows us that we encounter other types of numbers besides the natural numbers, because there may be temperatures which may go below 0. Thus we encounter naturally what we shall call **negative integers** which we call **minus 1, minus 2, minus 3**, . . . , and which we write as

$$-1, \ -2, \ -3, \ -4, \ldots .$$

We represent the negative integers on a line as being on the other side of 0 from the positive integers, like this:

Fig. 1-2

The positive integers, negative integers, and zero all together are called the **integers.** Thus $-9, 0, 10, -5$ are all integers.

If we view the line as a thermometer, on which a unit of temperature has been selected, say the degree Fahrenheit, then each integer represents a certain temperature. The negative integers represent temperatures below zero.

Our discussion is already typical of many discussions which will occur in this course, concerning mathematical objects and their applicability to physical situations. In the present instance, we have the integers as mathematical objects, which are essentially abstract quantities. We also have different applications for them, for instance measuring distance or temperatures. These are of course not the only applications. Namely, we can use the integers to measure time. We take the origin 0 to represent the year of the birth of Christ. Then the positive integers represent years after the birth of Christ (called AD years), while the negative integers can be used to represent BC years. With this convention, we can say that the year -500 is the year 500 BC.

Adding a positive number, say 7, to another number, means that we must move 7 units to the right of the other number. For instance,

$$5 + 7 = 12.$$

Seven units to the right of 5 yields 12. On the thermometer, we would of course be moving upward instead of right. For instance, if the temperature at a given time is 5° and if it goes up by 7°, then the new temperature is 12°.

Observe the very simple rule for addition with 0, namely

N1.
$$\boxed{0 + a = a + 0 = a}$$

for any integer a.

What about adding negative numbers? Look at the thermometer again. Suppose the temperature at a given time is 10°, and the temperature drops by 15°. The new temperature is then −5°, and we can write

$$10 - 15 = -5.$$

Thus −5 is the result of subtracting 15 from 10, or of adding −15 to 10.

In terms of points on a line, adding a negative number, say −3, to another number means that we must move 3 units to the left of this other number. For example,

$$5 + (-3) = 2$$

because starting with 5 and moving 3 units to the left yields 2. Similarly,

$$7 + (-3) = 4, \quad \text{and} \quad 3 + (-5) = -2.$$

Note that we have

$$3 + (-3) = 0 \quad \text{or} \quad 5 + (-5) = 0.$$

We can also write these equations in the form

$$(-3) + 3 = 0 \quad \text{or} \quad (-5) + 5 = 0.$$

For instance, if we start 3 units to the left of 0 and move 3 units to the right, we get 0. Thus, in general, we have the formulas (by assumption):

N2.
$$a + (-a) = 0 \quad \text{and also} \quad -a + a = 0.$$

In the representation of integers on the line, this means that a and $-a$ lie on opposite sides of 0 on that line, as shown on the next picture:

$$\quad\quad a \quad\quad\quad 0 \quad\quad\quad -a \quad\quad\quad\quad \textbf{Fig. 1-3}$$

Thus according to this representation we can now write

$$3 = -(-3) \quad \text{or} \quad 5 = -(-5).$$

In these special cases, the pictures are:

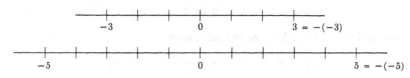

Fig. 1-4

Remark. We use the name

$$\textbf{minus } a \qquad \text{for} \qquad -a$$

rather than the words "negative a" which have found some currency recently. I find the words "negative a" confusing, because they suggest that $-a$ is a negative number. This is not true unless a itself is positive. For instance,

$$3 = -(-3)$$

is a positive number, but 3 is equal to $-a$, where $a = -3$, and a is a negative number.

Because of the property

$$a + (-a) = 0,$$

one also calls $-a$ the **additive inverse** of a.

The sum and product of integers are also integers, and the next sections are devoted to a description of the rules governing addition and multiplication.

§2. RULES FOR ADDITION

Integers follow very simple rules for addition. These are:

Commutativity. *If a, b are integers, then*

$$\boxed{a + b = b + a.}$$

For instance, we have

$$3 + 5 = 5 + 3 = 8,$$

or in an example with negative numbers, we have

$$-2 + 5 = 3 = 5 + (-2).$$

Associativity. *If a, b, c are integers, then*

$$\boxed{(a + b) + c = a + (b + c).}$$

In view of this, it is unnecessary to use parentheses in such a simple context, and we write simply

$$a + b + c.$$

For instance,

$$(3 + 5) + 9 = 8 + 9 = 17,$$
$$3 + (5 + 9) = 3 + 14 = 17.$$

We write simply

$$3 + 5 + 9 = 17.$$

Associativity also holds with negative numbers. For example,

$$(-2 + 5) + 4 = 3 + 4 = 7,$$
$$-2 + (5 + 4) = -2 + 9 = 7.$$

Also,

$$(2 + (-5)) + (-3) = -3 + (-3) = -6,$$
$$2 + (-5 + (-3)) = 2 + (-8) = -6.$$

The rules of addition mentioned above will not be proved, but we shall prove other rules from them.

To begin with, note that:

N3.

$$\boxed{\text{If } a + b = 0, \text{ then } b = -a \text{ and } a = -b.}$$

To prove this, add $-a$ to both sides of the equation $a + b = 0$. We get

$$-a + a + b = -a + 0 = -a.$$

Since $-a + a + b = 0 + b = b$, we find

$$b = -a$$

as desired. Similarly, we find $a = -b$. We could also conclude that

$$-b = -(-a) = a.$$

As a matter of convention, we shall write

$$a - b$$

instead of

$$a + (-b).$$

Thus a sum involving three terms may be written in many ways, as follows:

$$
\begin{aligned}
(a - b) + c &= (a + (-b)) + c \\
&= a + (-b + c) \quad &&\text{by associativity} \\
&= a + (c - b) \quad &&\text{by commutativity} \\
&= (a + c) - b \quad &&\text{by associativity,}
\end{aligned}
$$

and we can also write this sum as

$$
a - b + c = a + c - b,
$$

omitting the parentheses. Generally, in taking the sum of integers, we can take the sum in any order by applying associativity and commutativity repeatedly.

As a special case of **N3**, for any integer a we have

N4.
$$
\boxed{a = -(-a).}
$$

This is true because

$$
a + (-a) = 0,
$$

and we can apply **N3** with $b = -a$. Remark that this formula is true whether a is positive, negative, or 0. If a is positive, then $-a$ is negative. If a is negative, then $-a$ is positive. In the geometric representation of numbers on the line, a and $-a$ occur symmetrically on the line on opposite sides of 0. Of course, we can pile up minus signs and get other relationships, like

$$
-3 = -(-(-3)),
$$

or

$$
3 = -(-3) = -(-(-(-3))).
$$

Thus when we pile up the minus signs in front of a, we obtain a or $-a$ alternatively. For the general formula with the appropriate notation, cf. Exercises 5 and 6 of §4.

From our rules of operation we can now prove:

For any integers a, b we have

$$
-(a + b) = -a + (-b)
$$

or, in other words,

N5.

$$-(a + b) = -a - b.$$

Proof. Remember that if x, y are integers, then $x = -y$ and $y = -x$ mean that $x + y = 0$. Thus to prove our assertion, we must show that

$$(a + b) + (-a - b) = 0.$$

But this comes out immediately, namely,

$$
\begin{aligned}
(a + b) + (-a - b) &= a + b - a - b && \text{by associativity} \\
&= a - a + b - b && \text{by commutativity} \\
&= 0 + 0 \\
&= 0.
\end{aligned}
$$

This proves our formula.

Example. We have

$$
\begin{aligned}
-(3 + 5) &= -3 - 5 = -8, \\
-(-4 + 5) &= -(-4) - 5 = 4 - 5 = -1, \\
-(3 - 7) &= -3 - (-7) = -3 + 7 = 4.
\end{aligned}
$$

You should be very careful when you take the negative of a sum which involves itself in negative numbers, taking into account that

$$-(-a) = a.$$

The following rule concerning positive integers is so natural that you probably would not even think it worth while to take special notice of it. We still state it explicitly.

If a, b are positive integers, then $a + b$ is also a positive integer.

For instance, 17 and 45 are positive integers, and their sum, 62, is also a positive integer.

We assume this rule concerning positivity. We shall see later that it also applies to positive real numbers. From it we can prove:

If a, b are negative integers, then $a + b$ is negative.

Proof. We can write $a = -n$ and $b = -m$, where m, n are positive. Therefore

$$a + b = -n - m = -(n + m),$$

which shows that $a + b$ is negative, because $n + m$ is positive.

Example. If we have the relationship between three numbers

$$a + b = c,$$

then we can derive other relationships between them. For instance, add $-b$ to both sides of this equation. We get

$$a + b - b = c - b,$$

whence $a + 0 = c - b$, or in other words,

$$a = c - b.$$

Similarly, we conclude that

$$b = c - a.$$

For instance, if

$$x + 3 = 5,$$

then

$$x = 5 - 3 = 2.$$

If

$$4 - a = 3,$$

then adding a to both sides yields

$$4 = 3 + a,$$

and subtracting 3 from both sides yields

$$1 = a.$$

If
$$-2 - y = 5,$$
then
$$-7 = y \quad \text{or} \quad y = -7.$$

EXERCISES

Justify each step, using commutativity and associativity in proving the following identities.

1. $(a + b) + (c + d) = (a + d) + (b + c)$
2. $(a + b) + (c + d) = (a + c) + (b + d)$
3. $(a - b) + (c - d) = (a + c) + (-b - d)$
4. $(a - b) + (c - d) = (a + c) - (b + d)$
5. $(a - b) + (c - d) = (a - d) + (c - b)$
6. $(a - b) + (c - d) = -(b + d) + (a + c)$
7. $(a - b) + (c - d) = -(b + d) - (-a - c)$
8. $((x + y) + z) + w = (x + z) + (y + w)$
9. $(x - y) - (z - w) = (x + w) - y - z$
10. $(x - y) - (z - w) = (x - z) + (w - y)$
11. Show that $-(a + b + c) = -a + (-b) + (-c)$.
12. Show that $-(a - b - c) = -a + b + c$.
13. Show that $-(a - b) = b - a$.

Solve for x in the following equations.

14. $-2 + x = 4$ 15. $2 - x = 5$

16. $x - 3 = 7$ 17. $-x + 4 = -1$

18. $4 - x = 8$ 19. $-5 - x = -2$

20. $-7 + x = -10$ 21. $-3 + x = 4$

22. Prove the **cancellation law for addition:**

$$\boxed{\text{If } a + b = a + c, \text{ then } b = c.}$$

23. Prove: If $a + b = a$, then $b = 0$.

§3. RULES FOR MULTIPLICATION

We can multiply integers, and the product of two integers is again an integer. We shall list the rules which apply to multiplication and to its relations with addition.

We again have the rules of *commutativity* and *associativity:*

$$\boxed{ab = ba} \quad \text{and} \quad \boxed{(ab)c = a(bc).}$$

We emphasize that these apply whether a, b, c are negative, positive, or zero. Multiplication is also denoted by a dot. For instance

$$3 \cdot 7 = 21,$$

and

$$(3 \cdot 7) \cdot 4 = 21 \cdot 4 = 84,$$
$$3 \cdot (7 \cdot 4) = 3 \cdot 28 = 84.$$

For any integer a, the rules of multiplication by 1 and 0 are:

N6. $\qquad \boxed{1a = a} \quad \text{and} \quad \boxed{0a = 0.}$

Example. We have

$$(2a)(3b) = 2(a(3b))$$
$$= 2(3a)b$$
$$= (2 \cdot 3)ab$$
$$= 6ab.$$

In this example we have done something which is frequently useful, namely we have moved to one side all the explicit numbers like 2, 3 and put on the other side those numbers denoted by a letter like a or b. Using commutativity and associativity, we can prove similarly

$$(5x)(7y) = 35xy$$

or, with more factors,

$$(2a)(3b)(5x) = 30abx.$$

We suggest that you carry out the proof of this equality completely, using associativity and commutativity for multiplication.

Finally, we have the rule of *distributivity*, namely

$$a(b + c) = ab + ac$$

and also on the other side,

$$(b + c)a = ba + ca.$$

These rules will not be proved, but will be used constantly. We shall, however, make some comments on them, and prove other rules from them.

First observe that if we just assume distributivity on one side, and commutativity, then we can prove distributivity on the other side. Namely, assuming distributivity on the left, we have

$$(b + c)a = a(b + c) = ab + ac = ba + ca,$$

which is the proof of distributivity on the right.

Observe also that our rule $0a = 0$ can be proved from the other rules concerning multiplication and the properties of addition. We carry out the proof as an example. We have

$$0a + a = 0a + 1a = (0 + 1)a = 1a = a.$$

Thus

$$0a + a = a.$$

Adding $-a$ to both sides, we obtain

$$0a + a - a = a - a = 0.$$

The left-hand side is simply

$$0a + a - a = 0a + 0 = 0a,$$

so that we obtain $0a = 0$, as desired.

We can also prove

N7.

$$(-1)a = -a.$$

Proof. We have

$$(-1)a + a = (-1)a + 1a = (-1 + 1)a = 0a = 0.$$

By definition, $(-1)a + a = 0$ means that $(-1)a = -a$, as was to be shown.

We have

N8.

$$-(ab) = (-a)b.$$

Proof. We must show that $(-a)b$ is the negative of ab. This amounts to showing that

$$ab + (-a)b = 0.$$

But we have by distributivity

$$ab + (-a)b = (a + (-a))b = 0b = 0,$$

thus proving what we wanted.

Similarly, we leave to the reader the proof that

N9.

$$-(ab) = a(-b).$$

Example. We have

$$-(3a) = (-3)a = 3(-a).$$

Also,

$$4(a - 5b) = 4a - 20b.$$

Also,

$$-3(5a - 7b) = -15a + 21b.$$

In each of the above cases, you should indicate specifically each one of the rules we have used to derive the desired equality. Again, we emphasize that you should be especially careful when working with negative numbers and repeated minus signs. This is one of the most frequent sources of error when we work with multiplication and addition.

Example. We have

$$(-2a)(3b)(4c) = (-2) \cdot 3 \cdot 4abc$$
$$= -24abc.$$

Similarly,

$$(-4x)(5y)(-3c) = (-4)5(-3)xyc$$
$$= 60xyc.$$

Note that the product of two minus signs gives a plus sign.

Example. We have

$$(-1)(-1) = 1.$$

To see this, all we have to do is apply our rule

$$-(ab) = (-a)b = a(-b).$$

We find

$$(-1)(-1) = -(1(-1)) = -(-1) = 1.$$

Example. More generally, for any integers a, b we have

N10.

$$(-a)(-b) = ab.$$

We leave the proof as an exercise. From this we see that a product of two negative numbers is positive, because if a, b are positive and $-a$, $-b$ are therefore negative, then $(-a)(-b)$ is the positive number ab. For instance, -3 and -5 are negative, but

$$(-3)(-5) = -(3(-5)) = -(-(3 \cdot 5)) = 15.$$

Example. A product of a negative number and a positive number is negative. For instance, -4 is negative, 7 is positive, and

$$(-4) \cdot 7 = -(4 \cdot 7) = -28,$$

so that $(-4) \cdot 7$ is negative.

When we multiply a number with itself several times, it is convenient to use a notation to abbreviate this operation. Thus we write

$$aa = a^2,$$
$$aaa = a^3,$$
$$aaaa = a^4,$$

and in general if n is a positive integer,

$$a^n = aa \cdots a \quad \text{(the product is taken n times).}$$

We say that a^n is the **n-th power** of a. Thus a^2 is the second power of a, and a^5 is the fifth power of a.

If m, n are positive integers, then

N11.
$$\boxed{a^{m+n} = a^m a^n.}$$

This simply states that if we take the product of a with itself $m + n$ times, then this amounts to taking the product of a with itself m times and multiplying this with the product of a with itself n times.

Example
$$a^2 a^3 = (aa)(aaa) = a^{2+3} = aaaaa = a^5.$$

Example
$$(4x)^2 = 4x \cdot 4x = 4 \cdot 4xx = 16x^2.$$

Example
$$(7x)(2x)(5x) = 7 \cdot 2 \cdot 5xxx = 70x^3.$$

We have another rule for powers, namely

N12.
$$\boxed{(a^m)^n = a^{mn}.}$$

This means that if we take the product of a with itself m times, and then take the product of a^m with itself n times, then we obtain the product of a with itself mn times.

Example. We have

$$(a^3)^4 = a^{12}.$$

Example. We have

$$(ab)^n = a^n b^n$$

because

$$(ab)^n = abab \cdots ab \qquad \text{(product of } ab \text{ with itself } n \text{ times)}$$
$$= \underbrace{aa \cdots ab}_{n}\underbrace{b \cdots b}_{n}$$
$$= a^n b^n.$$

Example. We have

$$(2a^3)^5 = 2^5(a^3)^5 = 32a^{15}.$$

Example. The population of a city is 300 thousand in 1930, and doubles every 20 years. What will be the population after 60 years?

This is a case of applying powers. After 20 years, the population is $2 \cdot 300$ thousand. After 40 years, the population is $2^2 \cdot 300$ thousand. After 60 years, the population is $2^3 \cdot 300$ thousand, which is a correct answer. Of course, we can also say that the population will be 2 million 400 thousand.

The following three formulas are used constantly. They are so important that they should be thoroughly memorized by reading them out loud and repeating them like a poem, to get an aural memory of them.

$$\boxed{(a + b)^2 = a^2 + 2ab + b^2,} \qquad \boxed{(a - b)^2 = a^2 - 2ab + b^2,}$$

$$\boxed{(a + b)(a - b) = a^2 - b^2.}$$

Proofs. The proofs are carried out by applying repeatedly the rules for multiplication. We have:

$$(a + b)^2 = (a + b)(a + b) = a(a + b) + b(a + b)$$
$$= aa + ab + ba + bb$$
$$= a^2 + ab + ab + b^2$$
$$= a^2 + 2ab + b^2,$$

which proves the first formula.

$$(a - b)^2 = (a - b)(a - b) = a(a - b) - b(a - b)$$
$$= aa - ab - ba + bb$$
$$= a^2 - ab - ab + b^2$$
$$= a^2 - 2ab + b^2,$$

which proves the second formula.

$$(a + b)(a - b) = a(a - b) + b(a - b) = aa - ab + ba - bb$$
$$= a^2 - ab + ab - b^2$$
$$= a^2 - b^2,$$

which proves the third formula.

Example. We have

$$(2 + 3x)^2 = 2^2 + 2 \cdot 2 \cdot 3x + (3x)^2$$
$$= 4 + 12x + 9x^2.$$

Example. We have

$$(3 - 4x)^2 = 3^2 - 2 \cdot 3 \cdot 4x + (4x)^2$$
$$= 9 - 24x + 16x^2.$$

Example. We have

$$(-2a + 5b)^2 = 4a^2 + 2(-2a)(5b) + 25b^2$$
$$= 4a^2 - 20ab + 25b^2.$$

Example. We have

$$(4a - 6)(4a + 6) = (4a)^2 - 36$$
$$= 16a^2 - 36.$$

We have discussed so far examples of products of two factors. Of course, we can take products of more factors using associativity.

Example. Expand the expression

$$(2x + 1)(x - 2)(x + 5)$$

as a sum of powers of x multiplied by integers.

We first multiply the first two factors, and obtain

$$(2x + 1)(x - 2) = 2x(x - 2) + 1(x - 2)$$
$$= 2x^2 - 4x + x - 2$$
$$= 2x^2 - 3x - 2.$$

We now multiply this last expression with $x + 5$ and obtain

$$(2x + 1)(x - 2)(x + 5) = (2x^2 - 3x - 2)(x + 5)$$
$$= (2x^2 - 3x - 2)x + (2x^2 - 3x - 2)5$$
$$= 2x^3 - 3x^2 - 2x + 10x^2 - 15x - 10$$
$$= 2x^3 + 7x^2 - 17x - 10,$$

which is the desired answer.

EXERCISES

1. Express each of the following expressions in the form $2^m 3^n a^r b^s$, where m, n, r, s are positive integers.

   a) $8a^2b^3(27a^4)(2^5ab)$
   b) $16b^3a^2(6ab^4)(ab)^3$
   c) $3^2(2ab)^3(16a^2b^5)(24b^2a)$
   d) $24a^3(2ab^2)^3(3ab)^2$
   e) $(3ab)^2(27a^3b)(16ab^5)$
   f) $32a^4b^5a^3b^2(6ab^3)^4$

2. Prove:
$$(a + b)^3 = a^3 + 3a^2b + 3ab^2 + b^3,$$
$$(a - b)^3 = a^3 - 3a^2b + 3ab^2 - b^3.$$

3. Obtain expansions for $(a + b)^4$ and $(a - b)^4$ similar to the expansions for $(a + b)^3$ and $(a - b)^3$ of the preceding exercise.

Expand the following expressions as sums of powers of x multiplied by integers. These are in fact called polynomials. You might want to read, or at least look at, the section on polynomials later in the book (Chapter 13, §2).

4. $(2 - 4x)^2$
5. $(1 - 2x)^2$
6. $(2x + 5)^2$
7. $(x - 1)^2$

8. $(x + 1)(x - 1)$

9. $(2x + 1)(x + 5)$

10. $(x^2 + 1)(x^2 - 1)$

11. $(1 + x^3)(1 - x^3)$

12. $(x^2 + 1)^2$

13. $(x^2 - 1)^2$

14. $(x^2 + 2)^2$

15. $(x^2 - 2)^2$

16. $(x^3 - 4)^2$

17. $(x^3 - 4)(x^3 + 4)$

18. $(2x^2 + 1)(2x^2 - 1)$

19. $(-2 + 3x)(-2 - 3x)$

20. $(x + 1)(2x + 5)(x - 2)$

21. $(2x + 1)(1 - x)(3x + 2)$

22. $(3x - 1)(2x + 1)(x + 4)$

23. $(-1 - x)(-2 + x)(1 - 2x)$

24. $(-4x + 1)(2 - x)(3 + x)$

25. $(1 - x)(1 + x)(2 - x)$

26. $(x - 1)^2(3 - x)$

27. $(1 - x)^2(2 - x)$

28. $(1 - 2x)^2(3 + 4x)$

29. $(2x + 1)^2(2 - 3x)$

30. The population of a city in 1910 was 50,000, and it doubles every 10 years. What will it be (a) in 1970 (b) in 1990 (c) in 2,000?

31. The population of a city in 1905 was 100,000, and it doubles every 25 years. What will it be after (a) 50 years (b) 100 years (c) 150 years?

32. The population of a city was 200 thousand in 1915, and it triples every 50 years. What will be the population

   a) in the year 2215?   b) in the year 2165?

33. The population of a city was 25,000 in 1870, and it triples every 40 years. What will it be

   a) in 1990?   b) in 2030?

§4. EVEN AND ODD INTEGERS; DIVISIBILITY

We consider the positive integers 1, 2, 3, 4, 5, . . . , and we shall distinguish between two kinds of integers. We call

$$1, 3, 5, 7, 9, 11, 13, \ldots$$

the **odd integers,** and we call

$$2, 4, 6, 8, 10, 12, 14, \ldots$$

the **even integers.** Thus the odd integers go up by 2 and the even integers go up by 2. The odd integers start with 1, and the even integers start with 2. Another way of describing an even integer is to say that it is a positive integer which can be written in the form $2n$ for some positive integer n. For instance, we can write

$$2 = 2 \cdot 1,$$
$$4 = 2 \cdot 2,$$
$$6 = 2 \cdot 3,$$
$$8 = 2 \cdot 4,$$

and so on. Similarly, an odd integer is an integer which differs from an even integer by 1, and thus can be written in the form $2m - 1$ for some positive integer m. For instance,

$$1 = 2 \cdot 1 - 1,$$
$$3 = 2 \cdot 2 - 1,$$
$$5 = 2 \cdot 3 - 1,$$
$$7 = 2 \cdot 4 - 1,$$
$$9 = 2 \cdot 5 - 1,$$

and so on. Note that we can also write an odd integer in the form

$$2n + 1$$

if we allow n to be a natural number, i.e., allowing $n = 0$. For instance, we have

$$1 = 2 \cdot 0 + 1,$$
$$3 = 2 \cdot 1 + 1,$$
$$5 = 2 \cdot 2 + 1,$$
$$7 = 2 \cdot 3 + 1,$$
$$9 = 2 \cdot 4 + 1,$$

and so on.

> **Theorem 1.** *Let a, b be positive integers.*
> *If a is even and b is even, then $a + b$ is even.*
> *If a is even and b is odd, then $a + b$ is odd.*
> *If a is odd and b is even, then $a + b$ is odd.*
> *If a is odd and b is odd, then $a + b$ is even.*

Proof. We shall prove the second statement, and leave the others as exercises. Assume that a is even and that b is odd. Then we can write

$$a = 2n \qquad \text{and} \qquad b = 2k + 1$$

for some positive integer n and some natural number k. Then

$$a + b = 2n + 2k + 1$$
$$= 2(n + k) + 1$$
$$= 2m + 1 \qquad \text{(letting } m = n + k).$$

This proves that $a + b$ is odd.

   Theorem 2. *Let a be a positive integer. If a is even, then a^2 is even. If a is odd, then a^2 is odd.*

   Proof. Assume that a is even. This means that $a = 2n$ for some positive integer n. Then

$$a^2 = 2n \cdot 2n = 2(2n^2) = 2m,$$

where $m = 2n^2$ is a positive integer. Thus a^2 is even.

   Next, assume that a is odd, and write $a = 2n + 1$ for some natural number n. Then

$$a^2 = (2n + 1)^2 = (2n)^2 + 2(2n)1 + 1^2$$
$$= 4n^2 + 4n + 1$$
$$= 2(2n^2 + 2n) + 1$$
$$= 2k + 1, \qquad \text{where } k = 2n^2 + 2n.$$

Hence a^2 is odd, thus proving our theorem.

   Corollary. *Let a be a positive integer. If a^2 is even, then a is even. If a^2 is odd, then a is odd.*

   Proof. This is really only a reformulation of the theorem, taking into account ordinary logic. If a^2 is even, then a cannot be odd because the square of an odd number is odd. If a^2 is odd, then a cannot be even because the square of an even number is even.

   We can generalize the property used to define an even integer. Let d be a positive integer and let n be an integer. We shall say that d **divides** n, or that n **is divisible by** d if we can write

$$n = dk$$

for some integer k. Thus an even integer is a positive integer which is divisible by 2. According to our definition, the number 9 is divisible by 3 because

$$9 = 3 \cdot 3.$$

Also, 15 is divisible by 3 because

$$15 = 3 \cdot 5.$$

Also, -30 is divisible by 5 because

$$-30 = 5(-6).$$

Note that every integer is divisible by 1, because we can always write

$$n = 1 \cdot n.$$

Furthermore, every positive integer is divisible by itself.

EXERCISES

1. Give the proofs for the cases of Theorem 1 which were not proved in the text.

2. Prove: If a is even and b is any positive integer, then ab is even.

3. Prove: If a is even, then a^3 is even.

4. Prove: If a is odd, then a^3 is odd.

5. Prove: If n is even, then $(-1)^n = 1$.

6. Prove: If n is odd, then $(-1)^n = -1$.

7. Prove: If m, n are odd, then the product mn is odd.

Find the largest power of 2 which divides the following integers.

8. 16	9. 24	10. 32	11. 20
12. 50	13. 64	14. 100	15. 36

Find the largest power of 3 which divides the following integers.

16. 30	17. 27	18. 63	19. 99
20. 60	21. 50	22. 42	23. 45

24. Let a, b be integers. Define $a \equiv b$ (mod 5), which we read "a is **congruent to b modulo 5**", to mean that $a - b$ is divisible by 5. Prove: If $a \equiv b$ (mod 5) and $x \equiv y$ (mod 5), then

$$a + x \equiv b + y \quad (\text{mod } 5)$$

and

$$ax \equiv by \quad (\text{mod } 5).$$

25. Let d be a positive integer. Let a, b be integers. Define

$$a \equiv b \quad (\text{mod } d)$$

to mean that $a - b$ is divisible by d. Prove that if $a \equiv b$ (mod d) and $x \equiv y$ (mod d), then

$$a + x \equiv b + y \quad (\text{mod } d)$$

and

$$ax \equiv by \quad (\text{mod } d).$$

26. Assume that every positive integer can be written in one of the forms $3k$, $3k + 1$, $3k + 2$ for some integer k. Show that if the square of a positive integer is divisible by 3, then so is the integer.

§5. RATIONAL NUMBERS

By a **rational number** we shall mean simply an ordinary fraction, that is a quotient

$$\frac{m}{n} \quad \text{also written} \quad m/n,$$

where m, n are integers and $n \neq 0$. In taking such a quotient m/n, we emphasize that **we cannot divide by 0,** and thus we must always be sure that $n \neq 0$. For instance,

$$\frac{1}{4}, \frac{2}{3}, -\frac{3}{4}, -\frac{5}{7}$$

are rational numbers. Finite decimals also give us examples of rational numbers. For instance,

$$1.4 = \frac{14}{10} \quad \text{and} \quad 1.41 = \frac{141}{100}.$$

Just as we did with the integers, we can represent the rational numbers on the line. For instance, $\frac{1}{2}$ lies one-half of the way between 0 and 1, while

$\frac{2}{3}$ lies two-thirds of the way between 0 and 1, as shown on the following picture.

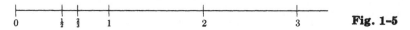

$$\text{Fig. 1-5}$$

The negative rational number $-\frac{3}{4}$ lies on the opposite side of 0 at a distance $\frac{3}{4}$ from 0. On the next picture, we have drawn $-\frac{3}{4}$ and $-\frac{5}{4}$.

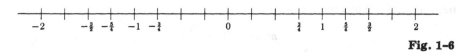

$$\text{Fig. 1-6}$$

There is no unique representation of a rational number as a quotient of two integers. For instance, we have

$$\frac{1}{2} = \frac{2}{4}.$$

We can interpret this geometrically on the line. If we cut up the segment between 0 and 1 into four equal pieces, and we take two-fourths of them, then this is the same as taking one-half of the segment. Picture:

$$\text{Fig. 1-7}$$

We need a general rule to determine when two expressions of quotients of integers give the same rational numbers. We assume this rule without proof. It is stated as follows.

Rule for cross-multiplying. Let m, n, r, s be integers and assume that n ≠ 0 and s ≠ 0. Then

$$\frac{m}{n} = \frac{r}{s} \qquad \text{if and only if} \qquad ms = rn.$$

The name "cross-multiplying" comes from our visualization of the rule in the following diagram:

$$\frac{m}{n} \times \frac{r}{s}.$$

Example. We have

$$\frac{1}{2} = \frac{2}{4}$$

because

$$1 \cdot 4 = 2 \cdot 2.$$

Also, we have

$$\frac{3}{7} = \frac{9}{21}$$

because

$$3 \cdot 21 = 9 \cdot 7$$

(both sides are equal to 63).

We shall make no distinction between an integer m and the rational number $m/1$. Thus we write

$$m = m/1 = \frac{m}{1}.$$

With this convention, we see that every integer is also a rational number. For instance, $3 = 3/1$ and $-4 = -4/1$.

Observe the special case of cross-multiplying when one side is an integer. For instance:

$$\frac{2n}{5} = \frac{6}{1}, \qquad \frac{2n}{5} = 6, \qquad 2n = 30, \qquad n = \frac{30}{2} = 15$$

are all equivalent formulations of a relation involving n.

Of course, cross-multiplying also works with negative numbers. For instance,

$$\frac{-4}{5} = \frac{8}{-10}$$

because

$$(-4)(-10) = 8 \cdot 5$$

(both sides are equal to 40).

Remark. For the moment, we are dealing with quotients of integers and describing how they behave. In the next section we shall deal with multiplicative inverses. There, you can see how the rule for cross-multiplication can in fact be proved from properties of such an inverse. Some people view this proof as the reason why cross-multiplication "works". However, in some contexts, one wants to *define* the multiplicative inverse by using the rule for cross-multiplication. This is the reason for emphasizing it here independently.

Cancellation rule for fractions. Let a be a non-zero integer. Let m, n be integers, $n \neq 0$. Then

$$\frac{am}{an} = \frac{m}{n}.$$

Proof. To test equality, we apply the rule for cross-multiplying. We must verify that

$$(am)n = m(an),$$

which we see is true by associativity and commutativity.

The examples which we gave are special cases of this cancellation rule. For instance

$$\frac{-4}{5} = \frac{(-2)(-4)}{(-2)5} = \frac{8}{-10}.$$

In dealing with quotients of integers which may be negative, it is useful to observe that

$$\boxed{\frac{-m}{n} = \frac{m}{-n}.}$$

This is proved by cross-multiplying, namely we must verify that

$$(-m)(-n) = mn,$$

which we already know is true.

The cancellation rule leads us to use the notion of divisibility already mentioned in §4. Indeed, suppose that d is a positive integer and m, n are divisible by d (or as we also say, that d is a **common divisor** of m and n). Then we can write

$$m = dr \qquad \text{and} \qquad n = ds$$

for some integers r and s, so that

$$\frac{m}{n} = \frac{dr}{ds} = \frac{r}{s}.$$

We see that our cancellation rule is applicable.

Example. We have

$$\frac{10}{15} = \frac{2 \cdot 5}{3 \cdot 5} = \frac{2}{3}$$

because 10 and 15 are both divisible by 5.

We say that a rational number is **positive** if it can be written in the form m/n, where m, n are positive integers. Let a be a positive rational number. We shall say that a is **expressed in lowest form** as a fraction

$$a = \frac{r}{s}$$

where r, s are positive integers if the only common divisor of r and s is 1.

Theorem 3. *Any positive rational number has an expression as a fraction in lowest form.*

Proof. First write a given positive rational number as a quotient of positive integers m/n. We know that 1 is a common divisor of m and n. Furthermore, any common divisor is at most equal to m or n. Thus among all common divisors there is a greatest one, which we denote by d. Thus we can write

$$m = dr \qquad \text{and} \qquad n = ds$$

with positive integers r and s. Our rational number is equal to

$$\frac{m}{n} = \frac{dr}{ds} = \frac{r}{s}.$$

All we have to do now is to show that the only common divisor of r and s is 1. Suppose that e is a common divisor which is greater than 1. Then we can write

$$r = ex \qquad \text{and} \qquad s = ey$$

with positive integers x and y. Hence

$$m = dr = dex \qquad \text{and} \qquad n = ds = dey.$$

Therefore de is a common divisor for m and n, and is greater than d since e is greater than 1. This is impossible because we assumed that d was the greatest common divisor of m and n. Therefore 1 is the only common divisor of r and s, and our theorem is proved.

Example. Any positive rational number can be expressed as a quotient m/n, where m, n are positive integers which are not both even, because if m/n is the expression of this rational number in lowest form, then 2 cannot divide both m and n, and therefore at least one of them must be odd.

Let

$$\frac{m}{n} \qquad \text{and} \qquad \frac{r}{s}$$

be rational numbers, expressed as quotients of integers. We can put these rational numbers over a common denominator ns by writing

$$\frac{m}{n} = \frac{ms}{ns} \qquad \text{and} \qquad \frac{r}{s} = \frac{nr}{ns}.$$

For instance, to put $3/5$ and $5/7$ over the common denominator $5 \cdot 7 = 35$, we write

$$\frac{3}{5} = \frac{3 \cdot 7}{5 \cdot 7} = \frac{21}{35} \quad \text{and} \quad \frac{5}{7} = \frac{5 \cdot 5}{7 \cdot 5} = \frac{25}{35}.$$

This leads us to the formula for the addition of rational numbers. Consider first a special case, when the rational numbers have a common denominator, for instance,

$$\frac{3}{5} + \frac{8}{5} = \frac{11}{5}.$$

This is reasonable just from the interpretation of rational numbers: If we have three-fifths of something, and add eight-fifths of that same thing, then we get eleven-fifths of that thing. In general, we can write the rule for addition when the rational numbers have a common denominator as

$$\boxed{\frac{a}{d} + \frac{b}{d} = \frac{a + b}{d}.}$$

Example. We have

$$\frac{-5}{8} + \frac{2}{8} = \frac{-3}{8}.$$

When the rational numbers do not have a common denominator, we get the formula for their addition by putting them over a common denominator. Namely, let $\frac{m}{n}$ and $\frac{r}{s}$ be rational numbers, expressed as quotients of integers m, n and r, s with $n \neq 0$ and $s \neq 0$. Then we have seen that

$$\frac{m}{n} = \frac{sm}{sn} \quad \text{and} \quad \frac{r}{s} = \frac{nr}{ns}.$$

Thus our rational numbers now have the common denominator sn, and thus the formula for addition in this general case is

$$\boxed{\frac{m}{n} + \frac{r}{s} = \frac{ms + rn}{ns}.}$$

Example. We have

$$\frac{3}{5} + \frac{4}{7} = \frac{3 \cdot 7 + 4 \cdot 5}{35} = \frac{21 + 20}{35} = \frac{41}{35}.$$

Example. We have

$$\frac{-5}{2} + \frac{3}{7} = \frac{(-5)\cdot 7 + 2 \cdot 3}{14} = \frac{-29}{14} \cdot$$

Example. We have

$$\frac{3}{-4} + \frac{5}{7} = \frac{21 - 20}{-28} = \frac{1}{-28} \cdot$$

Using our rule for adding rational numbers, we conclude at once:

The sum of positive rational numbers is also positive.

Observe that our number 0 has the property that

$$\boxed{\frac{0}{n} = \frac{0}{1} = 0}$$

for any integer $n \neq 0$. Indeed, applying our test for the equality of two fractions, we must verify that

$$0 \cdot 1 = 0 \cdot n,$$

and this is true because both sides are equal to 0.

For any rational number a, we have

$$\boxed{0 + a = a + 0 = a.}$$

This is easily seen using the analogous property for integers. Namely, write $a = m/n$, where m, n are integers, and $n \neq 0$. Then

$$0 + a = \frac{0}{n} + \frac{m}{n} = \frac{0 + m}{n} = \frac{m}{n} = a,$$

and similarly on the other side.

Let $a = m/n$ be a rational number, where m, n are integers and $n \neq 0$. Then we have

$$\frac{-m}{n} + \frac{m}{n} = \frac{-m + m}{n} = 0.$$

For this reason, we shall write

$$\boxed{\frac{-m}{n} = -\frac{m}{n}.}$$

By a previous remark, we also see that

$$\boxed{-\frac{m}{n} = \frac{m}{-n}.}$$

This shows how a minus sign can be moved around the various terms of a fraction without changing the value of the fraction.

A rational number which can be written as a fraction

$$-\frac{m}{n} = \frac{-m}{n} = \frac{m}{-n}$$

where m, n are positive integers will be called **negative**. For example, the number

$$\frac{3}{-5} = \frac{-3}{5} = -\frac{3}{5}$$

is negative. Using the definition of addition of rational numbers, you can easily verify for yourselves that a sum of negative rational numbers is negative.

Addition of rational numbers satisfies the properties of commutativity and associativity.

Just as we did for integers, the above statement will be accepted without proof. It is in fact a general property of much more general numbers, which will be restated again for these numbers in the next section.

In §2, we proved a number of properties of addition using only commutativity and associativity, together with the rules

$$0 + a = a \quad \text{and} \quad a + (-a) = 0.$$

These properties therefore remain valid for rational numbers. Similarly, all the exercises of §2 remain valid for rational numbers.

This remark will again be made later whenever we meet a similar situation. For instance, we see as before that

$$if\ a + b = 0,\ then\ b = -a.$$

We just add $-a$ to both sides of the equation $a + b = 0$. In words, we can say: To test whether a given rational number is equal to minus another, all we need to verify is that the sum of the numbers is equal to 0.

We shall now give the formula for **multiplication** of rational numbers. This formula is:

$$\frac{m}{n} \cdot \frac{r}{s} = \frac{mr}{ns}.$$

Thus to take the product of two rational numbers, we multiply their numerators and multiply their denominators. More precisely, the numerator of the product is the product of the numerators, and the denominator of the product is the product of the denominators.

Example. We have

$$\frac{3}{5} \cdot \frac{7}{8} = \frac{21}{40}.$$

Also,

$$\frac{2}{7} \cdot \frac{11}{16} = \frac{22}{112}.$$

We can write this last fraction in simpler form, namely

$$\frac{2}{7} \cdot \frac{11}{16} = \frac{2 \cdot 11}{7 \cdot 2 \cdot 8}.$$

We can then cancel 2 and get

$$\frac{2}{7} \cdot \frac{11}{16} = \frac{11}{56}.$$

This shows that sometimes it is best not to carry out a multiplication before looking at the possibility of cancellations.

Example. We have

$$\frac{-4}{5} \cdot \frac{7}{-3} = \frac{(-4)7}{5(-3)} = \frac{-28}{-15} = \frac{28}{15}.$$

Example. Let $a = m/n$ be a rational number expressed as a quotient of integers. Then

$$a^2 = \left(\frac{m}{n}\right)^2 = \frac{m}{n}\frac{m}{n} = \frac{m^2}{n^2}.$$

Similarly,

$$a^3 = \frac{m}{n}\frac{m}{n}\frac{m}{n} = \frac{m^3}{n^3}.$$

In general, for any positive integer k, we have

$$a^k = \left(\frac{m}{n}\right)^k = \frac{m^k}{n^k}.$$

Example. We have

$$\left(\frac{1}{2}\right)^3 = \frac{1}{2^3} = \frac{1}{8}.$$

Also,

$$\left(\frac{3}{5}\right)^4 = \frac{3^4}{5^4} = \frac{81}{525}.$$

Example. A chemical substance disintegrates in such a way that it gets halved every 10 min. If there are 20 grams (g) of the substance present at a given time, how much will be left after 50 min?

This is easily done. At the end of 10 min, we have $\frac{1}{2} \cdot 20$ g left. At the end of 20 min, we have $\frac{1}{2^2} \cdot 20$ g left, and so on; at the end of 50 min, we have

$$\frac{1}{2^5} \cdot 20 = \frac{20}{32}$$

grams left. This is a correct answer. If you want to put the fraction in lowest form, you may do so, and then you get the answer in the form $\frac{5}{8}$ g. You can also put it in approximate decimals, which we don't do here.

We ask: Is there a positive rational number a whose square is 2? The answer is at first not obvious. Such a number would be a square root of 2. Note that $1^2 = 1 \cdot 1 = 1$ and $2^2 = 4$. Thus the square of 1 is smaller than 2 and the square of 2 is bigger than 2. Any positive square root of 2 will therefore lie between 1 and 2 if it exists. We could experiment with various decimals to see whether they yield a square root of 2. For instance, let us try the decimal just in the middle between 1 and 2. We have

$$(1.5)^2 = 2.25,$$

which is bigger than 2. Thus 1.5 is not a square root of 2, and is too big to be one.

We could try more systematically, namely:

$$(1.1)^2 = 1.21 \quad \text{(too small)},$$
$$(1.2)^2 = 1.44 \quad \text{(too small)},$$
$$(1.3)^2 = 1.69 \quad \text{(too small)},$$
$$(1.4)^2 = 1.96 \quad \text{(too small but coming closer)}.$$

We know that 1.5 is too big, and hence we must go to the next decimal place to try out further.

$$(1.41)^2 = 1.9881 \quad \text{(too small)},$$
$$(1.42)^2 = 2.0164 \quad \text{(too big)}.$$

Thus we must go to the next decimal place for further experimentation. We try successively $(1.411)^2$, $(1.412)^2$, $(1.413)^2$, $(1.414)^2$ and find that they are too small. Computing $(1.415)^2$ we see that it is too big. We could keep on going like this. There are several things to be said about our procedure.

(1) It is very systematic, and could be programmed on a computer.

(2) It gives us increasingly good approximations to a square root of 2, namely it gives us rational numbers whose squares come closer and closer to 2.

However, to find a rational number whose square is 2, the procedure is a bummer because of the following theorem.

Theorem 4. *There is no positive rational number whose square is 2.*

Proof. Suppose that such a rational number exists. We can write it in lowest form m/n by Theorem 3. In particular, not both m and n can be even. We have

$$\left(\frac{m}{n}\right)^2 = \frac{m^2}{n^2} = 2.$$

Consequently, we obtain

$$m^2 = 2n^2,$$

and therefore m^2 is even. By the Corollary of Theorem 2 of §4, we conclude that m must be even, and we can therefore write

$$m = 2k$$

for some positive integer k. Thus we obtain

$$m^2 = (2k)^2 = 4k^2 = 2n^2.$$

We can cancel 2 from both sides of the equation

$$4k^2 = 2n^2,$$

and obtain

$$n^2 = 2k^2.$$

This means that n^2 is even, and as before, we conclude that n itself must be even. Thus from our original assumption that $(m/n)^2 = 2$ and m/n is in lowest form, we have obtained the impossible fact that both m, n are even. This means that our original assumption $(m/n)^2 = 2$ cannot be true, and concludes the proof of our theorem.

A number which is not rational is called **irrational**. From Theorem 4, we see that if a positive number a exists such that $a^2 = 2$, then a must be irrational. We shall discuss this further in the next section dealing with real numbers in general.

Multiplication of rational numbers satisfies the same basic rules as multiplication of integers. We state these once more:

For any rational number a we have $1a = a$ and $0a = 0$. Furthermore, multiplication is associative, commutative, and distributive with respect to addition.

As before, we *assume* these as properties of numbers. Moreover, we have the same remark for multiplication that we did for addition. All the properties of §3 which were proved using only the basic ones are therefore also valid for rational numbers. Thus the formulas which we had, like

$$(a + b)^2 = a^2 + 2ab + b^2,$$

are now seen to be valid for rational numbers as well. All the exercises at the end of §3 are valid for rational numbers.

Example. Solve for a in the equation

$$3a - 1 = 7.$$

We add 1 to both sides of the equation, and thus obtain

$$3a = 7 + 1 = 8.$$

We then divide by 3 and get

$$a = \frac{8}{3}.$$

Example. Solve for x in the equation

$$2(x - 3) = 7.$$

To do this, we use distributivity first, and get the equivalent equation

$$2x - 6 = 7.$$

Next we find

$$2x = 7 + 6 = 13,$$

whence

$$x = \frac{13}{2}.$$

Of course we could have given other arguments to find the answer. For instance, we could first get

$$x - 3 = \frac{7}{2},$$

whence

$$x = \frac{7}{2} + 3.$$

This is a perfectly correct answer. However, we can also give the answer in fraction form. We write $3 = \frac{6}{2}$, and find that

$$x = \frac{7}{2} + \frac{6}{2} = \frac{13}{2}.$$

Example. Solve for x in the equation

$$\frac{3x - 7}{2} + 4 = 2x.$$

We multiply both sides of the equation by 2 and obtain

$$3x - 7 + 8 = 4x.$$

We then add $-3x$ to both sides, to get

$$1 = 4x - 3x = x.$$

This solves our problem.

EXERCISES

1. Solve for a in the following equations.

   a) $2a = \dfrac{3}{4}$ b) $\dfrac{3a}{5} = -7$ c) $\dfrac{-5a}{2} = \dfrac{3}{8}$

2. Solve for x in the following equations.

   a) $3x - 5 = 0$ b) $-2x + 6 = 1$ c) $-7x = 2$

3. Put the following fractions in lowest form.

   a) $\dfrac{10}{25}$ b) $\dfrac{3}{9}$ c) $\dfrac{30}{25}$ d) $\dfrac{50}{15}$

   e) $\dfrac{45}{9}$ f) $\dfrac{62}{4}$ g) $\dfrac{23}{46}$ h) $\dfrac{16}{40}$

4. Let $a = m/n$ be a rational number expressed as a quotient of integers m, n with $m \neq 0$ and $n \neq 0$. Show that there is a rational number b such that $ab = ba = 1$.

5. Solve for x in the following equations.

   a) $2x - 7 = 21$ b) $3(2x - 5) = 7$ c) $(4x - 1)2 = \dfrac{1}{4}$

   d) $-4x + 3 = 5x$ e) $3x - 2 = -5x + 8$ f) $3x + 2 = -3x + 4$

   g) $\dfrac{4x}{3} + 1 = 3x$ h) $-\dfrac{3x}{2} + \dfrac{4}{3} = 5x$ i) $\dfrac{2x - 1}{3} + 4x = 10$

6. Solve for x in the following equations.

   a) $2x - \dfrac{3}{7} = \dfrac{x}{5} + 1$ b) $\dfrac{3}{4}x + 5 = -7x$ c) $\dfrac{-2}{13}x = 3x - 1$

   d) $\dfrac{4x}{3} + \dfrac{3}{4} = 2x - 5$ e) $\dfrac{4(1 - 3x)}{7} = 2x - 1$ f) $\dfrac{2 - x}{3} = \dfrac{7}{8}x$

7. Let n be a positive integer. By n **factorial,** written $n!$, we mean the product

$$1 \cdot 2 \cdot 3 \cdots n$$

of the first n positive integers. For instance,

$$2! = 2,$$
$$3! = 2 \cdot 3 = 6,$$
$$4! = 2 \cdot 3 \cdot 4 = 24.$$

   a) Find the value of $5!$, $6!$, $7!$, and $8!$.

b) Define $0! = 1$. Define the **binomial coefficient**

$$\binom{m}{n} = \frac{m!}{n!\,(m-n)!}$$

for any natural numbers m, n such that n lies between 0 and m. Compute the binomial coefficients

$$\binom{3}{0},\ \binom{3}{1},\ \binom{3}{2},\ \binom{3}{3},\ \binom{4}{0},\ \binom{4}{1},\ \binom{4}{2},\ \binom{4}{3},\ \binom{4}{4},$$

$$\binom{5}{0},\ \binom{5}{1},\ \binom{5}{2},\ \binom{5}{3},\ \binom{5}{4},\ \binom{5}{5}.$$

The binomial coefficient $\binom{m}{n}$ is equal to the number of ways n things can be selected out of m things. You may want to look at the discussion of Chapter 16, §1 at this time to see why this is so.

c) Show that

$$\binom{m}{n} = \binom{m}{m-n}.$$

d) Show that if n is a positive integer at most equal to m, then

$$\binom{m}{n} + \binom{m}{n-1} = \binom{m+1}{n}.$$

8. Prove that there is no positive rational number a such that $a^3 = 2$.

9. Prove that there is no positive rational number a such that $a^4 = 2$.

10. Prove that there is no positive rational number a such that $a^2 = 3$. You may assume that a positive integer can be written in one of the forms $3k$, $3k + 1$, $3k + 2$ for some integer k. Prove that if the square of a positive integer is divisible by 3, then so is the integer. Then use a similar proof as for $\sqrt{2}$.

11. a) Find a positive rational number, expressed as a decimal, whose square approximates 2 up to 3 decimals.

   b) Same question, but with 4 decimals accuracy instead.

12. a) Find a positive rational number, expressed as a decimal, whose square approximates 3 up to 2 decimals.

   b) Same question but with 3 decimals instead.

13. Find a positive rational number, expressed as a decimal, whose square approximates 5 up to

   a) 2 decimals,   b) 3 decimals.

14. Find a positive rational number whose cube approximates 2 up to
    a) 2 decimals,   b) 3 decimals.

15. Find a positive rational number whose cube approximates 3 to
    a) 2 decimals,   b) 3 decimals.

16. A chemical substance decomposes in such a way that it halves every 3 min. If there are 6 grams (g) of the substance present at the beginning, how much will be left
    a) after 3 min?   b) after 27 min?   c) after 36 min?

17. A chemical substance reacts in such a way that one third of the remaining substances decomposes every 15 min. If there are 15 g of the substance present at the beginning, how much will be left
    a) after 30 min?   b) after 45 min?   c) after 165 min?

18. A substance reacts in water in such a way that one-fourth of the undissolved part dissolves every 10 min. If you put 25 g of the substance in water at a given time, how much will be left after
    a) 10 min?   b) 30 min?   c) 50 min?

19. You are testing the effect of a noxious substance on bacteria. Every 10 min, one-tenth of the bacteria which are still alive are killed. If the population of bacteria starts with 10^6, how many bacteria are left after
    a) 10 min?   b) 30 min?   c) 50 min?
    d) Within which period of 10 min will half the bacteria be killed?
    e) Within which period of 10 min will 70% of the bacteria be killed?
    f) Within which period of 10 min will 80% of the bacteria be killed?
    [*Note:* If one-tenth of those alive are killed, then nine-tenths remain.]

20. A chemical pollutant is being emptied in a lake with 50,000 fishes. Every month, one-third of the fish still alive die from this pollutant. How many fish will be alive after
    a) 1 month?   b) 2 months?
    c) 4 months?   d) 6 months?
    (Give your answer to the nearest 100.)
    e) What is the first month when more than half the fish will be dead?
    f) During which month will 80% of the fish be dead?
    [*Note:* If one-third die, then two thirds remain.]

21. Every 10 years the population of a city is five-fourths of what it was 10 years before. How many years does it take
    a) before the population doubles?   b) before it triples?

§6. MULTIPLICATIVE INVERSES

Rational numbers satisfy one property which is not satisfied by integers, namely:

If a is a rational number $\neq 0$, then there exists a rational number, denoted by a^{-1}, such that

$$a^{-1}a = aa^{-1} = 1.$$

Indeed, if $a = m/n$ where m, n are integers $\neq 0$, then $a^{-1} = n/m$ because

$$\frac{m}{n}\frac{n}{m} = \frac{mn}{mn} = 1.$$

We call a^{-1} the **multiplicative inverse of** a.

Example. The multiplicative inverse of $\frac{1}{2}$ is $\frac{2}{1}$, or simply 2, because

$$2 \cdot \frac{1}{2} = 1.$$

The multiplicative inverse of $\frac{2}{3}$ is $\frac{3}{2}$. The multiplicative inverse of $-\frac{5}{7}$ is $-\frac{7}{5}$.

Observe that if a and b are rational numbers such that

$$ab = 1,$$

then

$$b = a^{-1}.$$

Proof. We multiply both sides of the relation $ab = 1$ by a^{-1}, and get

$$a^{-1}ab = a^{-1} \cdot 1 = a^{-1}.$$

Using associativity on the left, we find

$$a^{-1}ab = 1b = b,$$

so that we do find $b = a^{-1}$ as desired.

From the existence of an inverse for non-zero rational numbers, we deduce:

If $ab = 0$, then $a = 0$ or $b = 0$.

Proof. Suppose $a \neq 0$. Multiply both sides of the equation $ab = 0$ by a^{-1}. We get:

$$a^{-1}ab = 0a^{-1} = 0.$$

On the other hand, $a^{-1}ab = 1b = b$, so that we find $b = 0$, as desired.

We shall use the same notation as for quotients of integers in taking quotients of rational numbers. We write

$$\frac{a}{b} \quad \text{or} \quad a/b \quad \text{instead of} \quad b^{-1}a \quad \text{or} \quad ab^{-1}.$$

Example. Let $a = \frac{3}{4}$ and $b = \frac{5}{7}$. Then

$$\frac{3/4}{5/7} = \frac{3}{4}\left(\frac{5}{7}\right)^{-1} = \frac{3}{4}\frac{7}{5} = \frac{21}{20}.$$

Example. We have

$$\frac{1 + \frac{1}{2}}{2 - \frac{4}{3}} = \left(1 + \frac{1}{2}\right) \cdot \left(2 - \frac{4}{3}\right)^{-1}$$

$$= \frac{2 + 1}{2} \cdot \left(\frac{6 - 4}{3}\right)^{-1}$$

$$= \frac{3}{2}\left(\frac{2}{3}\right)^{-1} = \frac{3}{2}\frac{3}{2} = \frac{9}{4}.$$

Our rule for cross-multiplication which applied to quotients of integers applies as well when we want to cross-multiply rational numbers. We state it, and prove it using only the basic properties of addition, multiplication, and inverses.

Cross-multiplication. *Let a, b, c, d be rational numbers, and assume that $b \neq 0$ and $d \neq 0$.*

$$\text{If } \frac{a}{b} = \frac{c}{d}, \text{ then } ad = bc.$$

$$\text{If } ad = bc, \text{ then } \frac{a}{b} = \frac{c}{d}.$$

Proof. Assume that $a/b = c/d$. We can rewrite this relation in the form

$$b^{-1}a = d^{-1}c.$$

Multiply both sides by db (which is the same as bd). We obtain

$$dbb^{-1}a = bdd^{-1}c,$$

so that

$$da = bc$$

because $bb^{-1}a = 1a = a$, and similarly, $dd^{-1}c = 1c = c$.

Conversely, assume that $ad = bc$. Multiply both sides by $b^{-1}d^{-1}$, which is equal to $d^{-1}b^{-1}$. We find:

$$add^{-1}b^{-1} = d^{-1}b^{-1}bc,$$

whence

$$ab^{-1} = d^{-1}c.$$

This means that $a/b = c/d$, as desired.

Example. By cross-multiplying, we have

$$\frac{3}{x-1} = 2$$

if and only if

$$3 = 2(x-1) = 2x - 2,$$

which is equivalent to

$$3 + 2 = 2x.$$

Thus we can solve for x, and get $x = \frac{5}{2}$.

Example. By cross-multiplying we have

$$\frac{4+x}{\frac{1}{2}x} = 5$$

if and only if

$$4 + x = 5 \cdot \frac{1}{2}x = \frac{5x}{2}.$$

Again by cross-multiplication this is equivalent to

$$2(4 + x) = 5x,$$

or

$$8 + 2x = 5x.$$

Subtracting $2x$ from both sides of this equation, we solve for x, and get

$$x = \frac{8}{3}.$$

Cancellation law for multiplication. *Let a be a rational number $\neq 0$.*

If $ab = ac$, then $b = c$.

Proof. Multiply both sides of the equation $ab = ac$ by a^{-1}. We get

$$a^{-1}ab = a^{-1}ac,$$

whence $b = c$.

We also have a **cancellation law** similar to that for quotients of integers.

If a, b, c, d are rational numbers and $a \neq 0$, $c \neq 0$, then

$$\frac{ab}{ac} = \frac{b}{c}.$$

This can be verified, for instance, by cross-multiplication, because we have

$$abc = bac$$

(using commutativity and associativity).

Thus we can operate with fractions formed with rational numbers much as we could operate with fractions formed with integers.

Example. If a/b and c/d are two quotients of rational numbers (and $b \neq 0$, $d \neq 0$), then we can put them over a "common denominator" and write

$$\frac{a}{b} = \frac{ad}{bd}, \qquad \frac{c}{d} = \frac{bc}{bd}.$$

Example. If x, y, b are rational numbers and $b \neq 0$, then we can add quotients in a manner similar to the addition for quotients of integers, namely

$$\frac{x}{b} + \frac{y}{b} = b^{-1}x + b^{-1}y$$

$$= b^{-1}(x + y) \qquad \text{by distributivity}$$

$$= \frac{x + y}{b} \qquad \text{by definition.}$$

Combining this with the "common denominator" procedure of the preceding example, we find

$$\boxed{\frac{a}{b} + \frac{c}{d} = \frac{ad + bc}{bd}.}$$

This formula is entirely analogous to the formula expressing the sum of two rational numbers.

Example. Show that

$$\frac{1}{x - y} + \frac{1}{x + y} = \frac{2x}{x^2 - y^2}.$$

To do this, we add the two quotients on the left by our general formula which we just derived, and get:

$$\frac{1(x + y) + 1(x - y)}{(x - y)(x + y)} = \frac{x + y + x - y}{x^2 - y^2} = \frac{2x}{x^2 - y^2},$$

as was to be shown.

Remark. In the preceding example, the quotients $1/(x - y)$ and $1/(x + y)$ make no sense if $x - y = 0$ or $x + y = 0$. In such instances, we assume tacitly that x and y are such that $x - y \neq 0$ and $x + y \neq 0$. In the sequel we shall sometimes omit the explicit mention of such conditions if there is no danger of confusion.

Example. Solve for x in the equation

$$\frac{3x + 1}{2x - 5} = 4.$$

We cross-multiply. For $2x - 5 \neq 0$, i.e. $x \neq \frac{5}{2}$, we find the equivalent equation

$$3x + 1 = 4(2x - 5) = 8x - 20.$$

Hence

$$8x - 3x = 1 - (-20) = 1 + 20 = 21.$$

This yields finally

$$5x = 21,$$

whence

$$x = \frac{21}{5}.$$

Example. We give an example from the physical world. Suppose that an object is moving along a straight line at constant speed. Let s denote the speed, let d denote the distance traveled by the object, and let t denote the time taken to travel the distance d. Then in physics one verifies the formula

$$d = st.$$

Of course, we must select units of time and distance before we can associate numbers with these. For instance, suppose that the distance traveled is 5 mi, and the time taken is $\frac{1}{2}$ hr. Then the speed is

$$s = d/t = \frac{5 \text{ mi}}{\frac{1}{2} \text{ hr}} = 2 \cdot 5 \text{ mi/hr} = 10 \text{ mi/hr}.$$

Example. A person takes a trip and drives 8 hr, a distance of 400 mi. His average speed is 60 mph on the freeway, and 30 mph when he drives through a town. How long did the person drive through towns during his trip?

To solve this, let x be the length of time the person drives through towns. Then the length of time the person is on the freeway is $8 - x$. The distance driven through towns is therefore equal to $30x$, and the distance driven on freeways is $60(8 - x)$. Since the total distance driven is 400 mi, we have

$$30x + 60(8 - x) = 400.$$

This is equivalent to the equations

$$30x + 480 - 60x = 400$$

and

$$80 = 30x.$$

Thus we find

$$x = \frac{80}{30} = \frac{8}{3}.$$

Hence the person spent $\frac{8}{3}$ hrs driving through towns.

Example. The radiator of a car contains 8 qt of liquid, consisting of water and 40% antifreeze. How much should be drained and replaced by antifreeze if the resultant mixture should have 90% antifreeze?

Let x be the number of quarts which must be drained. After draining this amount, we are left with $(8 - x)$ qt of liquid, of which 40% is antifreeze. Thus we are left with

$$\frac{40}{100}(8 - x) \text{ qt}$$

of antifreeze. Since we now add x qt of antifreeze, we see that x satisfies

$$x + \frac{40}{100}(8 - x) = \frac{90}{100} \cdot 8.$$

From this we can solve for x, transforming this equation into equivalent equations as follows:

$$x + \frac{40}{100} \cdot 8 - \frac{40}{100} x = \frac{90}{100} \cdot 8,$$

which amounts to

$$\frac{60}{100} x = \frac{50}{100} \cdot 8,$$

whence

$$x = \frac{400}{60} = \frac{20}{3}.$$

This is a correct answer, but if you insist on putting the fraction in lowest form, then we can say that $6\frac{2}{3}$ qt should be replaced by antifreeze.

Remark. The above examples, and the exercises, can also be worked using two unknowns. Cf. the end of Chapter 2, §1.

EXERCISES

1. Solve for x in the following equations.

a) $\dfrac{2x - 1}{3x + 2} = 7$

b) $\dfrac{2 - 4x}{x + 1} = \dfrac{3}{4}$

c) $\dfrac{x}{x + 5} = \dfrac{5}{7}$

d) $2x + 5 = \dfrac{3x - 2}{7}$

e) $\dfrac{1 - 2x}{3x + 4} = -3$

f) $\dfrac{-2 - 5x}{-3x - 4} = \dfrac{4}{-3}$

g) $\dfrac{-2 - 7x}{4} + 1 = \dfrac{1 - x}{5}$

h) $\dfrac{3x + 1}{4 - 2x} + \dfrac{7}{3} = 0$

i) $\dfrac{-2 - 4x}{3} = \dfrac{x - 1}{4} + 5$

2. Prove the following relations. It is assumed that all values of x and y which occur are such that the denominators in the indicated fractions are not equal to 0.

a) $\dfrac{1}{x+y} - \dfrac{1}{x-y} = \dfrac{-2y}{x^2-y^2}$

   b) $\dfrac{x^3-1}{x-1} = 1 + x + x^2$

c) $\dfrac{x^4-1}{x-1} = 1 + x + x^2 + x^3$

d) $\dfrac{x^n-1}{x-1} = x^{n-1} + x^{n-2} + \cdots + x + 1.$ [*Hint:* Cross-multiply and cancel as much as possible.]

3. Prove the following relations.

a) $\dfrac{1}{2x+y} + \dfrac{1}{2x-y} = \dfrac{4x}{4x^2-y^2}$

b) $\dfrac{2x}{x+5} - \dfrac{3x+1}{2x+1} = \dfrac{x^2-14x-5}{2x^2+11x+5}$

c) $\dfrac{1}{x+3y} + \dfrac{1}{x-3y} = \dfrac{2x}{x^2-9y^2}$

d) $\dfrac{1}{3x-2y} + \dfrac{x}{x+y} = \dfrac{x+y+3x^2-2xy}{3x^2+xy-2y^2}$

For more exercises of this type, see Chapter 13, §2

4. Prove the following relations.

a) $\dfrac{x^3-y^3}{x-y} = x^2 + xy + y^2$

b) $\dfrac{x^4-y^4}{x-y} = x^3 + x^2y + xy^2 + y^3$

c) Let

$$x = \frac{1-t^2}{1+t^2} \quad \text{and} \quad y = \frac{2t}{1+t^2}.$$

Show that $x^2 + y^2 = 1$.

5. Prove the following relations.

a) $\dfrac{x^3+1}{x+1} = x^2 - x + 1$

b) $\dfrac{x^5 + 1}{x + 1} = x^4 - x^3 + x^2 - x + 1$

c) If n is an odd integer, prove that

$$\frac{x^n + 1}{x + 1} = x^{n-1} - x^{n-2} + x^{n-3} - \cdots - x + 1.$$

[*Hint:* Cross-multiply.]

6. Assume that a particle moving with uniform speed on a straight line travels a distance of $\frac{3}{4}$ ft at a speed of $\frac{2}{5}$ ft/sec. What time did it take the particle to do that?

7. If a solid has uniform density d, occupies a volume v, and has mass m, then we have the formula

$$m = vd.$$

Find the density if

a) $m = \frac{3}{10}$ lb and $v = \frac{2}{3}$ in^3,   b) $m = 6$ lb and $v = \frac{4}{3}$ in^3.

c) Find the volume if the mass is 15 lb and the density is $\frac{2}{3}$ lb/in^3.

8. Let F denote temperature in degrees Fahrenheit, and C the temperature in degrees centigrade. Then F and C are related by the formula

$$C = \tfrac{5}{9}(F - 32).$$

Find C when F is

a) 32,   b) 50,   c) 99,   d) 100,   e) -40.

9. Let F and C be as in Exercise 8. Find F when C is:

a) 0,   b) -10,   c) -40,   d) 37,   e) 40,   f) 100.

10. In electricity theory, one denotes the current by I, the resistance by R, and the voltage by E. These are related by the formula

$$E = IR$$

(with appropriate units). Find the resistance when the voltage and current are:

a) $E = 10,\ I = 3$;   b) $E = 220,\ I = 10$.

11. A solution contains 35% alcohol and 65% water. If you start with 12 kilograms of solution, how much water must be added to make the percentage of alcohol equal to

     a) 20%?   b) 10%?   c) 5%?

12. A plane travels 3,000 mi in 4 hr. When the wind is favorable, the plane averages 900 mph. When the wind is unfavorable, the plane averages 500 mph. During how many hours was the wind favorable?

13. Tickets for a performance sell at $5.00 and $2.00. The total amount collected was $4,100, and there are 1,300 tickets in all. How many tickets of each price were sold?

14. A salt solution contains 10% salt and weighs 80 g. How much pure water must be added so that the percentage of salt drops to

   a) 4%?   b) 6%?   c) 8%?

15. How many kilograms of water must you add to 6 kg of pure alcohol to get a mixture containing

   a) 25% alcohol?   b) 20% alcohol?   c) 15% alcohol?

16. A boat travels a distance of 500 mi, along two rivers, for 50 hr. The current goes in the same direction as the boat along one river, and then the boat averages 20 mph. The current goes in the opposite direction along the other river, and then the boat averages 8 mph. During how many hours was the boat on the first river?

17. How much water must evaporate from a salt solution weighing 2 lb and containing 25% salt, if the remaining mixture must contain

   a) 40% salt?   b) 60% salt?

18. The radiator of a car can contain 10 kg of liquid. If it is half full with a mixture having 60% antifreeze and 40% water, how much more water must be added so that the resulting mixture has only

   a) 40% antifreeze?   b) 10% antifreeze?

   Will it fit in the radiator?

2 Linear Equations

§1. EQUATIONS IN TWO UNKNOWNS

Suppose that we are given two equations like

(1) $$2x + y = 1,$$
(2) $$3x - 2y = 4.$$

We wish to solve these equations for x and y. We follow what is known as the elimination method. We try to get rid of x, say, so as to obtain only one equation in y. We observe that x is multiplied by 2 in the first equation and by 3 in the second. We want to multiply each one of these equations by a suitable number so that the coefficients of x become the same. Thus we multiply the first equation by 3 and the second by 2. We obtain

$$6x + 3y = 3,$$
$$6x - 4y = 8.$$

If we now subtract the second equation from the first, i.e. subtract each side of the second equation from the corresponding side of the first, we see that the $6x$ cancels, and we find:

$$3y - (-4y) = 3 - 8,$$

whence

$$3y + 4y = 7y = -5.$$

This yields

$$y = \frac{-5}{7}.$$

53

We can then solve for x, using (1), which gives $2x = 1 - y$. Thus

$$2x = 1 - \frac{-5}{7} = \frac{7 + 5}{7} = \frac{12}{7}.$$

Hence

$$x = \frac{12}{2 \cdot 7} = \frac{12}{14}.$$

Our answer is therefore:

$$y = \frac{-5}{7} \quad \text{and} \quad x = \frac{12}{14}.$$

If we want x in lowest form, we can always write $x = \frac{6}{7}$, but $\frac{12}{14}$ is quite correct.

As a variation, we could also have eliminated y first. Thus we multiply the first equation by 2, leave the second unchanged, and add the equations. We get:

$$4x + 2y = 2,$$
$$3x - 2y = 4.$$

Adding yields

$$4x + 3x = 6.$$

Thus $7x = 6$ and $x = \frac{6}{7}$, which is of course the same answer that we found above. We could then solve for y using the first equation, namely,

$$y = 1 - 2x,$$

so that

$$y = 1 - \frac{12}{7} = \frac{7 - 12}{7} = \frac{-5}{7}.$$

It may happen that a system of linear equations has no solutions. For instance the system

(3)
$$2x - y = 5,$$
$$2x - y = 7$$

obviously has no solution. The system

(4)
$$2x - y = 5,$$
$$6x - 3y = 7$$

has no solution either. Indeed, any solution of $6x - 3y = 7$ is also a solution of

$$2x - y = \tfrac{7}{3},$$

(divide the equation by 3), and it is again obvious that no simultaneous solution exists for the system of equations (4).

We do not want to overemphasize here the theory determining precisely the cases when a solution exists and when it does not. It "usually" exists, unless one has a case essentially like the examples above. Our purposes here are mainly to put you at ease with two simple equations in two unknowns, so that you have some simple approach to them. We don't intend to overburden you or give you any worries about them. On the other hand, you may wish to do Exercises 9 and 10 to get the general criterion indicating when a solution exists. These exercises make precise our meaning of "usually".

When you learn about coordinates, then you will see that the simultaneous equations we have been considering represent straight lines, and that finding their simultaneous solution gives the coordinates of the point of intersection of these lines. If you wish, you may look up coordinates right away, and the first section of Chapter 12 to see about this.

One final remark. Observe that our elimination procedure actually proves that if x, y are numbers satisfying the simultaneous equations, then they must have the value obtained by the method indicated. Conversely these values for x, y actually are solutions of the equations. This can be checked each time explicitly. To prove it in general is easy but requires setting up convenient notation and using general letters for the coefficients of the equation. See Exercise 11. Here we don't want to get bogged down in abstraction. Our purpose in this section was simply to teach you a simple and efficient way of finding the solutions of a simple system of equations.

Simultaneous equations like the above can be used to solve problems which we gave in the context of one variable at the end of Chapter 1. We give an example of this.

Example. A person takes a trip and drives 8 hr, a distance of 400 mi. His average speed is 60 mph on the freeway, and 30 mph when he drives through a town. How long did the person drive through towns during his trip?

To solve this, let x be the length of time driven on freeways, and let y be the length of time driven through towns. Then

$$x + y = 8.$$

This gives us a first equation. Furthermore, the distance driven on freeways is equal to $60x$, and the distance driven through towns is equal to $30y$. Hence we get a second equation

$$60x + 30y = 400.$$

We can now solve our pair of equations, by multiplying the first by 60 and subtracting the second. We get

$$60y - 30y = 480 - 400,$$

or more simply,

$$30y = 80.$$

Therefore

$$y = \frac{80}{30} = \frac{8}{3}$$

is our numerical answer, and the person drove $\frac{8}{3}$ hr through towns. This is of course the same answer that we found when working with only one variable.

You may now wish to work out the exercises at the end of Chapter 1, §6 by means of two unknowns, which may be easier to handle the problems.

EXERCISES

Solve the following systems of equations for x and y.

1. $2x - y = 3$
   $x + y = 2$

2. $-4x + 7y = -1$
   $x - 2y = -4$

3. $3x + 4y = -2$
   $-2x - 3y = 1$

4. $-3x + 2y = -1$
   $x - y = 2$

5. $-3x + y = 0$
   $x - y = 1$

6. $3x + 7y = 0$
   $x - y = 0$

7. $7x - y = 2$
   $2x + 2y = 4$

8. $-4x - 7y = 5$
   $2x + y = 6$

9. Let a, b, c, d be numbers such that $ad - bc \neq 0$. Solve the following systems of equations for x and y in terms of a, b, c, d.

a) $ax + by = 1$
   $cx + dy = 2$

b) $ax + by = 3$
   $cx + dy = -4$

c) $ax + by = -2$
   $cx + dy = 3$

d) $ax + by = 5$
   $cx + dy = 7$

10. Making the same assumptions as in Exercise 9, show that the solution of the system

$$ax + by = 0,$$
$$cx + dy = 0$$

must be $x = 0$ and $y = 0$.

11. Let a, b, c, d, u, v be numbers and assume that $ad - bc \neq 0$. Solve the following system of equations for x and y in terms of a, b, c, d, u, v:

$$ax + by = u,$$
$$cx + dy = v.$$

Verify that the answer you get is actually a solution.

§2. EQUATIONS IN THREE UNKNOWNS

We now want to solve a system of equations like

$$\begin{aligned} 3x + 2y + 4z &= 1, \\ -x + y + 2z &= 2, \\ x - 3y + z &= -1 \end{aligned}$$

(1)

for x, y, z. We follow the same pattern as before, eliminating successively x, y, and then solving for z. We choose the order of elimination so as to make it easier on ourselves. Thus adding the second and third equations already gets rid of x, so we do this, and get

$$y - 3y + 3z = 2 - 1$$

or

(2) $$-2y + 3z = 1.$$

We go back to (1), and eliminate x from the first two equations. We multiply the second by 3 and add it to the first. This yields

$$2y + 3y + 4z + 6z = 1 + 6$$

or

(3) $$5y + 10z = 7.$$

Then equations (2) and (3) form a pair of equations in two unknowns which can be solved as in the first section of this chapter. We multiply (2) by 5, we multiply (3) by 2, and add. This gets rid of y, and we obtain

$$15z + 20z = 5 + 14$$

which yields

$$35z = 19,$$

whence

$$z = \frac{19}{35}.$$

Having found the value for z, we can go back to (2) or (3) to find the value for y. Suppose we use (2). We get

$$2y = 3z - 1 = 3 \cdot \frac{19}{35} - 1$$
$$= \frac{57 - 35}{35}$$
$$= \frac{22}{35}.$$

Hence dividing by 2, we find the value for y, namely

$$y = \frac{11}{35}.$$

Finally we can solve for x using any one of the first three equations in (1). Say we use the third equation. We have

$$x = -1 + 3y - z$$
$$= -\frac{35}{35} + \frac{33}{35} - \frac{19}{35}$$
$$= -\frac{91}{35}.$$

Thus the solution to our problem is:

$$x = \frac{-21}{35},$$

$$y = \frac{11}{35},$$

$$z = \frac{19}{35}.$$

EXERCISES

Solve the following equations for x, y, z.

1. $2x - 3y + z = 0$
   $x + y + z = 1$
   $x - 2y - 4z = 2$

2. $2x - y + z = 1$
   $4x + y + z = 2$
   $x - y - 2z = 0$

3. $x + 4y - 4z = 1$
   $x + 2y + z = 2$
   $4x - 3y - 2z = 1$

4. $x + y + z = 0$
   $x - y + z = 0$
   $2x - y - z = 0$

5. $5x + 3y - z = 0$
   $x + 2y + 2z = 1$
   $x - 2y - 2z = 0$

6. $2x + 2y - 3z = 0$
   $x - 3y + z = 3$
   $2x + y - 4z = 0$

7. $4x - 2y + 5z = 1$
   $x + y + z = 0$
   $-x + y - 2z = 2$

8. $x + y + z = 0$
   $x - y - z = 1$
   $x + y - z = 1$

In the next exercises, you will find it easiest to clear denominators before solving.

9. $\frac{1}{2}x + y - \frac{3}{4}z = 1$
   $\frac{2}{3}x - \frac{1}{3}y + z = 2$
   $x - \frac{1}{5}y + 2z = 1$

10. $\frac{1}{2}x - y + z = 1$
    $x + \frac{1}{3}y - \frac{2}{3}z = 2$
    $x + y - z = 3$

11. $\frac{3}{4}x - y + z = 1$
    $x - \frac{1}{2}y + z = 0$
    $x + y - \frac{1}{3}z = 1$

12. $\frac{1}{2}x - \frac{2}{3}y + z = 1$
    $x - \frac{1}{5}y + z = 0$
    $2x - \frac{1}{3}y + \frac{2}{5}z = 1$

3 Real Numbers

§1. ADDITION AND MULTIPLICATION

The integers and rational numbers are part of a larger system of numbers. As we know, the integers and rational numbers correspond to some points on the line. The real numbers are those numbers which correspond to all points on the line. Another way of describing them is to say that they consist of all numbers which have a decimal expansion, possibly infinite. For instance

$$9.123145\ldots$$

is a real number. It is a rather long and tedious process to develop the theory of the real numbers systematically. Hence we shall first summarize some of the algebraic properties which they satisfy, and then, as we go along, if the need arises, we shall mention other properties. Unless otherwise specified, **"number"** will mean **"real number"**.

The sum and product of two numbers are numbers, and they satisfy the following properties, similar to those of rational numbers.

Properties of addition. Addition is commutative and associative, meaning that for all real numbers a, b, c we have

$$a + b = b + a \quad \text{and} \quad a + (b + c) = (a + b) + c.$$

Furthermore, we have

$$0 + a = a.$$

To each real number a there is associated a number denoted by $-a$ such that

$$a + (-a) = 0.$$

As with integers, or rational numbers, we represent a and $-a$ on opposite sides of 0 on the line.

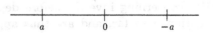

Fig. 3-1

61

The number $-a$ is called the **additive inverse** of a, as before. We also read it as **minus** a. If b is a number such that $a + b = 0$, then we must have $b = -a$, as we see by adding $-a$ to both sides of the equation $a + b = 0$. This is called the **uniqueness of the additive inverse.**

Properties of multiplication. Multiplication is commutative and associative, meaning that for all real numbers a, b, c we have

$$ab = ba \quad \text{and} \quad a(bc) = (ab)c.$$

Furthermore, we have

$$1a = a \quad \text{and} \quad 0a = 0.$$

Multiplication is distributive with respect to addition, meaning that

$$a(b + c) = ab + ac \quad \text{and} \quad (b + c)a = ba + ca.$$

So far, these properties are the same as those satisfied by the integers and rational numbers. *In particular, further properties which were proved using only these basic ones are now valid for the real numbers.* For instance, we recall the important formulas:

$$(a + b)^2 = a^2 + 2ab + b^2, \qquad (a - b)^2 = a^2 - 2ab + b^2,$$

$$(a + b)(a - b) = a^2 - b^2.$$

These are true if a, b are real numbers because their proofs used only commutativity and associativity.

Existence of the multiplicative inverse. If a is a real number $\neq 0$, then there exists a real number denoted by a^{-1} such that

$$a^{-1}a = aa^{-1} = 1.$$

As with rational numbers, this number a^{-1} is called the **multiplicative inverse of** a. Instead of writing a^{-1} we write $1/a$, and

$$\text{we write} \quad a/b \quad \text{instead of} \quad b^{-1}a \quad \text{or} \quad ab^{-1}.$$

The proofs of properties concerning inverses before depended only on the basic ones we have mentioned so far, and are thus applicable to the real numbers. Thus we have cross-multiplication, cancellation rules, etc. We have the **uniqueness of the multiplicative inverse.** Namely, if

$$ab = 1,$$

then multiplying both sides by a^{-1} shows that $b = a^{-1}$.

EXERCISES

1. Let E be an abbreviation for even, and let I be an abbreviation for odd. We know that:

$$E + E = E,$$
$$E + I = I + E = I,$$
$$I + I = E,$$
$$EE = E,$$
$$II = I,$$
$$IE = EI = E.$$

a) Show that addition for E and I is associative and commutative. Show that E plays the rôle of a zero element for addition. What is the additive inverse of E? What is the additive inverse of I?

b) Show that multiplication for E and I is commutative and associative. Which of E or I behaves like 1? Which behaves like 0 for multiplication? Show that multiplication is distributive with respect to addition.

Remark. The system consisting of E and I gives an example of a system with only two objects satisfying the basic properties of addition and multiplication. Thus real numbers are not the only system to satisfy these properties.

§2. REAL NUMBERS: POSITIVITY

We have the positive numbers, represented geometrically on the straight line by those numbers unequal to 0 and lying to the right of 0. If a is a positive number, we write $a > 0$. We shall list the basic properties of positivity from which others will be proved.

POS 1. *If a, b are positive, so are the product ab and the sum $a + b$.*

POS 2. *If a is a real number, then either a is positive, or $a = 0$, or $-a$ is positive, and these possibilities are mutually exclusive.*

If a number is not positive and not 0, then we say that this number is **negative**. By **POS 2**, if a is negative, then $-a$ is positive.

We know already that the number 1 is positive, but this could be proved from our two properties, and the basic rules for addition and multiplication. It may interest you to see the proof, which runs as follows and is very simple. By **POS 2,** we know that either 1 or -1 is positive. If 1 is not positive, then -1 is positive. By **POS 1,** it must follow that $(-1)(-1)$ is positive. But this product is equal to 1. Consequently, it must be 1, which is positive and not -1.

Using property **POS 1,** we can now conclude that $1 + 1 = 2$ is positive, that $2 + 1 = 3$ is positive, and so forth. Thus our calling

$$1, 2, 3, \ldots$$

the positive integers is compatible with our two rules **POS 1** and **POS 2**.

Other basic properties of positivity are easily proved from the two basic ones, and are left as exercises (Exercises 1, 2, 3), namely:

If a is positive and b is negative, then ab is negative.

If a is negative and b is negative, then ab is positive.

If a is positive, then $1/a$ is positive.

If a is negative, then $1/a$ is negative.

One of the properties of real numbers which we assume without proof is that every positive real number has a square root. This means:

If $a > 0$, then there exists a number b such that $b^2 = a$.

Because of this, and Theorem 4, §5 of Chapter 1, we now see that a number whose square is 2 is *irrational*, but exists as a real number.

It is a reasonable question to ask right away how many numbers there are whose squares are equal to a given positive number. For instance, what are all the real numbers x such that $x^2 = 2$?

This is easily answered. *There are precisely two such numbers.* One is positive, and the other is negative. Let us prove this. Let $b^2 = 2$, and let x be any real number such that $x^2 = 2$ also. We have

$$x^2 - b^2 = 0.$$

However, the left-hand side factors, and we find

$$(x + b)(x - b) = 0.$$

Hence we must have

$$x + b = 0 \quad \text{or} \quad x - b = 0$$

so that

$$x = -b \quad \text{or} \quad x = b.$$

On the other hand, the square of $-b$ is equal to 2, because

$$(-b)^2 = (-b)(-b) = b^2 = 2.$$

Thus we have proved our assertion.

Of the two numbers whose square is 2, we conclude from **POS 2** that precisely one of them is positive. We now adopt a convention, in force throughout all of mathematics. **We agree to call the square root of 2 only the positive number b whose square is 2.** This positive number will be denoted by

$$\sqrt{2}.$$

Therefore the two numbers whose square is 2 are

$$\sqrt{2} \quad \text{and} \quad -\sqrt{2},$$

and we have

$$\sqrt{2} > 0.$$

Exactly the same arguments show that given any positive number a, there exist precisely two numbers whose square is a. If b is one of them, then $-b$ is the other. Just replace 2 by a in the preceding arguments. Again by convention, we let

$$\sqrt{a}$$

denote the *unique positive number whose square is a.* The other number whose square is a is therefore $-\sqrt{a}$. We shall express this also by saying that the solutions of the equation $x^2 = a$ are

$$x = \pm\sqrt{a}.$$

We read this as "x equals plus or minus square root of a".

Another way of putting this is:

If x, y are numbers such that $x^2 = y^2$, then $x = y$ or $x = -y$.

But we cannot conclude that $x = y$. Furthermore, for any number x, the number

$$\sqrt{x^2}$$

is ≥ 0. Thus

$$\sqrt{(-3)^2} = \sqrt{9} = 3.$$

There is a special notation for this. We call $\sqrt{x^2}$ the **absolute value** of x, and denote it by

$$|x| = \sqrt{x^2}.$$

Thus we have

$$|-3| = 3 \qquad \text{and also} \qquad |-5| = 5.$$

Of course, for any positive number a, we have

$$|a| = a.$$

Thus

$$|3| = 3 \qquad \text{and} \qquad |5| = 5.$$

We won't work too much with absolute values in this book, and we do not want to overemphasize them here. Occasionally, we need the notion, and we need to know that the absolute value of -3 is 3. In that spirit, we give an example showing how to solve an equation with an absolute value in it, just to drive the definition home, but not to belabor the point.

Example. Find all values of x such that $|x + 5| = 2$.

To do this, we note that $|x + 5| = 2$ if and only if $x + 5 = 2$ or $x + 5 = -2$. Thus we have two possibilities, namely

$$x = 2 - 5 = -3 \qquad \text{and} \qquad x = -5 - 2 = -7.$$

This solves our problem.

Observe that

$$\frac{1}{\sqrt{2}} = \frac{\sqrt{2}}{2}.$$

This is because

$$2 = \sqrt{2}\,\sqrt{2},$$

and so our assertion is true because of cross-multiplication. It is a tradition in elementary schools to transform a quotient like

$$\frac{1}{\sqrt{2}}$$

into another one in which the square root sign does not appear in the denominator. As far as we are concerned, doing this is not particularly useful in general. It may be useful in special cases, but neither more nor less than other manipulations with quotients, to be determined *ad hoc* as the need arises. Actually, in many cases it is useful to have the square root in the denominator. We shall give two examples of how to transform an expression involving square roots in the numerator or denominator. The manipulations of these examples will be based on the old rule

$$(x + y)(x - y) = x^2 - y^2.$$

Example. Consider a quotient

$$\frac{3}{2 + \sqrt{5}}.$$

We wish to express it as a quotient where the denominator is a rational number. We multiply both numerator and denominator by

$$2 - \sqrt{5}.$$

This yields

$$\frac{3}{(2 + \sqrt{5})} \frac{(2 - \sqrt{5})}{(2 - \sqrt{5})} = \frac{6 - 3\sqrt{5}}{2^2 - (\sqrt{5})^2} = \frac{6 - 3\sqrt{5}}{-1} = -6 + 3\sqrt{5}.$$

Example. This example has the same notation as an actual case which arises in more advanced courses of calculus. Let x and h be numbers such that x and $x + h$ are positive. We wish to write the quotient

$$\frac{\sqrt{x + h} - \sqrt{x}}{h}$$

in such a way that the square root signs occur only in the denominator. We multiply numerator and denominator by $(\sqrt{x+h} + \sqrt{x})$. We obtain:

$$\frac{(\sqrt{x+h} - \sqrt{x})}{h} \frac{(\sqrt{x+h} + \sqrt{x})}{(\sqrt{x+h} + \sqrt{x})} = \frac{(\sqrt{x+h})^2 - (\sqrt{x})^2}{h(\sqrt{x+h} + \sqrt{x})}$$

$$= \frac{x+h-x}{h(\sqrt{x+h} + \sqrt{x})}$$

$$= \frac{h}{h(\sqrt{x+h} + \sqrt{x})}$$

$$= \frac{1}{\sqrt{x+h} + \sqrt{x}}.$$

Thus we find finally:

$$\frac{\sqrt{x+h} - \sqrt{x}}{h} = \frac{1}{\sqrt{x+h} + \sqrt{x}}.$$

In the first example, the procedure we have followed is called **rationalizing the denominator**. In the second example, the procedure is called **rationalizing the numerator**. In a quotient involving square roots, rationalizing the numerator means that we transform this quotient into another one, equal to the first, but such that no square root sign appears in the numerator. Similarly, rationalizing the denominator means that we transform this quotient into another one, equal to the first, but such that no square root appears in the denominator. Both procedures are useful in practice.

Square roots will be used when we discuss the Pythagoras Theorem, and the distance between points in Chapter 8, §2. You could very well look up these sections right now to see these applications, especially the section on distance.

EXERCISES

1. Prove:
   a) If a is a real number, then a^2 is positive.
   b) If a is positive and b is negative, then ab is negative.
   c) If a is negative and b is negative, then ab is positive.

2. Prove: If a is positive, then a^{-1} is positive.

3. Prove: If a is negative, then a^{-1} is negative.

4. Prove: If a, b are positive numbers, then

$$\sqrt{\frac{a}{b}} = \frac{\sqrt{a}}{\sqrt{b}}.$$

5. Prove that

$$\frac{1}{1 - \sqrt{2}} = -(1 + \sqrt{2}).$$

6. Prove that the multiplicative inverse of $2 + \sqrt{3}$ can be expressed in the form $c + d\sqrt{3}$, where c, d are rational numbers.

7. Prove that the multiplicative inverse of $3 + \sqrt{5}$ can be expressed in the form $c + d\sqrt{5}$, where c, d are rational numbers.

8. Let a, b be rational numbers. Prove that the multiplicative inverse of $a + b\sqrt{2}$ can be expressed in the form $c + d\sqrt{2}$, where c, d are rational numbers.

9. Same question as in Exercise 8, but replace $\sqrt{2}$ by $\sqrt{3}$.

10. Let x, y, z, w be rational numbers. Show that a product

$$(x + y\sqrt{5})(z + w\sqrt{5})$$

can be expressed in the form $c + d\sqrt{5}$, where c, d are rational numbers.

11. Generalize Exercise 10, replacing $\sqrt{5}$ by $\sqrt{a}$ for any positive integer a.

12. Rationalize the numerator in the following expressions.

a) $\dfrac{\sqrt{2x + 3} + 1}{4}$
b) $\dfrac{\sqrt{1 + x} - 3}{2}$

c) $\dfrac{\sqrt{x - h} - \sqrt{x}}{h}$
d) $\dfrac{\sqrt{x - h} + \sqrt{x}}{h}$

e) $\dfrac{\sqrt{x + h} + \sqrt{x}}{h}$
f) $\dfrac{\sqrt{x + 2h} - \sqrt{x}}{h}$

13. Find all possible numbers x such that
a) $|x - 1| = 2$,
b) $|x| = 5$,
c) $|x - 3| = 4$,
d) $|x + 1| = 6$,
e) $|x + 4| = 3$,
f) $|x - 2| = 1$.

14. Find all possible numbers x such that
a) $|2x - 1| = 3$,
b) $|3x + 1| = 2$,
c) $|2x + 1| = 4$,
d) $|3x - 1| = 1$,
e) $|4x - 5| = 6$.

15. Rationalize the numerator in the following expressions.

a) $\dfrac{\sqrt{x} + \sqrt{y}}{\sqrt{x} - \sqrt{y}}$

b) $\dfrac{\sqrt{x+y} - \sqrt{y}}{\sqrt{x+y} + \sqrt{y}}$

c) $\dfrac{\sqrt{x+1} + \sqrt{x-1}}{\sqrt{x+1} - \sqrt{x-1}}$

d) $\dfrac{\sqrt{x-3} + \sqrt{x}}{\sqrt{x-3} - \sqrt{x}}$

e) $\dfrac{\sqrt{x+y} - 1}{3 + \sqrt{x+y}}$

f) $\dfrac{\sqrt{x+y} + x}{\sqrt{x+y}}$

16. Rationalize the denominator in each one of the cases of Exercise 15.

17. Prove that there is no real number x such that

$$\sqrt{x-1} = 3 + \sqrt{x}.$$

[*Hint:* Start by squaring both sides.]

18. If $\sqrt{x-1} = 3 - \sqrt{x}$, prove that $x = \frac{25}{9}$.

19. Determine in each of the following cases whether there exists a real number x satisfying the indicated relation, and if there is, determine this number.

a) $\sqrt{x-2} = 3 + 2\sqrt{x}$ b) $\sqrt{x-2} = 3 - 2\sqrt{x}$

c) $\sqrt{x+3} = 1 + \sqrt{x}$ d) $\sqrt{x+3} = 1 - \sqrt{x}$

e) $\sqrt{x-4} = 3 + \sqrt{x}$ f) $\sqrt{x-4} = 3 - \sqrt{x}$

20. If a, b are two numbers, prove that $|a - b| = |b - a|$.

§3. POWERS AND ROOTS

Let n be a positive integer and let a be a real number. As before, we let

$$a^n$$

be the product of a with itself n times. The rule

$$a^{m+n} = a^m a^n$$

holds as before, if m, n are positive integers.

Let a be a positive number and let n be a positive integer. As part of the properties of real numbers, we assume, but do not prove, that *there exists a unique positive real number r such that*

$$r^n = a.$$

This number r is called the **n-th root** of a, and is denoted by

$$a^{1/n} \quad \text{or} \quad \sqrt[n]{a}.$$

The n-th root generalizes the existence and uniqueness of the square root discussed in the preceding section.

Theorem 1. *Let a, b be positive real numbers. Then*

$$(ab)^{1/n} = a^{1/n}b^{1/n}.$$

Proof. Let $r = a^{1/n}$ and $s = b^{1/n}$. This means that $r^n = a$ and $s^n = b$. Therefore

$$(rs)^n = r^n s^n = ab.$$

This means that rs is the n-th root of ab, and proves our theorem.

The n-th root can be further generalized to fractional powers. Let a be a positive real number. We shall assume without proof the following property of numbers.

Fractional powers. Let a be a positive number. To each rational number x we can associate a positive number denoted by a^x, which is the n-th power of a when x is a positive integer n, the n-th root of a when $x = 1/n$, and satisfying the following conditions:

POW 1. *For all rational numbers x, y we have*

$$a^{x+y} = a^x a^y.$$

POW 2. *For all rational numbers x, y we have*

$$(a^x)^y = a^{xy}.$$

POW 3. *If a, b are positive, then*

$$(ab)^x = a^x b^x.$$

We shall now derive consequences from these conditions.

First, we compute a^0. Let $b = a^0$. We have:

$$a = a^1 = a^{0+1} = a^0 a^1 = a^0 a.$$

Thus $a = ba$. Multiply both sides by a^{-1}. We get

$$1 = aa^{-1} = baa^{-1} = b.$$

Thus we find the important formula

$$\boxed{a^0 = 1.}$$

Next we shall see what a negative power is like. Let x be a positive rational number. Then

$$1 = a^0 = a^{x+(-x)} = a^x a^{-x}.$$

Thus the product of a^x and a^{-x} is 1. This means that a^{-x} is the multiplicative inverse of a^x, and thus that

$$\boxed{a^{-x} = \frac{1}{a^x}.}$$

This also justifies our notation, writing a^{-1} for the multiplicative inverse of a.

Example. Let $x = 3$. Then

$$a^{-3} = (a^3)^{-1} = \frac{1}{a^3}.$$

Similarly,

$$2^{-4} = (2^4)^{-1} = \frac{1}{2^4}.$$

Thus very roughly speaking, taking negative powers corresponds to taking a quotient.

Next, let $x = m/n$ be a positive rational number, expressed as a quotient of positive integers m, n. Then by **POW 2,** we find:

$$\boxed{a^{m/n} = (a^m)^{1/n} = (a^{1/n})^m.}$$

The fractional power can be decomposed into an ordinary power with an integer and an n-th root.

Example. We have

$$8^{2/3} = (8^{1/3})^2$$
$$= 2^2 = 4.$$

Example. We have

$$(\sqrt{2})^{3/4} = (\sqrt{2}^{1/4})^3$$
$$= (2^{1/8})^3 = 2^{3/8}.$$

Example. We have

$$(\sqrt{2})^3 = \sqrt{2}\,\sqrt{2}\,\sqrt{2}$$
$$= 2\sqrt{2} = 2^{3/2}.$$

Example. We have

$$\left(\frac{25}{9}\right)^{3/2} = \frac{25^{3/2}}{9^{3/2}} = \frac{125}{27}.$$

We would also like to take powers with irrational exponents, i.e. we would like to define numbers like

$$2^{\sqrt{2}}.$$

This is much more difficult, but it can be done in such a way that the two conditions **POW 1** and **POW 2** are satisfied. We shall not need this, and therefore shall postpone a systematic development for a more advanced course, although we shall make some further comments on the situation in the chapter on functions. However, we are led to make a final comment concerning the real numbers, as distinguished from the rational numbers.

Note that the properties of addition, multiplication, and positivity hold for rational numbers. What distinguishes the real numbers from the rationals is the existence of more numbers, like square roots, n-th roots, general exponents, etc. To make this "etc." precise is a more complicated undertaking. We can ask: Is there a neat way (besides stating that the real numbers consist of all infinite decimals) of expressing a property of the reals which guarantees that any number which we want to exist intuitively can be proved to exist using just this property? The answer is yes, but belongs to a much more advanced course. Thus throughout this course and throughout elementary calculus, whenever we wish a real number to exist so that we can carry out a certain discussion, our policy is to assume its existence and to postpone the proof to more advanced courses.

EXERCISES

1. Express each one of the following in the form $2^k 3^m a^r b^s$, where k, m, r, s are integers.

    a) $\dfrac{1}{8} a^3 b^{-4} 2^5 a^{-2}$

    b) $3^{-4} 2^5 a^3 b^6 \cdot \dfrac{1}{2^3} \cdot \dfrac{1}{a^4} \cdot b^{-1} \cdot \dfrac{1}{9}$

    c) $\dfrac{3a^3 b^4}{2a^5 b^6}$

    d) $\dfrac{16 a^{-3} b^{-5}}{9 b^4 a^7 2^{-3}}$

2. What integer is $81^{1/4}$ equal to?

3. What integer is $(\sqrt{2})^6$ equal to?

4. Is $(\sqrt{2})^5$ an integer?

5. Is $(\sqrt{2})^{-5}$ a rational number? Is $(\sqrt{2})^5$ a rational number?

6. In each case, the expression is equal to an integer. Which one?
    a) $16^{1/4}$ 　　b) $8^{1/3}$ 　　c) $9^{3/2}$ 　　d) $1^{5/4}$
    e) $8^{4/3}$ 　　f) $64^{2/3}$ 　　g) $25^{3/2}$

7. Express each of the following expressions as a simple decimal.
    a) $(.09)^{1/2}$ 　　　　　　b) $(.027)^{1/3}$
    c) $(.125)^{2/3}$ 　　　　　　d) $(1.21)^{1/2}$

8. Express each of the following expressions as a quotient m/n, where m, n are integers > 0.

    a) $\left(\dfrac{8}{27}\right)^{2/3}$

    b) $\left(\dfrac{4}{9}\right)^{1/2}$

    c) $\left(\dfrac{25}{16}\right)^{3/2}$

    d) $\left(\dfrac{49}{4}\right)^{3/2}$

9. Solve each of the following equations for x.
    a) $(x - 2)^3 = 5$ 　　　　b) $(x + 3)^2 = 4$
    c) $(x - 5)^{-2} = 9$ 　　　d) $(x + 3)^3 = 27$
    e) $(2x - 1)^{-3} = 27$ 　　f) $(3x + 5)^{-4} = 64$

    [*Warning*: Be careful with possible minus signs when extracting roots.]

§4. INEQUALITIES

We recall that we write

$$a > 0$$

if a is positive. If a, b are two real numbers, we shall write

$$a > b \quad \text{instead of} \quad a - b > 0.$$

We shall write

$$a < 0 \quad \text{instead of} \quad -a > 0$$

and also

$$b < a \quad \text{instead of} \quad a > b.$$

Example. We have $3 > 2$ because $3 - 2 = 1 > 0$. We have

$$-1 > -2$$

because

$$-1 + 2 = 1 > 0.$$

In the geometric representation of numbers on the line, the relation $a > b$ means that a lies to the right of b. We see that -1 lies to the right of -2 in our example.

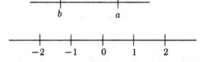

Fig. 3-2

We shall write

$$a \geqq b$$

to mean a is **greater than or equal to** b. Thus

$$3 \geqq 2 \quad \text{and} \quad 3 \geqq 3$$

are both true inequalities.

Using only our two properties **POS 1** and **POS 2**, we shall prove rules for dealing with inequalities. In what follows, we let a, b, c be real numbers.

IN 1. *If $a > b$ and $b > c$, then $a > c$.*

IN 2. *If $a > b$ and $c > 0$, then $ac > bc$.*

IN 3. *If $a > b$ and $c < 0$, then $ac < bc$.*

Rule **IN 2** expresses the fact that an inequality which is multiplied by a positive number is preserved. Rule **IN 3** tells us that if we multiply both sides of an inequality by a negative number, then the inequality gets *reversed*. For instance, we have the inequality

$$1 < 3.$$

But -2 is negative, and if we multiply both sides by -2 we get

$$-2 > -6.$$

This is represented geometrically by the fact that -2 lies to the right of -6 on the line.

Let us now prove the rules for inequalities.

To prove **IN 1**, suppose that $a > b$ and $b > c$. By definition, this means that

$$a - b > 0$$

and

$$b - c > 0.$$

Using property **POS 1**, we conclude that

$$a - b + b - c > 0.$$

Canceling b gives us

$$a - c > 0,$$

which means that $a > c$, as was to be shown.

To prove **IN 2**, suppose that $a > b$ and $c > 0$. By definition,

$$a - b > 0.$$

Hence using **POS 1** concerning the product of positive numbers, we conclude that

$$(a - b)c > 0.$$

The left-hand side of this inequality is equal to $ac - bc$ by distributivity. Therefore

$$ac - bc > 0,$$

which means that

$$ac > bc,$$

thus proving **IN 2**.

We shall leave the proof of **IN 3** as an exercise.

Other properties which can easily be proved from the three basic ones will be left as exercises (see Exercises 2 through 5). They will be used constantly without further reference. In particular, we use some of them in the next examples.

Example. We wish to show that the inequality

$$2x - 4 > 5$$

is equivalent to an inequality of type $x > a$ or $x < b$. Indeed, it is equivalent to

$$2x > 5 + 4 = 9,$$

which is equivalent to

$$x > \frac{9}{2}.$$

Example. Suppose that x is a number such that

(1) $$\frac{3x + 5}{x - 4} < 2.$$

We wish to find equivalent conditions under which this is true, expressed by simpler inequalities like $x > a$ or $x < b$. Note that the quotient on the left makes no sense if $x = 4$. Thus it is natural to consider the two cases separately, $x > 4$ and $x < 4$. Suppose that $x > 4$. Then $x - 4 > 0$ and hence, in this case, our inequality (1) is equivalent to

$$3x + 5 < 2(x - 4) = 2x - 8.$$

This in turn is equivalent to

$$3x - 2x < -8 - 5$$

or, in other words,

$$x < -13.$$

However, in our case $x > 4$, so that $x < -13$ is impossible. Hence there is no number $x > 4$ satisfying (1).

Now assume that $x < 4$. Then $x - 4 < 0$ and $x - 4$ is negative. We multiply both sides of our inequality (1) by $x - 4$ and reverse the inequality. Thus inequality (1) is equivalent in the present case to

(2) $$3x + 5 > 2(x - 4) = 2x - 8.$$

Furthermore, this inequality is equivalent to

(3) $3x - 2x > -8 - 5$

or, in other words,

(4) $x > -13.$

However, in our case, $x < 4$. Thus in this case, we find that the numbers x such that $x < 4$ and $x > -13$ are precisely those satisfying inequality (1). This achieves what we wanted to do. Note that the preceding two inequalities holding simultaneously can be written in the form

$$-13 < x < 4.$$

The set of numbers x satisfying such inequalities is called an **interval.** The numbers -13 and 4 are called the **endpoints** of the interval. We can represent the interval as in the following figure.

Fig. 3-3

Example. The set of numbers x such that $3 < x < 7$ is an interval, shown in the next figure.

Fig. 3-4

Example. The set of numbers x such that $3 \leq x \leq 7$ is also called an interval. In this case, we include the endpoints, 3 and 7, in the interval. The word "interval" applies to both cases, whether or not we admit the endpoints. We represent the interval with the endpoints in the next figure.

Fig. 3-5

In general, let a, b be numbers with $a \leq b$. Any one of the following sets of numbers is called an **interval.**

The set of numbers x such that $a < x < b$, called an **open interval.**
The set of numbers x such that $a \leq x \leq b$, called a **closed interval.**
The set of numbers x such that $a \leq x < b$.
The set of numbers x such that $a < x \leq b$.

The last two intervals are called **half open** or **half closed.**

Example. Again by convention, it is customary to say that the set of all numbers x such that $x > 7$ is an **infinite interval.** Similarly, the set of all numbers x such that $x < -3$ is an infinite interval. In general, if a is a number, the set of numbers x such that $x > a$ is an infinite interval, and so is the set of numbers x such that $x < a$. Again by convention, we may wish to include the endpoint. For instance, the set of numbers x such that $x \geq 7$ is also called an infinite interval. The set of numbers x such that $x \leq -3$ is called an infinite interval. We illustrate some of these intervals in the next figure.

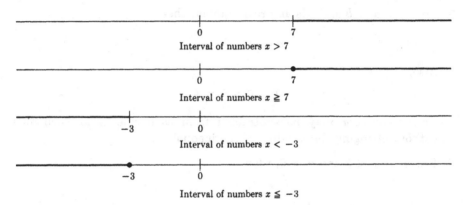

Interval of numbers $x > 7$

Interval of numbers $x \geq 7$

Interval of numbers $x < -3$

Interval of numbers $x \leq -3$

Fig. 3–6

EXERCISES

1. Prove **IN 3.**

2. Prove: If $0 < a < b$, if $c < d$, and $c > 0$, then

$$ac < bd.$$

3. Prove: If $a < b < 0$, if $c < d < 0$, then

$$ac > bd.$$

4. a) If $x < y$ and $x > 0$, prove that $1/y < 1/x$.
   b) Prove a rule of **cross-multiplication of inequalities:** If a, b, c, d are numbers and $b > 0$, $d > 0$, and if

$$\frac{a}{b} < \frac{c}{d},$$

   prove that

$$ad < bc.$$

   Also prove the converse, that if $ad < bc$, then $a/b < c/d$.

5. Prove: If $a < b$ and c is any real number, then

$$a + c < b + c.$$

   Also,

$$a - c < b - c.$$

   Thus a number may be subtracted from each side of an inequality without changing the validity of the inequality.

6. Prove: If $a < b$ and $a > 0$, then

$$a^2 < b^2.$$

   More generally, prove successively that

$$a^3 < b^3,$$
$$a^4 < b^4,$$
$$a^5 < b^5.$$

   Proceeding stepwise, we conclude that

$$a^n < b^n$$

   for every positive integer n. To make this stepwise argument formal, one must state explicitly a property of integers which is called induction, and is discussed later in the book.

7. Prove: If $0 < a < b$, then $a^{1/n} < b^{1/n}$. [*Hint:* Use Exercise 6.]

8. Let a, b, c, d be numbers and assume $b > 0$ and $d > 0$. Assume that

$$\frac{a}{b} < \frac{c}{d}.$$

a) Prove that

$$\frac{a}{b} < \frac{a+c}{b+d} < \frac{c}{d}.$$

(There are two inequalities to be proved here, the one on the left and the one on the right.)

b) Let r be a number > 0. Prove that

$$\frac{a}{b} < \frac{a+rc}{b+rd} < \frac{c}{d}.$$

c) If $0 < r < s$, prove that

$$\frac{a+rc}{b+rd} < \frac{a+sc}{b+sd}.$$

9. If $3x - 1 > 0$, prove that $x > \frac{1}{3}$.

10. If $4x + 5 < 0$, prove that $x < -\frac{5}{4}$.

In each of the following cases, find the intervals of numbers x satisfying the stated inequalities.

11. $5x + 2 > -3$

12. $-2x + 1 < 4$

13. $3x + 2 < 1$

14. $-3x - 2 > 5$

15. $3x - 1 < 4x + 5$

16. $2x + 7 > -x + 3$

17. $-3x - 1 > 5x$

18. $2x + 1 < -3x - 2$

19. $\dfrac{3x - 1}{x - 2} < 1$

20. $\dfrac{-2x + 5}{x + 3} < 1$

21. $\dfrac{2 - x}{2x + 1} > 2$

22. $\dfrac{3 - x}{x - 5} > 4$

23. $\dfrac{3 - 4x}{3x - 1} > 2$

24. $\dfrac{3x + 1}{2x - 6} < 3$

25. $x^2 < 1$

26. $x^2 < 2$

27. $x^2 < 3$

28. $x^2 < 4$

29. $x^2 > 1$

30. $x^2 > 2$

31. $x^2 > 3$

32. $x^2 > 4$

4 Quadratic Equations

We know how to solve an equation like

$$3x - 2 = 0.$$

In such an equation, x appears only in the first power. We shall now consider the next most difficult case, when x appears to the second power. We first deal with examples.

Example 1. Consider the equation

(1) $$x^2 - 3x + 1 = 0.$$

We wish to solve for x, that is, determine all values for x which satisfy this equation. We shall ultimately derive a general formula for this. Before deriving the formula, we carry out on this special example the method used to derive the general formula.

Solving our equation amounts to solving

(2) $$x^2 - 3x = -1.$$

We wish to add a number to both sides of this equation so that the left-hand side becomes a square, of the form $(x - s)^2$. We know that

$$(x - s)^2 = x^2 - 2sx + s^2.$$

Thus we need $2s = 3$, or $s = \frac{3}{2}$. Consequently, adding $(\frac{3}{2})^2$ to each side of equation (2), we find

$$x^2 - 3x + \frac{9}{4} = -1 + \frac{9}{4} = \frac{5}{4}.$$

The left-hand side has been adjusted so that it is a square, namely

$$x^2 - 3x + \frac{9}{4} = \left(x - \frac{3}{2}\right)^2,$$

and hence solving this equation amounts to solving

$$\left(x - \frac{3}{2}\right)^2 = \frac{5}{4}.$$

We can now take the square root, and we find that x is a solution if and only if

$$x - \frac{3}{2} = \pm\sqrt{\frac{5}{4}}.$$

Therefore finally we find two possible values for x, namely

$$x = \frac{3}{2} \pm \sqrt{\frac{5}{4}}.$$

This is an abbreviation for the two values

$$x = \frac{3}{2} + \sqrt{\frac{5}{4}} \quad \text{and} \quad x = \frac{3}{2} - \sqrt{\frac{5}{4}}.$$

Example 2. We wish to solve the equation

(3) $$x^2 + 2x + 2 = 0.$$

We apply the same method as before. We must solve

$$x^2 + 2x = -2.$$

We add 1 to both sides, so that we are able to express the left-hand side in the form

$$x^2 + 2x + 1 = (x + 1)^2.$$

Solving equation (3) is equivalent to solving

$$(x + 1)^2 = -2 + 1 = -1.$$

But a negative real number cannot possibly be a square of a real number, and we conclude that our equation does not have a solution in real numbers.

Example 3. We wish to solve the equation

(4) $$2x^2 - 3x - 5 = 0.$$

This amounts to solving

$$2x^2 - 3x = 5.$$

This time, we see that x^2 is multiplied by 2. To reduce our problem to one similar to those already considered, we divide the whole equation by 2, and solving (4) is equivalent to solving

(5)
$$x^2 - \frac{3}{2} x = \frac{5}{2}.$$

We can now complete the square on the left as we did before. We need to find a number s such that

$$x^2 - \frac{3}{2} x = x^2 - 2sx.$$

This means that $s = \frac{3}{4}$. Adding s^2 to both sides of (5), we find

$$x^2 - \frac{3}{2} x + \frac{9}{16} = \frac{5}{2} + \frac{9}{16} = \frac{49}{16}.$$

Expressing the left-hand side as a square, this is equivalent to

$$\left(x - \frac{3}{4} \right)^2 = \frac{49}{16}.$$

We can now solve for x, getting

$$x - \frac{3}{4} = \pm \sqrt{\frac{49}{16}}$$

or equivalently,

$$x = \frac{3}{4} \pm \sqrt{\frac{49}{16}},$$

which is our answer. Although this answer is correct, it is sometimes convenient to watch for possible simplifications. In the present case, we note that

$$\sqrt{\frac{49}{16}} = \frac{7}{4},$$

and hence

$$x = \frac{3}{4} \pm \frac{7}{4}.$$

Therefore

$$x = \frac{10}{4} \quad \text{and} \quad x = \frac{-4}{4} = -1$$

are the two possible solutions of our equation.

We are now ready to deal with the general case.

Theorem. *Let a, b, c be real numbers and $a \neq 0$. The solutions of the quadratic equation*

(6)
$$ax^2 + bx + c = 0$$

are given by the formula

$$x = \frac{-b \pm \sqrt{b^2 - 4ac}}{2a}$$

provided that $b^2 - 4ac$ is positive, or 0. If $b^2 - 4ac$ is negative, then the equation has no solution in the real numbers.

Proof. Solving our equation amounts to solving

$$ax^2 + bx = -c.$$

Dividing by a, we see that this is equivalent to solving

(7) $$x^2 + \frac{b}{a}x = -\frac{c}{a}.$$

To complete the square on the left, we need

$$x^2 + \frac{b}{a}x = x^2 + 2sx,$$

and hence $s = b/2a$. We therefore add $s^2 = b^2/4a^2$ to both sides of (7), and find the equivalent equation

$$\left(x + \frac{b}{2a}\right)^2 = -\frac{c}{a} + \frac{b^2}{4a^2}$$

$$= \frac{b^2 - 4ac}{4a^2}.$$

If $b^2 - 4ac$ is negative, then the right-hand side

$$\frac{b^2 - 4ac}{4a^2}$$

is negative, and hence cannot be the square of a real number. Thus our equation has no real solution. If $b^2 - 4ac$ is positive, or 0, then we can take the square root, and we find

$$x + \frac{b}{2a} = \pm \frac{\sqrt{b^2 - 4ac}}{2a}.$$

Solving for x now yields

$$x = -\frac{b}{2a} \pm \frac{\sqrt{b^2 - 4ac}}{2a},$$

which can be rewritten as

$$x = \frac{-b \pm \sqrt{b^2 - 4ac}}{2a}.$$

This proves our theorem.

Remark. If $b^2 - 4ac = 0$, then we get precisely one solution for the quadratic equation, namely

$$x = \frac{-b}{2a}.$$

If $b^2 - 4ac > 0$, then we get precisely two solutions, namely

$$x = \frac{-b + \sqrt{b^2 - 4ac}}{2a}$$

and

$$x = \frac{-b - \sqrt{b^2 - 4ac}}{2a}.$$

The quadratic formula is so important that it should be memorized. Read it out loud like a poem, to get an aural memory of it:

"x equals minus b plus or minus square root of b squared minus four ac over two a."

Example 4. Solve the equation

$$3x^2 - 2x + 1 = 0.$$

We use the formula this time, and get

$$x = \frac{-(-2) \pm \sqrt{(-2)^2 - 4 \cdot 3}}{2 \cdot 3}$$

$$= \frac{4 \pm \sqrt{-8}}{6}.$$

In this case, we see that the expression $b^2 - 4ac$ under the square root sign is negative, and thus our equation has no solution in the real numbers.

Example 5. Solve the equation

$$2x^2 + 3x - 4 = 0.$$

Again, use the formula, to get

$$x = \frac{-3 \pm \sqrt{9 - 4 \cdot 2 \cdot (-4)}}{2 \cdot 2}$$

$$= \frac{-3 \pm \sqrt{9 + 32}}{4}$$

$$= \frac{-3 \pm \sqrt{41}}{4}.$$

This is our answer, and we get the usual two values for x, namely

$$x = \frac{-3 + \sqrt{41}}{4} \quad \text{and} \quad x = \frac{-3 - \sqrt{41}}{4}.$$

Remark. In the proof of our theorem concerning the solutions of the quadratic equation, we needed to operate with addition, multiplication, and square roots. If we knew that the real numbers could be extended to a larger system of numbers in which these operations were valid, *including the possibility of taking square roots of negative real numbers*, then our formula would be valid in this bigger system of numbers, and would again give the solutions of the equation in all cases. We shall see in the chapter on complex numbers how to get such a system.

EXERCISES

Solve the following equations. If there is no solution in the real numbers, say so, and give your reasons why. In each case, however, give the values for x which would solve the equation in a larger system of numbers where negative numbers have a "square root". Use the formula.

1. $x^2 + 3x - 2 = 0$

2. $x^2 - 3x - 2 = 0$

3. $x^2 - 4x + 5 = 0$

4. $x^2 - 4x - 5 = 0$

5. $3x^2 + 2x - 1 = 0$

6. $3x^2 - 4x + 1 = 0$

7. $3x^2 + 3x - 4 = 0$

8. $-2x^2 - 5x = 7$

9. $-2x^2 - 5x = -7$

10. $4x^2 + 5x = 6$

11. $x^2 - \sqrt{2}\,x + 1 = 0$ 12. $x^2 + \sqrt{2}\,x - 1 = 0$

13. $x^2 + 3x - \sqrt{2} = 0$ 14. $x^2 - 3x - \sqrt{5} = 0$

15. $x^2 - 3x + \sqrt{5} = 0$ 16. $x^2 - 2x - \sqrt{3} = 0$

You will solve more quadratic equations when you do the exercises in Chapter 12, finding the intersection of a straight line with a circle, parabola, ellipse, or hyperbola.

Interlude On Logic
and Mathematical Expressions

§1. ON READING BOOKS

This part of the book can really be read at any time. We put it in the middle because that's as good as any place to start reading a book. Very few books are meant to be read from beginning to end, and there are many ways of reading a book. One of them is to start in the middle, and go simultaneously backwards and forward, looking back for the definitions of any terms you don't understand, while going ahead to see applications and motivation, which are very hard to put coherently in a systematic development. For instance, although we must do algebra first, it is quite appealing to look simultaneously at the geometry, in which we use algebraic tools to systematize our geometric intuition.

In writing the book, the whole subject has to be organized in a totally ordered way, along lines and pages, which is not the way our brain works naturally. But it is unavoidable that some topics have to be placed before others, even though our brain would like to perceive them simultaneously. This simultaneity cannot be achieved in writing, which thus gives a distortion of the subject. It is clear, however, that I cannot substitute for you in perceiving various sections of this book together. You must do that yourself. The book can only help you, and must be organized so that any theorem or definition which you need can be easily found.

Another way of reading this book is to start at the beginning, and then skip what you find obvious or skip what you find boring, while going ahead to further sections which appeal to you more. If you meet some term you don't understand, or if you need some previous theorem to push through the logical development of that section, you can look back to the proper reference, which now becomes more appealing to you because you need it for something which you already find appealing.

Finally, you may want to skim through the book rapidly from beginning to end, looking just at the statements of theorems, or at the discussions between theorems, to get an overall impression of the whole subject. Then you can go back to cover the material more systematically.

Any of these ways is quite valid, and which one you follow depends on your taste. When you take a course, the material will usually be covered in the same order as the book, because that is the safest way to keep going logically. Don't let that prevent you from experimenting with other ways.

§2. LOGIC

We always try to keep clearly in mind what we assume and what we prove. By a "proof" we mean a sequence of statements each of which is either assumed, or follows from the preceding statements by a rule of deduction, which is itself assumed. These rules of deduction are essentially rules of common sense.

We use "If ..., then" sentences when one statement implies another. For instance, we use sentences like:

$$(1) \qquad\qquad \text{If } 2x = 5, \text{ then } x = \frac{5}{2}.$$

This is a true statement, patterned after the general sentence structure:

$$\text{If } A, \text{ then } B.$$

The **converse** of this statements is given by:

$$\text{If } B, \text{ then } A.$$

Thus the converse of our assertion (1) is:

$$(2) \qquad\qquad \text{If } x = \frac{5}{2}, \text{ then } 2x = 5.$$

We see that the converse is also true.

Whenever we meet such a situation, we can save ourselves space, and simply say:

$$(3) \qquad\qquad 2x = 5 \text{ if and only if } x = \frac{5}{2}.$$

Thus

$$\text{``}A \text{ only if } B\text{''} \qquad \text{means} \qquad \text{``If } A, \text{ then } B\text{''}.$$

However, using "only if" by itself rather than in the context of "if and only if" always sounds a little awkward. Because of the structure of the English language, one has a tendency to interpret "A only if B" to mean "if B, then A". Consequently, we shall never use the phrase "only if" by itself, only as part of the full phrase "A if and only if B".

Example. The assertion: "If $x = -3$, then $x^2 = 9$" is a true statement. Its converse:

$$\text{``If } x^2 = 9, \text{ then } x = -3\text{''}$$

is a false statement, because x may be equal to 3. Thus the statement:

$$\text{``}x^2 = 9 \text{ if and only if } x = -3\text{''}$$

is a false statement.

Example. The statement:

"If two lines are perpendicular, then they have a point in common"

is a true statement. Its converse:

"If two lines have a point in common, then they are perpendicular"

is a false statement.

Example. The statement:

"Two circles are congruent if and only if they have the same radius"

is a true statement.

We often give proofs by what is called the "method of **contradiction**". We want to prove that a certain statement A is true. To do this, we suppose that A is false, and then by logical reasoning starting from the supposition that A is false, we arrive at an absurdity, or at a contradiction of a true statement. We then conclude that our supposition "A is false" cannot hold, whence A must be true. An example of this occurred when we proved that

$\sqrt{2}$ is not a rational number. We did this by assuming that $\sqrt{2}$ is rational, then expressing it as a fraction in lowest form, and then showing that in fact, both numerator and denominator of this fraction must be even. This contradicted the hypothesis that $\sqrt{2}$ could be a fraction, in lowest form, whence we concluded that $\sqrt{2}$ is not a rational number.

Some assertions are true, some are false, and some are meaningless. Sometimes a set of symbols is meaningless because some letters, like x, or a, appear without being properly qualified. We give examples of this. When we write an equation like $2x = 5$, as in (1), the context is supposed to make it clear that x denotes a number. However, if there is any chance of doubt, this should always be specified. Thus a more adequate formulation of (1) would be:

(4)   If x is a real number and $2x = 5$, then $x = \dfrac{5}{2}$.

Similarly, a more adequate formulation of (2) would be:

(5)   Let x be a real number. Then $2x = 5$ if and only if $x = \dfrac{5}{2}$.

The symbols

$$2x = 5$$

by themselves are called an equation. As it stands, this equation simply indicates a possible relationship, but to give it meaning we must say something more about x. For instance:

    a) There exists a number x such that $2x = 5$.

    b) For all numbers x, we have $2x = 5$.

    c) There is no number x such that $2x = 5$.

    d) If x is a real number and $2x = 5$, then $x < 7$.

Of these statements, (a) is true, (b) is false, (c) is false, and (d) is true. We can also use the symbols "$2x = 5$" in a context like:

    e) Determine all numbers x such that $2x = 5$.

This sentence is actually a little ambiguous, because of the word "determine". In a sense, the equation itself, $2x = 5$, determines such numbers x. We have tried to avoid such ambiguities in this book. However, the context of a chapter can make the meaning of this sentence clear to us as follows:

    f) Express all rational numbers x such that $2x = 5$ in the form m/n, where m, n are integers, $n \neq 0$.

This is what we would understand when faced with sentence (e), or with a similar sentence like:

g) Solve for x in the equation $2x = 5$.

In writing mathematics, it is essential that complete sentences be used. Many mistakes occur because you allow incomplete symbols like

$$2x = 5$$

to occur, without the proper qualifications, as in sentences (a), (b), (c), (d), (e), (f), (g).

Example. The symbols "$x^2 = 2$" by themselves are merely an equation. The sentence:

"There exists a rational number x such that $x^2 = 2$."

is false. The sentence:

"There exists a real number x such that $x^2 = 2$."

is true.

Equality

We shall use the word "equality" between objects to mean that they are the same object. Thus when we write

$$2 + 3 = 6 - 1,$$

we mean that the number obtained by adding 2 and 3 is the same number as that obtained by subtracting 1 from 6. It is the number 5.

We use the word "equivalent" in several contexts. First, if A and B are assertions (which may be true or false), we say that they are **equivalent** to mean:

A is true if and only if B is true.

For instance, the following two assertions are equivalent in this sense:

The number x satisfies the equation $2x + 5 = 3$.

The number x is equal to -1.

We shall use the word "equivalent" in other contexts, but will explain these as the need arises.

We DO NOT USE THE WORD "EQUALITY" AS IT IS SOME-TIMES USED, for instance in elementary geometry. The following two triangles are not equal:

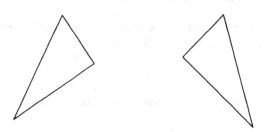

Fig. I-1

They are, however, congruent, and under a suitable definition of equivalence for triangles, we might even say that they are equivalent. Note that the *areas* of these triangles are *equal*. In the same vein, the following line segments are not equal:

Fig. I-2

However, their lengths are equal.

The mathematics which we discuss in this book, like most mathematics, has many applications and counterparts in the physical world. For instance, numbers can be used to measure length, area, speed, density, etc. For clarity, we try to use language in such a way that the mathematical notions are not usually identified with their physical counterparts. Thus we use words like *"correspond"*, or *"represent"*, when we wish to associate a physical quantity with a mathematical one. In line with this, we can deal with mathematical objects on two levels: the purely logical level of axioms, deductions, and proofs; and the mixed physical level. Often, it is quite tedious and not necessarily illuminating to insist that we follow only the strictly logical procedures. It is useful and perhaps more pleasant to follow our physical intuition for certain arguments. We shall see examples of both types of arguments when we discuss geometry in its intuitive setting and its analytical setting.

§3. SETS AND ELEMENTS

Following mathematical terminology, a collection of objects is called a **set**. The objects in this set are called the **elements** of the set.

The set of all real numbers is denoted by **R**. To say:

"x is an element of **R**"

means the same thing as to say

"x is a real number".

Let S and T be sets. We say that S is a **subset** of T if every element of S is also an element of T. For instance:

The set of rational numbers is a subset of the set of real numbers.

The set of integers is a subset of the set of rational numbers.

The set of integers is a subset of the set of rational numbers. It is also a subset of **R** (i.e. a subset of the real numbers).

The set of boys is a subset of the set of all children.

The set of all real numbers x such that $2x + 3 < 5$ is a subset of the real numbers.

As a matter of convention, we allow a subset of a set S to be all of S. Thus any set is a subset of itself. The sentence:

"For any set S, S is a subset of S"

is a true sentence.

A set is often described by stating the conditions under which something is an element of the set. Sometimes we state such conditions so that there are no elements in the set.

Example. There is no element in the set of all numbers x which satisfy the conditions

$$x < 0 \quad \text{and} \quad x > 0.$$

There is no element in the set of all positive numbers x which satifsy the conditions

$$\frac{2x}{x+1} > 1 \quad \text{and} \quad x < \frac{1}{2}.$$

Whenever this happens, that a set has no elements, we say that the set is **empty.** Thus the set of numbers x such that $2x > 1$ and $x < -3$ is empty.

Let S, S' be sets. Often, to prove that $S = S'$, we prove that S is a subset of S' and that S' is a subset of S.

Example. Let S be the set of numbers x such that $1 \leq x \leq 2$. Let T be the set of all numbers $5x$ with all x in S. We contend that T is the set of numbers y with $5 \leq y \leq 10$. First note that if x is in S, then $5x$ satisfies the inequalities

$$5 \leq 5x \leq 10.$$

Hence if T' is the set of all numbers y satisfying $5 \leq y \leq 10$, we see that T is contained in T'. Conversely, let y be a point of T', i.e. assume that

$$5 \leq y \leq 10.$$

Let $x = y/5$. Then x is in S and $y = 5x$. Hence T' is contained in T. This proves that $T = T'$.

§4. INDICES

In a sentence like

"Let x, y be numbers"

it is a convention of mathematical language to allow the possibility that $x = y$. Similarly, if we say

"Let P, Q be points in the plane"

we do not exclude the possibility that $P = Q$. If we wish to exclude this possibility then we say so explicitly. For instance, we would say:

"Let x, y be distinct numbers"

or

"Let x, y be numbers, $x \neq y$"

or

"Let P, Q be points, such that $P \neq Q$".

Similarly, we may wish to speak of several numbers instead of two numbers like x, y. Thus we might say

"Let x, y, z be numbers",

without excluding the possibility that some of these numbers may be equal to each other. It is clear that we would soon run out of letters of the alphabet in enumerating numbers just with letters, and hence we use a notation with subscripts, as exemplified in the following sentences.

"Let x_1, x_2 be numbers."

"Let x_1, x_2, x_3 be numbers."

"Let x_1, x_2, x_3, x_4 be numbers."

Finally, in the most general case we have the corresponding sentence:

"Let $x_1, \ldots, x_n$ be numbers."

We repeat that in such a sentence, it is possible that $x_i = x_j$ for some pair of subscripts i, j such that $i \neq j$. Such subscripts are also called **indices**.

Objects indexed by integers from 1 to n (or sometimes from 0 to n) are called a **sequence** of objects or, more precisely, a finite sequence. Thus in a finite sequence of numbers, denoted by

$$\{x_1, \ldots, x_n\}$$

we associate a number x_j to each integer j satisfying $1 \leq j \leq n$. Thus, considering a sequence as above amounts to considering a first number x_1, a second number x_2, and so forth, up to an n-th number x_n.

Example. For each integer j we let $x_j = (-1)^j$. Then

$$x_1 = -1, \quad x_2 = 1, \quad x_3 = -1, \quad x_4 = 1, \quad \ldots, \quad x_n = (-1)^n.$$

Observe how in this sequence the numbers x_j take on the values 1 or -1.

Example. We shall study polynomials later, and we shall write a polynomial in the form

$$a_n x^n + a_{n-1} x^{n-1} + \cdots + a_0.$$

The sequence of coefficients is the sequence

$$\{a_0, a_1, \ldots, a_n\}.$$

For instance, the sequence of coefficients of the polynomial

$$4x^3 - 2x^2 + 4x - 5$$

is the sequence $\{-5, 4, -2, 4\}$. We have

$$a_0 = -5, \quad a_1 = 4, \quad a_2 = -2, \quad a_3 = 4.$$

§5. NOTATION

The notation used in giving an account of a mathematical theory is important. It is very useful that the printed page should look appealing visually as well as mathematically.

It is also important that notation be fairly consistent, namely that certain symbols be used only to denote certain objects. For instance, we use lower case letters like a, b, c, x, y, z to denote numbers. We use capital letters like A, B, P, Q, X, Y to denote mostly points, although we also use A, B, to denote angles. When we do that, we reserve P, Q, for points. Within any given section, we try not to mix the two. Although one should of course always specify what a letter stands for, it is convenient if the use of letters follows a pattern, so that one knows at one glance what certain letters represent.

Notation can be slick. You will see in the chapters on coordinates that the notation of points and vectors is fairly slick. Sometimes, notation can be too slick. I hope that this is not the case in those chapters.

We usually reserve letters like f, g, F, G for functions or mappings. We cannot observe complete uniformity in this respect; otherwise, we would run out of letters very soon. For instance, we use (a), (b), (c), ... to denote a sequence of exercises, not numbers.

We use m, n to denote integers, except in cases where they are used as abbreviations for words. For instance, m is sometimes used as an abbreviation to denote the measure of an angle A, which we write $m(A)$.

In any book, it is impossible to avoid some mistakes, some confusion, some incorrectness of language, and some misuse of notation. If you find any such things in the present book, then correct them or improve them for yourself, or write your own book. This is still the best way to learn a subject, aside from teaching it.

Part Two
INTUITIVE GEOMETRY

This part is concerned with the geometry of the plane. We assume some basic properties, which we always try to state explicitly, including properties of straight lines, segments, angles, distance, etc. We then prove other facts from these.

There are two basic aspects of geometry: the Pythagoras theorem, and the notion of congruence. Classical treatments obscure these because of undue emphasis on "constructions", and ever more complicated diagrams involving triangles, parallelograms, etc. I am trying to combat this by returning to an exposition based directly on these two aspects. You will see how easy some of the usual properties of triangles are to prove, if one starts with the Pythagoras theorem. On the other hand, the briefest glance at other expositions will convince you how unnatural and complicated plane geometry can be otherwise. For instance, the theorem that the shortest distance between a point P and points on a line is given by the perpendicular segment becomes obvious from Pythagoras, so obvious that we leave it as an exercise.

Taking the Pythagoras theorem and the general properties of isometries as our starting point has other advantages: it is precisely this approach which fits more advanced mathematics best. It provides an exceedingly nice introduction to mappings. It provides the intuitive basis for the study of perpendicularity (and ultimately of orthogonality in vector spaces in subsequent courses in linear algebra).

In the next part, we shall indicate systematically how we can give algebraic definitions for most of the concepts handled in intuitive geometry, and how we can prove the results of geometry using only properties of numbers, but in a way which most often parallels exactly the geometric arguments. This program involves giving algebraic definitions for points, lines, angles, triangles, reflections, dilations, translations, congruence, etc.

A proof which is based only on properties of real numbers is called an **analytic** proof. A proof which is based on properties of the plane involving our intuition as in Part II is called a **geometric** proof.

For the logical development of the subject, it is not absolutely necessary to read all of this part. The reader could skip it, and refer to it as needed later. But I don't think it is desirable to inhibit your geometric intuition (as a recent trend among some educational schools would do). Don't be afraid of arguing geometrically.

Giving analytic foundations for geometry, however, does not serve solely, or even principally, the purpose of making such foundations more "rigorous". It provides computational means which are not available in the intuitive context. For instance, given two lines which intersect, how do you compute their point of intersection? There is little problem if you have an analytic definition for a line. If you don't, the question doesn't even make that much sense. For another example, try to see geometrically why reflection through an arbitrary point can be obtained as a composite of reflection through the origin and a translation (to understand this, read the appropriate sections). It isn't nearly as clear as when you have the analytic definitions for translations and reflections given in Chapters 8 and 9. Thus analytic foundations for the subject should not be viewed as an invention to make the subject arid. Feel free to mix the two approaches—geometric and analytic—to get the best feeling for the subject. One reason why the Greeks did not get further in their mathematics is that they suffered from the inhibition of using numbers to deal with geometric objects. Essentially, mathematics had to wait till Descartes to overcome this inhibition. It is equally pointless to fall now into the opposite inhibition.

5 Distance and Angles

§1. DISTANCE

The notion of distance is perhaps the most basic one concerning the plane.

We shall assume basic properties of distance without proof. We denote the distance between points P, Q in the plane by $d(P, Q)$. It is a number, which satisfies the following properties.

DIST 1. *For any points P, Q, we have $d(P, Q) \geq 0$. Furthermore,*

$$d(P, Q) = 0 \quad \text{if and only if} \quad P = Q.$$

DIST 2. *For any points P, Q we have*

$$d(P, Q) = d(Q, P).$$

DIST 3. *Let P, Q, M be points. Then*

$$d(P, M) \leq d(P, Q) + d(Q, M).$$

This third property is called the **triangle inequality**. The reason is that it expresses the geometric fact that the length of one side of a triangle is at most equal to the sum of the lengths of the other two sides, as illustrated in Fig. 5–1.

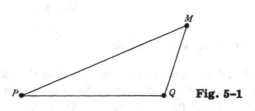

Fig. 5–1

107

We assume the basic fact that two distinct points P, Q lie on one and only one line, denoted by L_{PQ}. The portion of this line lying between P and Q is called the line **segment** between P and Q, and is denoted by $\overline{PQ}$. If units of measurement are selected, then the length of this segment is equal to the distance $d(P, Q)$. The straight line and the segment determined by P and Q are illustrated in Fig. 5–2.

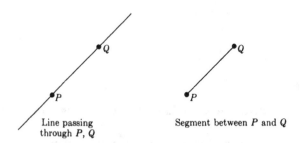

Line passing
through P, Q Segment between P and Q **Fig. 5–2**

We shall assume two facts relating line segments with the notion of distance. The first one is:

SEG 1. *Let P, Q, M be points. We have*

$$d(P, M) = d(P, Q) + d(Q, M)$$

if and only if Q lies on the segment between P and M.

This property **SEG 1** certainly fits our intuition of line segments, and is illustrated in Fig. 5–3, where Q lies on the segment $\overline{PM}$.

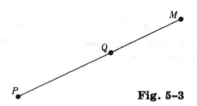

Fig. 5–3

The second fact which we assume is:

SEG 2. *Let P, M be points in the plane, and let $d = d(P, M)$. If c is a number such that $0 \leq c \leq d$, then there exists a unique point Q on the segment $\overline{PM}$ such that $d(P, Q) = c$.*

Again, this property fits our intuition perfectly.

Let r be a positive number, and let P be a point in the plane. We define the **circle** of center P and radius r to be the set of all points Q whose distance from P is r. We define the **disc** of center P and radius r to be the set of all points Q whose distance from P is $\leq r$. The circle and the disc are drawn on Fig. 5–4.

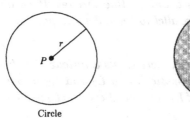

Circle   Disc   **Fig. 5–4**

Note. In many books you will find that the word "circle" is used to denote both what we call the circle and also the disc. This is not good terminology and leads to confusion, because it is always best if a single word is not used to denote two different objects or concepts. Geometrically speaking, we see that the circle is the boundary of the disc.

Remark. In our preceding discussion, we have made the usual tacit convention that a unit of length has been fixed. For instance, just to speak of the distance between two points as a number, we must have agreed already on a unit of distance. Interpreting the distance between P and Q as the length of the segment between P and Q again presupposes that such a unit of measurement has been fixed.

A similar remark applies to later discussions, for instance concerning area. When a unit of distance is selected, then it determines a unit of area. For instance, if our unit of distance is the inch, then the unit area is the square inch. We then commit a simplification of language by referring to distance or area as numbers. We say that the area of a square of side a is a^2. Having fixed the unit of measurement as the inch, this means that the length of each side is a in., and that the area is a^2 in^2. For simplicity of language, we agree once for all that a unit of measurement is fixed throughout our discussion, and then omit the units when speaking of length (distance) or area. Sometimes we also speak of the "numerical value" of the length, or area, with respect to such a choice of units, to emphasize that we are dealing with a number. For instance, the numerical value of the area of a square whose area is 9 in^2 is the number 9.

§2. ANGLES

We base our discussion of geometry on our concept of the plane. We are willing to assume some standard geometric facts, but for the convenience of the reader, we have reproved many facts from more basic ones. We assume the following facts about straight lines.

Two distinct points P, Q lie on one and only one line, denoted by L_{PQ}. Two lines which are not parallel meet in exactly one point. Given a line L and a point P, there exists a unique line through P parallel to L. If L_1, L_2, L_3 are lines, if L_1 is parallel to L_2 and L_2 is parallel to L_3, then L_1 is parallel to L_3.

Given a line L and a point P, there exists a unique line through P perpendicular to L. If L_1 is perpendicular to L_2 and L_2 is parallel to L_3, then L_1 is perpendicular to L_3. If L_1 is perpendicular to L_2 and L_2 is perpendicular to L_3, then L_1 is parallel to L_3.

Two points P and Q also determine two **rays**, one starting from P and the other starting from Q, as shown in Fig. 5–5. Each of these rays stops at P, but extends infinitely in one direction.

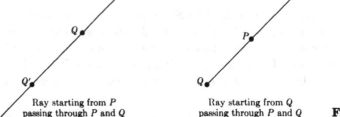

Ray starting from P Ray starting from Q
passing through P and Q passing through P and Q **Fig. 5–5**

A ray starting from P is simply a half line, consisting of all points on a line through P lying to one side of P. The ray starting from P and passing through another point Q will be denoted by R_{PQ}. If Q' is another point on this ray, distinct from P, then of course we have

$$R_{PQ} = R_{PQ'}.$$

In other words, a ray is determined by its starting point and by any other point on it.

If a ray starts at a point P, we also call P the **vertex** of the ray.

Consider two rays R_{PQ} and R_{PM} starting from the same point P. These rays separate the plane into two regions, as shown in Fig. 5–6.

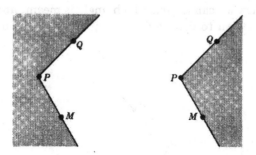

Each one of these regions will be called an **angle** determined by the rays. Thus the rays R_{PQ} and R_{PM} determine two angles.

Remark on terminology. There is some divergence in the way an angle is defined in other books. For instance, an angle is sometimes defined as the union of two rays having a common vertex, rather than the way we have defined it. I have chosen a different convention for several reasons. First, people do tend to think of one or the other side of the rays when they meet two rays like this:

Fig. 5–7

They do not think neutrally. Second, and more importantly, when we want to measure angles later, and assign a number to an angle, as when we shall say that an angle has 30 degrees, or 270 degrees, adopting the definition of an angle as the union of two rays would provide insufficient information for such purposes, and we would need to give additional information to determine the associated measure. Thus it is just as well to incorporate this information in our definition of an angle. Finally, the manner in which the measure of an angle will be found will rely on area, and is therefore natural, starting with our definition.

If we just draw these rays as shown in Fig. 5–8, like this, without any other indication, then we cannot tell which angle is meant, and thus we need some additional notation to distinguish one angle from the other, which we now describe.

Fig. 5–8

Recall that given a point P and a positive number r, the circle of radius r and center P is the collection of all points whose distance from P is equal to r.

Let R_{PQ} and R_{PM} be rays with vertex P. If C is a circle centered at P (of positive radius), then our two rays separate the circle into two arcs, as shown in Fig. 5–9.

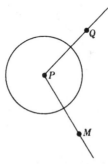

Fig. 5–9

Each arc lies within one of the angles, and thus to characterize each angle it suffices to draw the corresponding arc. The two parts of Fig. 5–10 thus show the usual way in which we draw the two angles formed by the rays.

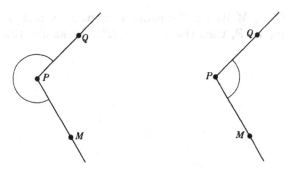

Fig. 5-10

Just knowing the two rays is not enough information to be able to distinguish one angle from the other. However, if the rays are given in an ordered fashion, selecting one of them as the first and the other as the second, then we do have enough information to determine one specific angle. This is done as follows.

Let R_{PQ} be the first ray and R_{PM} the second one. Then one of the angles determined by R_{PQ} and R_{PM} contains the arc going from the first ray to the second in the **counterclockwise** direction. We denote that angle by $\angle QPM$. The other angle contains the arc from R_{PM} to R_{PQ} in the **counterclockwise** direction, and we therefore denote this other angle by $\angle MPQ$. Thus the order in which Q, M occur is very important in this notation. We represent the angles $\angle QPM$ and $\angle MPQ$ by putting a little arrow on the arc, to indicate the counterclockwise direction, as in Fig. 5-11.

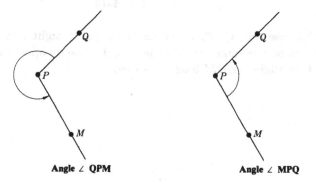

Angle $\angle QPM$ Angle $\angle MPQ$

Fig. 5-11

Example. If Q, P, M lie on the same straight line, and Q, M lie on the same ray starting at P, then the angle $\angle QPM$ looks like this:

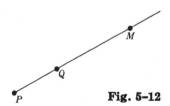

Fig. 5-12

In this case, the arc of a circle between the two rays is just a point, and we say that the angle $\angle QPM$ is the **zero angle**.

Note that when we deal with this degenerate case in which the two rays coincide, one of the angles is the zero angle but the other angle is the whole plane, and is called the **full** angle. However, with our conventions, we do not write this full angle with the notation $\angle QPM$. We do, however, represent it by an arrow going all the way around as follows:

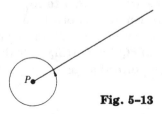

Fig. 5-13

Example. Suppose that Q, P, M lie on the same straight line but that Q and M do not lie on the same ray, that is, Q and M lie on opposite sides of P on the line. Our angle $\angle QPM$ looks like this:

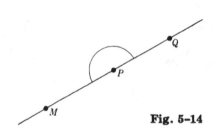

Fig. 5-14

In this case, we say that the angle $\angle QPM$ is a **straight angle**. Observe that we draw the angle $\angle MPQ$ with a different arc, namely:

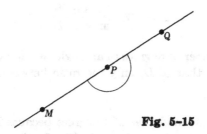

Fig. 5-15

Thus in this case, $\angle MPQ$ is different from $\angle QPM$, and both are straight angles because the three points P, Q, M lie on the same straight line.

Given an angle A with vertex P, let D be a disc centered at P. That part of the angle which also lies in the disc is called the **sector** of the disc determined by the angle. Picture:

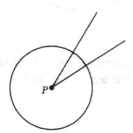

Fig. 5-16

The shaded part represents the sector S.

Just as we used numbers to measure distance, we can now use them to measure angles, provided that we select a unit of measurement first. This can be done in several ways. Here we discuss the most elementary way (but we shall return later to this question, and discuss another unit, which turns out to be more convenient in most mathematics).

The unit of measurement which we select here is the degree, such that the full angle has 360 degrees. Let A be an angle centered at P and let S be the sector determined by A in the disc D centered at P. Let x be a number between 0 and 360. We shall say that

A has x degrees

to mean that

$$\frac{\text{area of } S}{\text{area of } D} = \frac{x}{360}.$$

Thus

$$x = 360 \cdot \frac{\text{area of } S}{\text{area of } D}.$$

In computing the number of degrees in an angle, we do not have to determine the area of S or even that of D, only the ratio between the two. We shall now give examples.

Example. The straight angle has 180 degrees because it separates the disc into two sectors of equal area, as shown in Fig. 5–17.

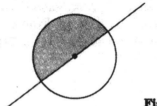

Fig. 5–17

Example. An angle whose measure is half that of the straight angle is called a **right angle**, and has 90 degrees, as in the Fig. 5–18.

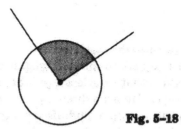

Fig. 5–18

Example. In Fig. 5–19, we have drawn the sectors determined by angles of 45 degrees and 30 degrees. The one with 30 degrees has one-third the measure of a right angle. In the picture of an angle of 45 degrees, we have drawn a dotted line to suggest the angle of 90 degrees. In the picture of an angle of 30 degrees, we have drawn two dotted lines to suggest the angles of 90 degrees

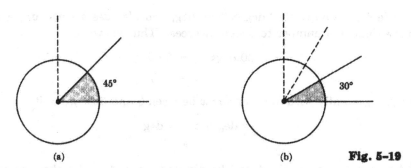

(a)   (b)   **Fig. 5–19**

and 60 degrees, respectively, showing how the angle of 90 degrees gets divided
into three parts having equal measures.

Example. In Fig. 5-19(c) we have drawn the sector lying between the angles
of 30° and 45°, and inside the circle of radius 2.

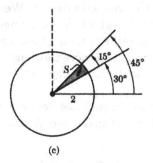

(c)   **Fig. 5–19 (cont.)**

We can compute the area of this sector using the definition of degrees. Let
us assume the fact that the area of the disc of radius r is πr^2, where π is approximately equal to 3.14159 . . . . (The decimal for π can be determined as
accurately as you wish, but we don't go into this here.) The area of the disc
of radius 2 is therefore equal to 4π. The sector S in Fig. 5-19(c) has 15°
(because 15 = 45 − 30), and hence

$$\text{area of } S = \frac{15}{360} \cdot 4\pi = \frac{\pi}{6} .$$

This is the numerical value of the area, in whatever units we are dealing with.
You can put this answer into decimal form, using tables for π, and a computer, but we prefer to leave it as $\pi/6$.

Similarly, the area of the sector lying between the angles of 30° and 45°,
and inside the circle of radius 5, is given by

$$\frac{15}{360} \cdot 25\pi = \frac{25}{24} \pi.$$

We shall abbreviate "degree" by deg, and also use a small upper circle to the right of a number to denote degrees. Thus we write

$$30 \text{ degrees} = 30 \text{ deg}$$
$$= 30°,$$

or more generally with any number x between 0 and 360, we write

$$x \text{ degrees} = x \text{ deg}$$
$$= x°.$$

The measure of an angle A will be denoted by $m(A)$. For the moment we shall deal with the measure in degrees. Thus to say that an angle A has 50° means the same thing as

$$m(A) = 50°.$$

Remark addressed to those who like to ask questions. In defining the number of degrees of an angle, we used a disc D. We did not specify the radius. It should be intuitively clear that when we change the disc, and hence the sector at the same time, the ratio of their areas remains the same. We shall assume this for now, and return to a more thorough discussion of this question later when we discuss area, and similar figures.

It is convenient to write inequalities between angles. Let A and B be angles. Suppose that A has x degrees and B has y degrees, where x, y are numbers satisfying

$$0 \leqq x \leqq 360$$

and

$$0 \leqq y \leqq 360.$$

We shall say that A is **smaller than or equal to** B if $x \leqq y$. For instance, an angle of 37° is smaller than an angle of 52°.

EXERCISES

1. Let R_{PQ} be a ray as drawn, horizontally.

$$\underset{P}{\bullet}\qquad\underset{Q}{\bullet}\qquad\qquad\qquad\qquad R_{PQ}$$

Draw a second ray R_{PM} such that the angle $\angle QPM$ has:

a) 60°   b) 120°   c) 135°   d) 160°

e) 210°   f) 225°   g) 240°   h) 270°

Let D be a disc centered at P and assume that its area is 60 in². In each one of the above cases find the area of the sector in the disc cut out by the two rays.

2. Assume that the area of a disc of radius 1 is equal to the number π (approximately equal to 3.14159...) and that the area of a disc of radius r is equal to πr^2.

   a) What is the area of a sector in the disc of radius r lying between angles of θ_1 and θ_2 degrees, as shown in Fig. 5–20(a)?

   b) What is the area of the band lying between two circles of radii r_1 and r_2 as shown in Fig. 5–20(b)?

   c) What is the area in the region bounded by angles of θ_1 and θ_2 degrees and lying between circles of radii r_1 and r_2 as shown in Fig. 5–20(c)?

   Give your answers in terms of π, θ_1, θ_2, r_1, r_2.

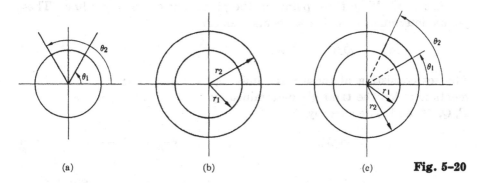

(a)   (b)   (c)   **Fig. 5–20**

Work out numerical examples of Exercise 2 as follows. In each case, express your answer in terms of rational multiples of π.

3. What is the area of a sector in the disc of radius 2 lying between angles of:

   a) 35° and 75°,   b) 15° and 60°,

   c) 80° and 110°,   d) 130° and 250°?

4. What is the area of the band lying between two circles having the same center, and having the following radii?

   a) 2 and 5   b) 3 and 4

   c) 2 and 6   d) 1 and 5

5. What is the area of the region bounded by angles of θ_1 and θ_2 degrees, and lying between circles of radii 3 and 5 when θ_1 and θ_2 have the values of Exercise 3(a), (b), (c), and (d)?

6. What is the area of the region lying between two circles having the same center, of radii 3 and 4, and bounded by angles of:

   a) 60° and 70°,   b) 110° and 270°,

   c) 65° and 120°,   d) 240° and 310°?

§3. THE PYTHAGORAS THEOREM

Let P, Q, M be three points in the plane, not on the same line. These points determine three line segments, namely

$$\overline{PQ}, \qquad \overline{QM}, \qquad \overline{PM}.$$

The set consisting of these three line segments is called the **triangle** determined by P, Q, M, and is denoted by

$$\triangle PQM.$$

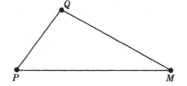

Fig. 5–21

Remark on terminology. We adopt here a convention which seems the most widespread. However, there is some pervasive ambiguity about the notion of a triangle, similar to the ambiguity which we have already mentioned about "circles". The word "triangle" is also used to denote the region bounded by the three line segments. There is no convenient word like "disc" which I could think of here to serve in a similar capacity. Nobody will accept "trisc". (Mathematicians have a good time thinking up words like that.) "Triangular region" is the expression that seems the most natural to use. There is a mathematical word, "simplex", which is used for triangular regions and their analogs in higher dimensions (pyramids, tetrahedrons, etc.). For this book, we shall be satisfied with "triangle" as we defined it. On the other hand, we shall commit a slight abuse of language, and speak of the "area of a triangle", when we mean the "area of the triangular region bounded by a triangle". This is current usage, and although slightly incorrect, it does not really lead to serious misunderstandings.

Each pair of sides of the triangle determines an angle. We shall say that a triangle is a **right triangle** if one of these angles is a right angle. The sides of the triangle which determine this angle are then called the **legs** of the right triangle. A right triangle looks like this. (Fig. 5-22.)

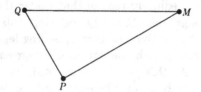

Fig. 5-22

The legs of the right triangle in Fig. 5-22 are the sides $\overline{PQ}$ and $\overline{PM}$.

In our development of geometry, we adopt the following attitude. We take for granted certain basic properties about lines (mentioned before), perpendicularity, and figures like right triangles and rectangles whose main features have to do with perpendicularity. These will be stated as explicitly as possible, to make the situation psychologically satisfactory, I hope. We then prove properties about other geometric figures from these.

In subsequent sections and chapters of the book, we shall show how such foundational materials can be further understood (e.g. by our discussion of congruences, and by coordinates).

One basic fact which we take for granted about right triangles is:

RT. *If two right triangles $\triangle PQM$ and $\triangle P'Q'M'$ have legs $\overline{PQ}$, $\overline{PM}$ and $\overline{P'Q'}$, $\overline{P'M'}$, respectively of equal lengths, that is,*

$$\text{length } \overline{PQ} = \text{length } \overline{P'Q'}$$
$$\text{length } \overline{PM} = \text{length } \overline{P'M'},$$

then: (a) the corresponding angles of the triangles have equal measure, (b) their areas are equal, and (c) the length of $\overline{QM}$ is equal to the length of $\overline{Q'M'}$.

Figure 5–23 illustrates such right triangles as in our axiom **RT**.

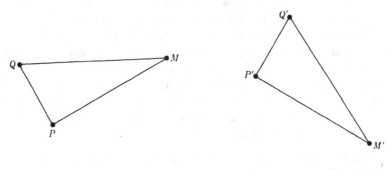

Fig. 5–23

Some of you may already know about the notion of congruence, and if you do then you will immediately realize that under the hypotheses of **RT**, the two triangles are congruent. Roughly speaking, this means that you can "move" one over the other so that the corresponding legs lie over each other. Later we shall deal formally with the notion of congruence, and develop the theory systematically. At this point, we are only concerned with getting one basic theorem, the Pythagoras Theorem, and we don't want to burden ourselves with a whole theory just to get to it, especially since our axiom **RT** is psychologically very satisfactory (to me, and I hope to you). We summarized in **RT** just what we need for our immediate purposes.

The choice of what is assumed in a theory and what is "proved" depends on many requirements. In a subject like geometry, what we assume must in some sense be intuitively obvious. We do not wish to assume so much that we feel uncomfortable about it, and feel that we have really cheated on the theory. On the other hand, we do not wish to assume so little that it becomes very difficult or involved to prove statements which our mind perceives at once as "obvious". We wish to minimize the basic assumptions, and to maximize what we can deduce easily from them. If our system of assumptions is very small, and corresponds to geometric properties which we regard as "obvious", and if we can then deduce easily and fast many properties which we do not regard as "obvious", then we have gone a long way towards finding a satisfactory set of assumptions. If I didn't think that the choice I have made about this was reasonably successful, I wouldn't have written a book . . . . In addition to that, however, experience with more advanced topics shows that the notion of perpendicularity is all-pervasive in mathematics, and that it turns out to be always worth while to have taken it as a fundamental notion, and to have taken as axioms some of its basic properties.

We shall see how some properties of triangles can be reduced to properties of rectangles.

Before defining rectangles, we mention explicitly a property relating parallel lines and distance. Let L, L' be parallel lines, and let P, Q be points on L. Let K_P be the line perpendicular to L passing through P, and let P' be the intersection of K_P with L'. Similarly, let K_Q be the line through Q perpendicular to L, and let Q' be the intersection of K_Q with L'. We shall assume:

PD. *The lengths of the segments $\overline{PP'}$ and $\overline{QQ'}$ are the same. In other words,*

$$d(P, P') = d(Q, Q').$$

This is illustrated by Fig. 5-24(a). We may call this length the **distance** between the two lines.

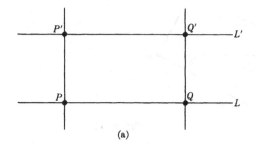
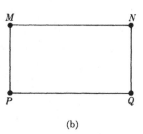

(a)   (b)

Fig. 5-24

Suppose now that P, Q, M, N are four points, such that the segments $\overline{PQ}, \overline{QN}, \overline{NM},$ and $\overline{MP}$ form a four-sided figure. Suppose that the opposite sides $\overline{PQ}, \overline{NM}$ are parallel, and also the opposite sides $\overline{QN}, \overline{MP}$ are parallel; suppose also that the adjacent sides are perpendicular, that is: $\overline{PQ}, \overline{QN}$ are perpendicular and $\overline{NM}, \overline{MP}$ are perpendicular. Then we shall call the set consisting of the four segments

$$\overline{PQ}, \ \overline{QN}, \ \overline{NM}, \ \overline{MP}$$

the **rectangle** determined by P, Q, N, M. This rectangle is illustrated in Fig. 5-24(b). Observe that according to our property **PD**, it follows that the opposite sides of the rectangle have the same length.

If a, b are the lengths of the sides of the rectangle, then we assume that the area of the rectangle is equal to ab. (*Comment:* We are committing here the same abuse of language by speaking of the area of the rectangle that we did with triangles. We mean, of course, the area of the region bounded by the rectangle.) As usual, a **square** is a rectangle all of whose sides have the same length. If this length is equal to a, then the area of the square is a^2.

We are interested in the area of a right triangle. Consider a right triangle $\triangle QPM$ such that $\angle MPQ$ is its right angle, as shown in Fig. 5–25. Then the segments $\overline{PQ}$ and $\overline{PM}$ are perpendicular. We let N be the point of intersection of the line through M parallel to $\overline{PQ}$, and the line through Q perpendicular to $\overline{PQ}$. Then N is the fourth corner of the rectangle whose three other corners are Q, P, M. Then the sides $\overline{QN}$ and $\overline{PM}$ have equal lengths. The sides $\overline{QP}$ and $\overline{NM}$ have equal lengths.

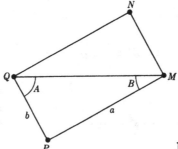

Fig. 5–25

Let A, B be the angles of the right triangle, other than the right angle, as shown in Fig. 5–25. It follows from **RT** that $\angle NQM$ has the same measure as B. Since $\angle NQP$ is a right angle, and since A and $\angle NQM$ together form this right angle $\angle NQP$, it follows that:

Theorem 1. *If A, B are the angles of a right triangle other than the right angle, then*

$$m(A) + m(B) = 90°.$$

Let a, b be the lengths of the sides of the rectangle. Then a, b are also the lengths of the legs of the right triangle $\triangle MPQ$. We assumed that the area of the rectangle is equal to ab. Again by **RT**, we conclude that the two triangles $\triangle QPM$ and $\triangle QNM$ which form this rectangle have the same area.

Hence we find:

Theorem 2. *The area of a right triangle whose legs have lengths a, b, is equal to*

$$\frac{ab}{2}.$$

The third side of a right triangle, which is not one of the legs, is called the **hypotenuse**. The next theorem gives us the relation between the length of the hypotenuse and the lengths of the other two sides.

Pythagoras Theorem. *Let a, b be the lengths of the two legs of a right triangle, and let c be the length of the hypotenuse. Then*

$$a^2 + b^2 = c^2.$$

Proof. Let us draw the right triangle with the leg of length a horizontally as shown in Fig. 5-26. Let the triangle be $\triangle PQM$, with right angle at P, as shown. Let P_1 be the point on the line through P, M at a distance b from M

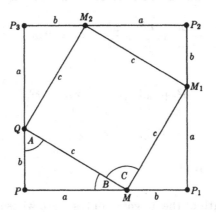

Fig. 5-26

and distance $a + b$ from P. We draw the segment $\overline{P_1P_2}$, perpendicular to $\overline{PP_1}$, on the same side of $\overline{PP_1}$ as the triangle, and of length $a + b$. We then draw the other two sides of the square whose sides have length $a + b$, as shown. The point P_1 is the vertex of a right angle, and we can form a right triangle one of whose legs is the segment $\overline{MP_1}$, of length b, and the other leg is the vertical segment $\overline{P_1M_1}$ of length a. We can now repeat this construction, forming a third right triangle $\triangle M_1P_2M_2$, and then a fourth right triangle $\triangle M_2P_3Q$. Each one of these right triangles has legs of lengths a and b, respectively. Consequently, by **RT**, the sides of the four-sided figure inside the big square have the same length, equal to c. Let A, B be the angles of our right triangle other than the right angle. Let C be any one of the angles of the four-sided figure, say $\angle M_1MQ$. By **RT** we know that $\angle M_1MP_1$ has the same measure as A, and therefore

$$m(B) + m(C) + m(A) = 180°.$$

But we have by Theorem 1,

$$m(A) + m(B) = 90°,$$

so that $m(C) = 90°$. Hence the four-sided figure inside the big square is a square, whose sides have length c.

We now compute areas. The area of the big square is

$$(a + b)^2 = a^2 + 2ab + b^2.$$

This area is equal to the sum of the areas of the four triangles, and the area of the square whose sides have length c. Thus it is also equal to

$$\frac{ab}{2} + \frac{ab}{2} + \frac{ab}{2} + \frac{ab}{2} + c^2 = 2ab + c^2.$$

This yields

$$a^2 + 2ab + b^2 = 2ab + c^2,$$

whence

$$a^2 + b^2 = c^2,$$

and the theorem is proved.

Example. The length of the diagonal of a square whose sides have length 1, as in Fig. 5–27(a), is equal to

$$\sqrt{1^2 + 1^2} = \sqrt{2}.$$

The length of the diagonal of a rectangle whose sides have lengths 3 and 4, as in Fig. 5–27(b), is equal to

$$\sqrt{3^2 + 4^2} = \sqrt{9 + 16}$$
$$= \sqrt{25} = 5.$$

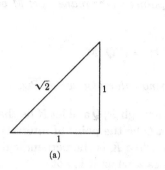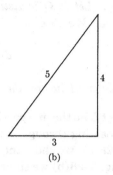

(a)   (b)   **Fig. 5-27**

Example. One leg of a right triangle has length 10 in., and the hypotenuse has length 15 in. What is the length of the other side?

This is easily done. Let b be this length. Then by Pythagoras, we get

$$10^2 + b^2 = 15^2,$$

whence

$$b^2 = 15^2 - 10^2$$
$$= 225 - 100 = 125.$$

Hence $b = \sqrt{125}$.

Let P, Q be distinct points in the plane. We recall that the **perpendicular bisector** of the segment $\overline{PQ}$ is the line perpendicular to $\overline{PQ}$ passing through the point which lies on $\overline{PQ}$, halfway between P and Q. Observe that if O is any point on the line passing through P and Q which is such that

$$d(O, P) = d(O, Q),$$

then O is necessarily on the segment between P and Q. Proof: If this were not the case, then either P would be on the segment $\overline{OQ}$ or Q would be on the segment $\overline{OP}$ (draw the picture). Say P is on the segment $\overline{OQ}$. Then

$$d(O, P) + d(P, Q) = d(O, Q),$$

whence $d(P, Q) = 0$ and $P = Q$, contrary to our assumption that P and Q are distinct. The case when Q might be on the segment $\overline{OP}$ is proved similarly.

The next result is an important consequence of the Pythagoras theorem.

Corollary. *Let P, Q be distinct points in the plane. Let M be also a point in the plane. We have*

$$d(P, M) = d(Q, M)$$

if and only if M lies on the perpendicular bisector of $\overline{PQ}$.

Proof. Let L be the line passing through P, Q and let K be the line perpendicular to L, passing through M. Let O be the point of intersection of K and L. In Fig. 5–28(a), we show the case when K is the perpendicular bisector of $\overline{PQ}$, and in Fig. 5–28(b) we show the case when it is not.

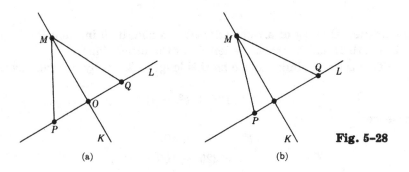

(a)   (b)   **Fig. 5–28**

Assume first that $d(P, M) = d(Q, M)$. By Pythagoras, we have

$$\begin{aligned} d(P, O)^2 + d(O, M)^2 &= d(P, M)^2 \\ &= d(Q, M)^2 \\ &= d(Q, O)^2 + d(O, M)^2. \end{aligned}$$

It follows that

$$d(P, O)^2 = d(Q, O)^2,$$

whence $d(P, O) = d(Q, O)$, and M is on the perpendicular bisector of $\overline{PQ}$.

Conversely, assume that $d(P, O) = d(Q, O)$. Similar steps show that

$$d(P, M) = d(Q, M),$$

thus proving our corollary.

EXERCISES

1. What is the length of the diagonal of a square whose sides have length

   a) 2,   b) 3,   c) 4,   d) 5,   e) r?

2. What is the length of the diagonal of a rectangle whose sides have lengths

   a) 1 and 2,   b) 3 and 5,   c) 4 and 7,

   d) r and $2r$,   e) $3r$ and $5r$,   f) $4r$ and $7r$?

3. What is the length of the diagonal of a cube whose sides have length

   a) 1,   b) 2,   c) 3,   d) 4,   e) r?

   [*Hint:* First compute the square of the length. Consider the diagonal of the base square of the cube, and apply the Pythagoras theorem twice.]

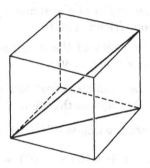

Fig. 5–29

4. What is the length of the diagonal of a rectangular solid whose sides have lengths

   a) 3, 4, 5;   b) 1, 2, 4;   c) 2, 3, 5;   d) 1, 3, 4;   e) 1, 3, 5?

5. What is the length of the diagonal of a rectangular solid whose sides have lengths a, b, c? What if the sides have lengths ra, rb, rc?

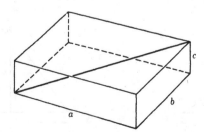

Fig. 5–30

6. You stand at a distance of 500 ft from a tower, and the tower is 100 ft high. What is the distance between you and the top of the tower?

7. a) In a right triangle, one side has length 7 ft and the hypotenuse has length 10 ft. What is the length of the other side?

   b) Same question if one side has length 11 ft and the hypotenuse has length 17 ft.

   c) Same question if one side has length 6 ft and the hypotenuse has length 13 ft.

8. a) You are flying a kite. Assume that the string between you and the kite forms a straight line segment. Suppose that the string has length 70 ft. A friend of yours stands exactly below the kite, and is at a distance of 30 ft from you. How high is the kite?

   b) Same question if the length of the string is 50 ft, and if the distance between you and your friend is 30 ft.

   c) Same question if the length of the string is 110 ft, and the distance between you and your friend is 40 ft.

9. Write down in detail the "similar steps" left to the reader in the proof of the corollary to the Pythagoras theorem.

10. Prove that if A, B, C are the angles of an arbitrary triangle, then

$$m(A) + m(B) + m(C) = 180°$$

by the following method: From any vertex draw the perpendicular to the line of the opposite side. Then use the result already known for right triangles. Distinguish the two pictures in Fig. 5–31.

11. Show that the area of an arbitrary triangle of height h whose base has length b is $bh/2$. [*Hint:* Decompose the triangle into two right triangles. Distinguish between the two pictures in Fig. 5–31. In one case the area of the triangle is the difference of the areas of two right triangles, and in the other case, it is the sum.]

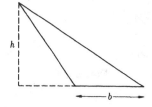
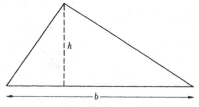

Fig. 5–31

12. a) Show that the length of the hypotenuse of a right triangle is $\geq$ the length of a leg.

b) Let P be a point and L a line. Show that the smallest value for the distances $d(P, M)$ between P and points M on the line is the distance $d(P, Q)$, where Q is the point of intersection between L and the line through P, perpendicular to L.

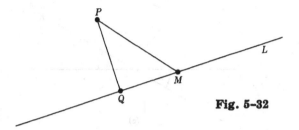

Fig. 5-32

13. This exercise asks you to derive some standard properties of angles from elementary geometry. They are used very commonly. We refer to the following figures.

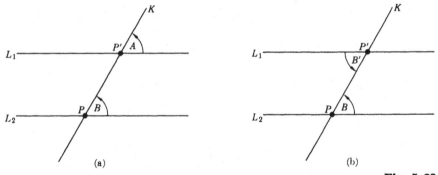

(a)   (b)

Fig. 5-33

a) In Fig. 5-33(a), you are given two parallel lines L_1, L_2 and a line K which intersects them at points P and P' as shown. Let A and B then be angles which K makes with L_1 and L_2 respectively, as shown. Prove that

$$m(A) = m(B).$$

[Hint: Draw a line from a point of K above L_1 perpendicular to L_1 and L_2. Then use the fact that the sum of the angles of a right triangle has 180°.]

b) In Fig. 5-33(b), you are given L_1, L_2 and K again. Let B and B' be the alternate angles formed by K and L_1, L_2 respectively, as shown. Prove that $m(B) = m(B')$. (Actually, all you need to do here is refer to the appropriate portion of the text. Which is it?)

c) Let K, L be two lines as shown on Fig. 5-33(c). Prove that the opposite angles A and A' as shown have equal measure.

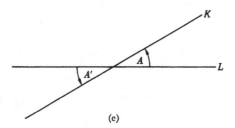

(c)

Fig. 5–33 (cont.)

14. Let $\triangle PQM$ be a triangle. Let L_1 be the perpendicular bisector of $\overline{PQ}$ and let L_2 be the perpendicular bisector of $\overline{QM}$. Let O be the point of intersection of L_1 and L_2. Show that $d(O, P) = d(O, M)$, and hence that O lies on the perpendicular bisector of $\overline{PM}$. Thus the perpendicular bisectors of the sides of the triangle meet in a point.

6 Isometries

§1. SOME STANDARD MAPPINGS OF THE PLANE

We need to define the notion of congruence. For instance, given two discs of the same radius as in Fig. 6–1, we want to say that they are congruent.

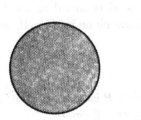

Fig. 6–1

Similarly, given two triangles as in Fig. 6–2, we also want to say that they are congruent.

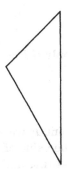
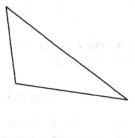

Fig. 6–2

Roughly speaking, this means that one figure can be laid over the other. To discuss the notion of congruence properly, it is convenient to define first a slightly more general notion, namely isometry. To do that, we must define still a more general notion, namely mapping. All these notions are quite common, and you will see that they include, as special cases, things which you can easily visualize, like reflections, rotations, stretching, etc., given as examples. We discuss these first, and then take up congruences in the last section.

By a **mapping** (or a **map**) of the plane into itself, we shall mean an association, which to each point of the plane associates another point of the plane. If P is a point and P' is the point associated with P by the mapping, then we denote this by the special arrow

$$P \mapsto P'.$$

The point P' associated with P is called the **value** of the mapping at P. We also say that P' **corresponds** to P under the mapping, or that P is **mapped** on P'.

Just as we used letters to denote numbers, it is useful to use letters to denote mappings. Thus if F is a mapping of the plane into itself, we denote the value of F at P by the symbols

$$F(P).$$

We shall also say that the value $F(P)$ of F at P is the **image** of P under F. If $F(P) = P'$, then we also say that F **maps** P on P'.

By definition, if F, G are mappings of the plane into itself, we have

$$F = G$$

if and only if, for every point P,

$$F(P) = G(P).$$

In other words, a map F is equal to a map G if and only if F and G have the same value at every point P.

Constant mapping

Let O be a given point in the plane. To each point P we associate this given point O. Then we obtain a mapping, and O is the value of the mapping at every point P. We say that this mapping is **constant,** and that O is its constant value.

Identity

To each point P we associate P itself. This is a rather simple mapping, which is called the **identity**, and is denoted by I. Thus we have

$$I(P) = P$$

for every point P.

Reflection through a line

Let L be a line. If P is any point, let

$$L' = L'_P$$

be the line through P perpendicular to L. Let O be the point of intersection of L and L'. Let P' be the point on L' which is at the same distance from O as P, but in the opposite direction. The association

$$P \mapsto P'$$

is called **reflection through** L, and could be denoted by R_L. Picture:

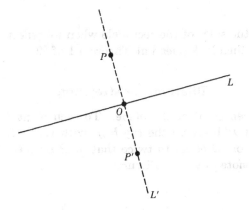

Fig. 6–3

Reflection through a point

Let O be a given point of the plane. To each point P of the plane we associate the point P' lying on the line passing through P and O, on the other side of O from P, and at the same distance from O as P. This mapping is called the **reflection through** O. Picture:

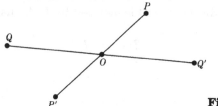

Fig. 6–4

We have drawn a point P and its value P' under the mapping, and also a point Q and its value Q' under the mapping.

For instance, we may reflect the four corners of a rectangle through the midpoint O of the rectangle. Each corner is mapped on the opposite corner, as in Fig. 6–5.

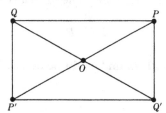

Fig. 6–5

What happens to the sides of the rectangle when we reflect them through O? Figure it out, and then look ahead at Theorem 1 of §2.

Dilations, or stretching

Let O be a given point of the plane. To each point P of the plane we associate the point P' lying on the ray R_{OP}, with vertex O, passing through P, at a distance from O equal to twice that of P from O. The point P' in this case is also denoted by $2P$. Picture:

Fig. 6–6

This particular mapping is called **dilation** by 2, or **stretching** by 2, **relative to** O. According to our notation, if F is dilation by 2, relative to O, then we have

$$F(P) = 2P.$$

Let r be any positive number. We can of course define the dilation by r just as we define dilation by 2, relative to a given point O. Namely, we define **dilation by** r to be the mapping F_r such that for any point P the value $F_r(P)$ is the point on the ray with vertex O, passing through P, at a distance from O equal to r times the distance between O and P. It is convenient to write rP instead of $F_r(P)$.

In Fig. 6–7(a) we have drawn the points

$$Q \text{ and } F_3(Q).$$

In Fig. 6–7(b) we have drawn the points

$$P, \ F_{1/3}(P), \ F_{2/3}(P).$$

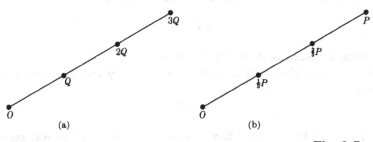

(a)   (b)

Fig. 6-7

A dilation is also sometimes called a **similarity transformation,** but the word dilation is the shortest and best term to be used to denote the concept.

Rotation

Let O be a given point in the plane, and let A be an angle. Let P be a point at distance d from O. Let C be the circle of radius d centered at O. Let P' be the point on this circle such that the angle $\angle POP'$ has the same number of degrees as A. The mapping which associates P' to P is called **rotation** (counterclockwise) by A, with respect to O, or relative to O. If we denote this mapping by G_A, then we can illustrate it on Fig. 6–8(a), where we have drawn O, P, and $G_A(P)$.

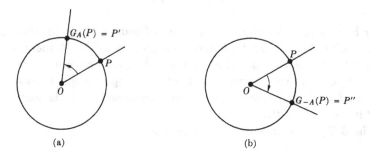

(a)   (b)

Fig. 6–8

Unless otherwise specified, a rotation by an angle A will always mean counterclockwise rotation. Of course, we can also define clockwise rotation by A, which we denote by G_{-A}. It associates with each point P the point P'' at the same distance from O as P, and such that the angle

$$\angle P''OP$$

has the same measure as A, as on Fig. 6–8(b).

Observe that a given angle A is not necessarily the same angle as that formed by

$$\angle POP'$$

(cf. Fig. 6–9) even though they have the same measure. However, experience shows that it is harmless for this type of discussion to indicate $\angle POP'$ as the

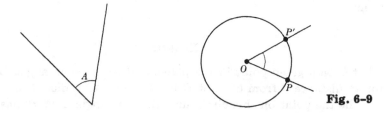

Fig. 6–9

angle A, and it is often convenient to do this to suggest what's going on. Still, it is not always harmless to do this. For instance, in a triangle like that in

Fig. 6–10 where the two bottom angles A and B have equal measures, we would not think of denoting them by the same letter.

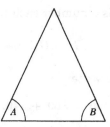

Fig. 6-10

Observe that rotation by 180° with respect to O is none other than reflection through O. Thus if R denotes a reflection through O, we have

$$G_{180°} = R.$$

We also have

$$G_{-180°} = R,$$

because the values of each one of these mappings at a point P is the same point P'. Even though these mappings are described by conditions which appear different, the mappings are nevertheless equal. Recall that, by definition, mappings F, G are equal if and only if

$$F(P) = G(P)$$

for all points P.

It is convenient to associate a rotation with a number rather than an angle. If x is a number between 0 and 360, we let

$$G_x$$

be the rotation by an angle of x degrees. Observe that

$$G_0 = G_{360} = I$$

is none other than the identity. (To be absolutely correct, we should also indicate the point O in our notation, but throughout our discussion we deal with the same given point, and thus omit it. If we wanted to indicate it explicitly, we could write for instance

$$G_{O,x}$$

for the rotation by an angle of x degrees relative to the given point O.)

Let x be an arbitrary number. We write x in the form

$$x = 360n + w,$$

where n is an integer, and w is a number such that $0 \leq w < 360$. We define G_x to be

$$G_x = G_w.$$

Example. Let $x = 500$. We write

$$500 = 360 + 140.$$

Then by definition

$$G_x = G_{140}$$

is rotation by 140°.

Example. Let $x = -210$. We write

$$-210 = -360 + 150.$$

Then

$$G_{-210} = G_{150}$$

is rotation by 150°.

Example. If x and y are numbers such that

$$x = y + 360m$$

for some integer m, then

$$G_x = G_y.$$

Namely, if $x = 360n + w$ with some integer n, and $0 \leq w < 360$, then

$$y + 360m = 360n + w,$$

and hence

$$y = 360(n - m) + w.$$

According to our definition, we have

$$G_x = G_w = G_y.$$

We can interpret counterclockwise rotation by a negative number as clockwise rotation by a positive number.

Example. Let $x = -90$. Then

$$-90 = -360 + 270.$$

Thus

$$G_{-90} = G_{270}.$$

We visualize this as saying that clockwise rotation by 90° is the same as counterclockwise rotation by 270°.

We shall use the convention of saying that the rotation G_x is rotation by x degrees, or even by an angle of x degrees, even though x may be greater than 360, or may be negative. This is convenient language, and reflects our geometric intuition without leading to great confusion.

Translations

Let us select a direction in the plane, and a distance d. We can represent these by an arrow as in Fig. 6–11. The arrow is simply an ordered pair of points (O, M), where O is its beginning point and M is its end point.

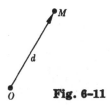

Fig. 6–11

The arrow points in the given direction, and the length of the arrow is equal to our given distance. To each point P, we associate the point P' which is at a distance d from P in the given direction. This is a mapping, which is called the **translation** (determined by the given direction and the given distance). In Fig. 6–12, letting T be this translation, we have drawn two points P, Q and their images under T.

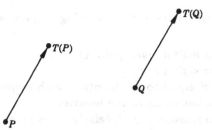

Fig. 6–12

The translation determined by an ordered pair of points (O, M) will be denoted by

$$T_{OM}.$$

Observe that if $T = T_{OM}$, then $T(O) = M$, and that T_{OO} is the identity.

Example. Let T_l be the translation by one inch to the left, T_r the translation by one inch to the right, and similarly T_u and T_d the translations by one inch upward and one inch downward respectively.

In Fig. 6–13 we have drawn a flower $\mathfrak{F}$ and its translations by T_l, T_r, T_d and T_u.

Fig. 6–13

EXERCISES

Let F be a mapping of the plane into itself. We define a **fixed point** for F to be a point P such that $F(P) = P$.

1. Describe the fixed points of the following mappings.

   a) The identity.
   b) Reflection through a given point O.
   c) Reflection through a line.
   d) A rotation not equal to the identity, with respect to a given point O.
   e) A translation not equal to the identity.
   f) Dilation by a number $r > 0$, relative to a given point O.

2. Write each one of the following numbers in the form

$$360n + w$$

with an integer n, and $0 \leq w < 360$.

a) -30 b) -90 c) -180 d) -270
e) -45 f) -225 g) 120 h) 540
i) -400 j) 600 k) 720 l) 450

3. Let the point P be as illustrated in Fig. 6–14. In each one of the cases of Exercise 2, draw the image of P under the corresponding rotation, by the number of degrees given in (a) through (l). You can use a protractor, or just make an approximate estimate of the position of this image.

$$\overset{\bullet}{O} \text{———} \overset{\bullet}{P} \qquad \text{Fig. 6–14}$$

If you want to make things look better, draw the image of a flower instead of a point.

§2. ISOMETRIES

Let F be a mapping of the plane into itself. We say that F **preserves distances,** or is **distance preserving,** if and only if for every pair of points P, Q in the plane, the distance between P and Q is the same as the distance between $F(P)$ and $F(Q)$. Such a mapping is also called an **isometry.** ("Iso" means same, and "metry" means measure. It is useful to have one word instead of two for this notion.)

Example. The constant mapping which to each point P associates a given point O is not an isometry. Dilation by 2 is not an isometry. Why?

Example. *Let F be any one of the following maps.*

Reflection through a point
Reflection through a line
Rotation
Translation

Then F is an isometry.

This will be assumed without proof. Later when we give definitions for these mappings depending on coordinates, we shall be able to prove that these maps are isometries very simply. In §5 we shall prove that any isometry can be obtained by simple combinations of the examples given above, in a sense which will be made precise.

Remark. Let F be an isometry. If P, Q are distinct points, then $F(P)$ and $F(Q)$ must be distinct, because the distance between P and Q is not 0, and hence the distance between $F(P)$ and $F(Q)$ cannot be 0 either. Cf. property **DIST 1** of distances.

Let S be a set of points in the plane and let F be a mapping of the plane into itself. The set of points consisting of all points $F(P)$, for all P in S, is called the **image of** S under F, and is denoted by $F(S)$.

Example. Let F be the reflection through a line L. Let S be the line segment between two points P and Q. Then the image of S under F is the line segment between $F(P)$ and $F(Q)$. Picture:

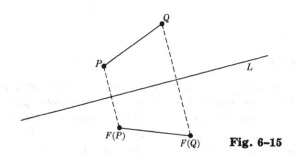

Fig. 6–15

With our notation, we have

$$F(\overline{PQ}) = \overline{F(P)F(Q)}.$$

The preceding example is but a special case of a general property of isometries, which we state in the next theorem.

Theorem 1. Let F be an isometry. The image of a line segment under F is a line segment. In fact, the image of the line segment $\overline{PQ}$ under F is the line segment between $F(P)$ and $F(Q)$.

Proof. (See Fig. 6–16.) Let X be a point on $\overline{PQ}$. For simplicity we denote $F(X)$ by X'. Since F preserves distances, we know that

$$d(P, X) = d(P', X'), \qquad d(X, Q) = d(X', Q').$$

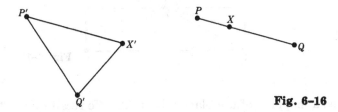

Fig. 6–16

By **SEG 1**, we have

$$d(P, Q) = d(P, X) + d(X, Q).$$

By assumption on F, we have

$$d(P', Q') = d(P', X') + d(X', Q').$$

Again by **SEG 1**, we conclude that X' must lie on the segment between P' and Q', thus proving that the image of $\overline{PQ}$ is contained in the segment $\overline{P'Q'}$.

We must still prove that every point of the segment $\overline{P'Q'}$ can be expressed as the image under F of a point on $\overline{QP}$. Let X' be a point on $\overline{P'Q'}$ at distance r from P'. Let X be the point on $\overline{PQ}$ at distance r from P. Then $F(X)$ is at distance r from $F(P) = P'$. It follows that $F(X) = X'$.

Remark. In this proof, we want to show that two sets of points are equal. We have followed a standard pattern, namely we have proved that each one is part of the other. This pattern will be repeated later.

> **Corollary.** *An isometry preserves straight lines. In other words, if L is a straight line in the plane and F is an isometry, then $F(L)$ (the image of L under F) is also a straight line. If L is the line passing through two distinct points P, Q, then $F(L)$ is the line passing through $F(P)$ and $F(Q)$.*

> *Proof.* We leave the proof as an exercise.

Example. Let O be the point of intersection of the diagonals of a rectangle. If we reflect through O, then the opposite corners are mapped on each other, and hence the opposite sides are mapped on each other, as illustrated in Fig. 6–17.

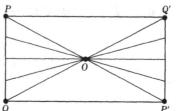

Fig. 6–17

Let F be a mapping of the plane into itself. We recall that a **fixed point** P of F is a point such that $F(P) = P$. Fixed points of isometries will now play a very important role in describing all isometries. You should definitely do Exercise 2 of the preceding section if you have not already done it. We shall investigate systematically isometries with no fixed point, one fixed point, two fixed points, and three fixed points, but in reverse order. In this last case, we shall see that the isometry must be the identity. Then we consider each case with one less fixed point, and analyze it by composing the given isometry with a reflection, a rotation, or a translation to get the ultimate result that any isometry must be a composite of these. (For the definition of a composite isometry, read ahead in §3.)

Theorem 2. *Let F be an isometry. Let P, Q be two distinct points in the plane. Assume that they are fixed points, in other words*

$$F(P) = P \qquad and \qquad F(Q) = Q.$$

Then every point on the line through P, Q is a fixed point of F.

Proof. We shall distinguish cases. Let M be a point on the line passing through P and Q. We wish to show that $F(M) = M$.

Case 1. The point M lies on the segment $\overline{PQ}$. Let $M' = F(M)$. Since F preserves distances, we have

$$d(P, M) = d(P, M'),$$
$$d(M', Q) = d(M, Q).$$

Hence

$$d(P, M') + d(M', Q) = d(P, Q).$$

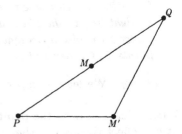

Fig. 6–18

By **SEG 1,** this means that M' lies on the segment between P and Q. Since

$$d(P, M) = d(P, M'),$$

it follows that $M = M'$.

Case 2. Suppose that M does not lie on the segment $\overline{PQ}$. Suppose that M lies on the ray having vertex P and passing through Q, but at a distance from P greater than that of Q, as in Fig. 6–19.

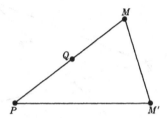

Fig. 6–19

Then

$$d(P, M') = d(P, M) = d(P, Q) + d(Q, M)$$
$$= d(P, Q) + d(Q, M').$$

By **SEG 1,** this means that Q lies on the segment between P and M'. Hence P, Q, M' lie on the same straight line, and therefore M' lies on the straight line passing through P, Q. Since Q lies on the segment between P and M', we conclude that M' lies on the ray having vertex P passing through Q. Since

$$d(P, M) = d(P, M'),$$

it follows that $M = M'$.

Case 3. This case is similar to Case 2, when M lies on the other side of P from Q. In this case, the role of P and Q is reversed, and the proof goes on as in Case 2, interchanging P and Q. This concludes the proof of Theorem 2.

Theorem 3. *Let F be an isometry. Let P, Q, M be three distinct points which do not lie on a straight line. Assume that P, Q, M are fixed points of F; that is,*

$$F(P) = P, \qquad F(Q) = Q, \qquad F(M) = M.$$

Then F is the identity.

Proof. Let L_{PQ} and L_{QM} be the lines passing through P, Q and Q, M, respectively. Let X be a point. We must show that $F(X) = X$. We can find a line L passing through X which intersects L_{PQ} in a point Z, and intersects L_{QM} in a point Y such that $Y \neq Z$. (For instance, pick a point Z on L_{QM} which is distinct from M, Q, X, and such that the line L_{XZ} is not parallel to L_{PQ}. Let $L = L_{XZ}$, and let Y be the point of intersection of L_{XZ} and L_{PQ}.) The situation is illustrated in Fig. 6–20.

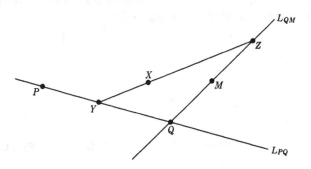

Fig. 6–20

By Theorem 2, every point on the lines L_{PQ} and L_{QM} are fixed. Therefore we have

$$F(Y) = Y \qquad \text{and} \qquad F(Z) = Z.$$

Again by Theorem 2, every point on the line L_{YZ} is fixed, and we conclude that $F(X) = X$. This proves our theorem.

Remark. A very important corollary of this theorem will be stated when we have the notion of inverse of an isometry, in §4.

EXERCISES

1. Draw the image of a line segment under
   a) reflection through a point,
   b) reflection through a line,
   c) rotation by 90°,

d) translation,

e) rotation by 180°.

2. For which values of r is dilation by r an isometry?

3. Draw the image of a circle of radius r, center P under

a) reflection through its center,

b) reflection through a line L outside of the circle as illustrated in Fig. 6–21,

c) rotation by 90° with respect to a point O outside of the circle,

d) rotation by 180° with respect to O,

e) rotation by 270° with respect to O,

f) translation.

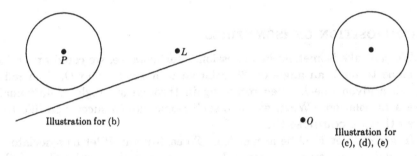

Illustration for (b)

Illustration for (c), (d), (e)

Fig. 6–21

4. Let L, K be two parallel lines, and let F be an isometry. Prove that $F(L)$ and $F(K)$ are parallel.

5. Let K, L be perpendicular lines, and let F be an isometry. Prove that $F(K)$ and $F(L)$ are perpendicular. [*Hint:* Use the corollary of the Pythagoras theorem.]

6. Visualize 3-dimensional space. We also have the notion of distance in space, satisfying the same basic properties as in a plane. We can therefore define an isometry of 3-space in the same way that we defined an isometry of the plane. It is a mapping of 3-space into itself which is distance preserving. Are Theorems 1 and 2 valid in 3-space? How would you formulate Theorem 3? (Consider the plane in which the three points lie.) Now formulate a theorem in 3-space about an isometry being the identity provided that it leaves enough points fixed. Describe a proof for such a theorem, similar to the proof of Theorem 3. Make a list of

what you need to assume to make such a proof go through. Write all of this up as if you were writing a book. Aside from learning mathematical substance, you will also learn how to think more clearly, and how to write mathematics in the process.

§3. COMPOSITION OF ISOMETRIES

We can take isometries in succession. For instance, we could first rotate the plane through an angle of 30° relative to a given point O; then reflect through a given line L; then rotate again through an angle of 45°; finally make a translation. When we take such isometries in succession like that, we say that we **compose** them.

In general, let F, G be isometries. To each point P let us associate the point $F(G(P))$, obtained by first taking the image of P under G, and then the image of this latter point under F. Then we obtain an association

$$P \mapsto F(G(P)),$$

which is a mapping. In fact, this mapping is an isometry, because the distance between two points P, Q is the same as the distance between $G(P)$, $G(Q)$ (because G is an isometry), and is the same as the distance between $F(G(P))$, $F(G(Q))$ (because F is an isometry). The association

$$P \mapsto F(G(P))$$

is called the **composite** of G and F, and is denoted by the symbols

$$F \circ G.$$

Thus we have

$$(F \circ G)(P) = F(G(P)).$$

Example. Let O be a given point. Let F be rotation by 90° with respect to O, and let G be reflection through O. Then $G \circ F$ is rotation by 270°. We also see that $F \circ F$ is rotation by 180°, and thus we may write

$$F \circ F = G.$$

Example. Let F be any isometry and let I be the identity. Then

$$F \circ I = I \circ F$$
$$= F.$$

Thus I behaves like multiplication by 1.

Composition of rotations. If F, G are rotations, relative to the same point O, then $F \circ G$ is also a rotation, relative to O.

Let O be a given point and let r be a number > 0. Let P, Q be points different from O, at the same distance r from O. Then there exists a unique rotation G_{PQ} relative to O, which maps P on Q (i.e. such that the value of G_{PQ} at P is Q).

We shall assume these statements without proof. They are both intuitively clear. Using them, we can write down a nice formula for the composite. Let P, Q, M be points at distance r from O. Then

$$\boxed{G_{QM} \circ G_{PQ} = G_{PM} \quad \text{and} \quad G_{PP} = I.}$$

Proof. The image of P under the composite $G_{QM} \circ G_{PQ}$ is

$$G_{QM}(G_{PQ}(P)) = G_{QM}(Q)$$
$$= M,$$

which is the same as the image of P under G_{PM}. By assumption there is only one rotation having this effect on P. Hence we get the first formula. The second is proved similarly, and even more simply.

In terms of numbers, we shall also assume without proof the following fact, which is intuitively clear.

Let x, y be numbers. Let G_x be the rotation associated with x as in §1. Then

$$G_x \circ G_y = G_{x+y}.$$

For example,

$$G_{45} \circ G_{45} = G_{90}.$$

This means that a rotation by 45° followed by a rotation by 45° is the same as a rotation by 90°. Also, as in Fig. 6–22, we have

$$G_{180} \circ G_{270} = G_{450} = G_{90}.$$

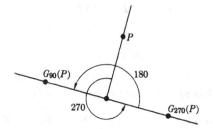

Fig. 6-22

Composition of translations. If F, G are translations, then the composite F∘G is also a translation.

Given points P, Q, there exists a unique translation T_{PQ} such that the image of P under T_{PQ} is Q.

Again, we assume these two statements without proof. As an exercise, prove the following formulas.

$$T_{QM} \circ T_{PQ} = T_{PM} \quad \text{and} \quad T_{PP} = I.$$

Associativity of isometries. Let F, G, H be isometries. Then we have

$$(F \circ G) \circ H = F \circ (G \circ H).$$

Proof. For any point P, we have

$$((F \circ G) \circ H)(P) = (F \circ G)(H(P)) = F(G(H(P)))$$
$$(F \circ (G \circ H))(P) = F((G \circ H)(P)) = F(G(H(P))).$$

This proves our remark, because the two maps $(F \circ G) \circ H$ and $F \circ (G \circ H)$ have the same value at P, and this is true for every point P.

We shall use the same notation with isometries that we used with numbers for multiplication. If F is an isometry,

$$\text{we denote } F \circ F \text{ by } F^2,$$
$$\text{we denote } F \circ F \circ F \text{ by } F^3,$$

and so on. We denote by F^n the isometry obtained by iterating F with itself n times.

Example. Let G be the reflection through a given point O. Then we see that

$$G^2 = \text{identity} = I.$$

This is like the relation $(-1)^2 = 1$. Note that we have:

$$G^3 = G$$
$$G^4 = I$$
$$\vdots$$

again in analogy with powers of -1.

Example. Let F be rotation by $90°$. Then:

$$F^2 = \text{rotation by } 180°,$$
$$F^3 = \text{rotation by } 270°,$$
$$F^4 = \text{rotation by } 360° = \text{identity},$$
$$F^5 = \text{rotation by } 90° = F.$$

Note this interesting cyclical nature of F, that $F^5 = F$.

If F is an isometry, we define

$$F^0 = I.$$

Then for any natural numbers m, n we have the old relation

$$F^{m+n} = F^m \circ F^n.$$

Thus composition behaves like a multiplication.

Example. Let T be translation by 1 in. to the right, and let P be a point. Then P, $T(P)$, $T^2(P)$, $T^3(P)$, ... are points on a horizontal line, and $T^{n+1}(P)$ is 1 in. to the right of $T^n(P)$. Draw one picture, with $0 \leq n \leq 5$.

EXERCISES

1. Let F be reflection through a line L. What is the smallest positive integer n such that $F^n = I$?

   In Exercises 2 and 3, let O be a given point in the plane. Let K be a vertical line and L a horizontal line intersecting at O. Let H be reflection through L and V reflection through K. Let G_x be rotation with respect to O, by an angle of x degrees.

2. Let P be the point shown in Fig. 6–23. Draw the image of P under the following isometries.

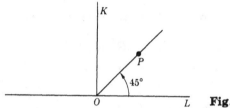

Fig. 6–23

   a) $H \circ G_{90}$ b) $G_{90} \circ H$ c) $V \circ G_{90}$ d) $G_{90} \circ V$

   e) $H \circ V \circ G_{90}$ f) $V \circ H \circ G_{90}$ g) $H \circ G_{180}$ h) $G_{180} \circ H$

   i) $G_{180} \circ V$ j) $V \circ G_{180}$ k) $H \circ V \circ G_{180}$ l) $V \circ H \circ G_{180}$

3. Let Q be the point shown in Fig. 6–24. Draw the image of Q under each one of the mappings (a) through (l) of the preceding exercise.

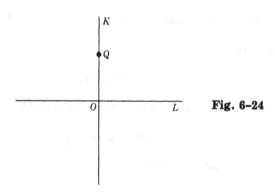

Fig. 6–24

4. Give an example of two isometries F_1, F_2 such that

$$F_1 \circ F_2 \neq F_2 \circ F_1.$$

5. Let G be rotation by
   a) 90°,   b) 60°,   c) 45°,
   d) 30°,   e) 15°.

   In each case, determine the smallest positive integer k such that $G^k = I$.

6. Draw a small flower. Let T be translation by 1 in. to the right, and let U be translation by 1 in. vertically upward. Draw the image of the flower under T, T^2, T^3, T^4, U, U^2, U^3, U^4, $T \circ U$, $T^2 \circ U$, $T^3 \circ U$, $T \circ U^2$, $T \circ U^3$, $T^2 \circ U^2$. Admire your pattern. Draw other images of the flower under isometries to make up other beautiful patterns.

§4. INVERSE OF ISOMETRIES

Let F be an isometry. By an **inverse** (isometry) for F we shall mean an isometry G such that

$$F \circ G = G \circ F = I.$$

Suppose that G and H are inverses for F. Then

$$H \circ F \circ G = H \circ I = H.$$

By associativity, the left-hand side is equal to

$$H \circ F \circ G = I \circ G = G.$$

Thus we find

$$G = H.$$

This is the same type of proof which we used before to prove the uniqueness of the inverse, and we see that it applies to our present setting with isometries. We denote the inverse of F by F^{-1} if it exists. Assume that the inverse exists. If P, Q are points, then the relations

$$P = F(Q) \qquad \text{and} \qquad Q = F^{-1}(P)$$

are equivalent. Indeed, if $P = F(Q)$, then applying F^{-1} we obtain

$$F^{-1}(P) = F^{-1}(F(Q)) = Q,$$

and similarly for the converse. If P is the image of Q under F, then we also say that Q is the **inverse image** of P under F.

Example. Let F be reflection through a given line L. Since

$$F^2 = F \circ F = I,$$

we conclude that

$$F^{-1} = F.$$

Example. For each number x let G_x be the associated rotation through x degrees. Then

$$G_x^{-1} = G_{-x},$$

because $G_{-x} \circ G_x = G_0 = I$. For instance,

$$G_{90}^{-1} = G_{270} = G_{-90}.$$

Also observe that if $G = G_{90}$, then

$$G^{-1} = G^3.$$

Example. Let T_{OM} denote the translation determined by the ordered pair of points (O, M). This is the translation in the direction of the ray with vertex O, passing through M, and such that the image of a point P lies at distance from P equal to the length of the segment $\overline{OM}$. Then T_{OM} has an inverse, which is none other than T_{MO}, namely the translation going in the opposite direction, but the same distance, because we have

$$T_{MO} \circ T_{OM} = I.$$

Example. Let F, G be isometries, having inverses F^{-1} and G^{-1}, respectively. Then the composite $F \circ G$ has an inverse, namely

$$(F \circ G)^{-1} = G^{-1} \circ F^{-1}.$$

This is easily seen. All we have to do is to verify that the right-hand side composed with $F \circ G$ on either side yields the identity. But we have

$$G^{-1} \circ F^{-1} \circ F \circ G = G^{-1} \circ I \circ G = G^{-1} \circ G = I,$$

and similarly on the other side. This proves our assertion.

Let n be a negative integer, say $n = -k$, where k is positive. We define F^n to be the composite of F^{-1} with itself k times, i.e.

$$F^{-k} = (F^{-1})^k.$$

Also we define $F^0 = I$ (identity). Then we have the formula

$$F^{m+n} = F^m \circ F^n$$

valid for any values of m, n as integers. This relation is analogous to that holding for powers of numbers. We omit the proof, which is in any case easy.

Example. If T is the translation by 1 in. to the right, then T^{-1} is the translation by 1 in. to the left. If U is the translation by 1 in. upward, then U^{-1} is the translation by 1 in. downward. Also, T^{-5} is the translation by 5 in. to the left, and U^{-6} is the translation by 6 in. downward.

Using inverses, we can now prove a very useful corollary of Theorem 3 which tells us when two isometries are equal.

Corollary of Theorem 3. Let P, Q, M be three distinct points which do not lie on the same line. Let F, G be isometries such that

$$F(P) = G(P), \qquad F(Q) = G(Q), \qquad F(M) = G(M).$$

Assume that F^{-1} exists. Then $F = G$.

Proof. The proof is very easy and will be left as an exercise.

Example. Let K be a vertical line and let L be a horizontal line. Let H be reflection with respect to L and let V be reflection with respect to K. Then $H \circ V = V \circ H$. To see this, we have only to verify that $H \circ V$ and $V \circ H$ have the same effect on three corners of a square centered at the point of intersection of the lines, and this is clear.

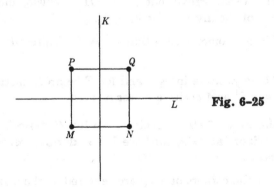

Fig. 6-25

Remark. It will be proved later that every isometry has an inverse.

EXERCISES

1. a) Let F be an isometry which has an inverse F^{-1}. Let S be a circle of radius r, and center P. Show that the image of S under F is a circle. [*Hint:* Let S' be the circle of center $F(P)$ and radius r. Show that $F(S)$ is contained in S' and that every point of S' is the image under F of a point in S.]

   b) Let F be an isometry which has an inverse F^{-1}. Let D be a disc of radius r and center P. Show that the image of D under F is a disc.

2. Let P, Q, P', Q' be points such that

$$d(P, Q) = d(P', Q').$$

Prove that there exists an isometry F such that $F(P) = P'$ and $F(Q) = Q'$. You may assume the same statements we have assumed in this section.

3. Let F, G, H be isometries and assume that F has an inverse. If

$$F \circ G = F \circ H,$$

prove that $G = H$ (**cancellation law** for isometries).

4. a) Let F be an isometry such that $F^2 = I$ and $F^3 = I$. Prove that $F = I$.

   b) Let F be an isometry such that $F^4 = I$ and $F^7 = I$. Prove that $F = I$.

   c) Let F be an isometry such that $F^5 = I$ and $F^8 = I$. Prove that $F = I$.

5. Write out the proof of the corollary to Theorem 3. (Consider $F^{-1} \circ G$.)

6. Let $F \circ G \circ H$ be the composite of three isometries. Assume that F^{-1}, G^{-1}, H^{-1} exist. Prove that $(F \circ G \circ H)^{-1}$ exists, and express this inverse in terms of the inverses for F, G, H.

7. Let F be an isometry such that $F^7 = I$. Express F^{-1} as a positive power of F.

8. Let n be a positive integer and let F be an isometry such that $F^n = I$. Express F^{-1} as a positive power of F.

   For the rest of the exercises, we let H denote reflection through the horizontal axis, and we let V denote reflection through the vertical axis.

9. Consider the corners of a square centered at the origin. For convenience of notation, number these corners 1, 2, 3, 4 as in Fig. 6–26.

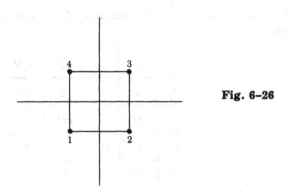

Fig. 6–26

Write the image of each one of these corners under the isometries H, V, $H \circ V$ and $V \circ H$. Just to show you an easy notation to do this, we write down the images of these corners under rotation by 90° in the following form:

$$\begin{bmatrix} 1 & 2 & 3 & 4 \\ 2 & 3 & 4 & 1 \end{bmatrix}.$$

This notation means that if G is rotation by 90°, then $G(1) = 2$, $G(2) = 3$, $G(3) = 4$, and $G(4) = 1$.

10. Let G be rotation by 90° so that $G^4 = I$. Express $H \circ G \circ H$ as a power of G. For what positive integer n do we have

$$H \circ G = G^n \circ H?$$

Write down the images of the corner of the square as in the preceding exercise, under the maps I, G, G^2, G^3, H, $H \circ G$, $H \circ G^2$, $H \circ G^3$, $G \circ H$, $G^2 \circ H$, $G^3 \circ H$.
Compare with the section on permutations in Chapter 14, §3.

Multiplication tables

Let us simplify the notation and write FG instead of $F \circ G$, to make the analogy with multiplication more striking. If H, V are the reflections along the horizontal line and vertical line, respectively, as above, then we can make a "multiplication table" for the products of the four elements I, H, V, HV, as follows.

	I	H	V	HV
I	I	H	V	HV
H	H	I	HV	V
V	V	HV	I	H
HV	HV	V	H	I

This multiplication table is to be read like a multiplication table for numbers. Where a row intersects a column, we have the value of the product of an element on the far left in the row, multiplied by the element on the top of each column. For instance, the product of H and HV is

$$HHV = V,$$

because

$$H^2 = I.$$

Similarly, the product of HV and HV is

$$HVHV = HHVV$$
$$= H^2V^2$$
$$= I.$$

A multiplication table for numbers would look like this.

	1	3	5	17
1	1	3	5	17
3	3	9	15	51
5	5	15	25	85
17	17	51	85	289

11. Let G be rotation by 90°, so that $G^4 = I$. Fill out the multiplication table given below. Write each entry in the form HG^k or G^m for suitable integers k, m between 0 and 3.

	I	G	G^2	G^3	H	HG	HG^2	HG^3
I								
G								
G^2								
G^3								
H								
HG								
HG^2								
HG^3								

12. Let G be rotation by 90°. Make up and fill out the multiplication table for the elements I, G, G^2, G^3, V, VG, VG^2, VG^3. Again, express each entry of the table as one of these elements.

13. Consider a triangle whose three sides have equal length and whose three angles have the same measure, 60°, as in Fig. 6–27.

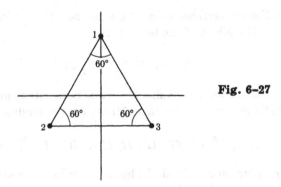

Fig. 6–27

The vertices of the triangle are numbered 1, 2, 3. Let G be rotation by 120° and let V, as usual, be reflection through the vertical axis.

a) Give the effect of the six isometries I, G, G^2, V, VG, VG^2 on the vertices, using the same notation as in Exercise 9.

b) Make up the multiplication table for these six isometries.

14. Let G be rotation by 60°. Find a positive integer k such that $HG = G^k H$. What is the smallest positive integer m such that $G^m = I$?

15. Consider a hexagon, i.e., a six-sided figure, whose sides all have the same length and whose angles have the same measure, as shown in Fig. 6–28.

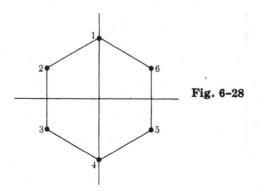

Fig. 6–28

a) What is the measure of these angles?
b) Let G be rotation by 60° and let H, V be as before, reflection through the horizontal and vertical axes, respectively. Give the effect of the 12 isometries

$$I, G, G^2, G^3, G^4, G^5, H, HG, HG^2, HG^3, HG^4, HG^5$$

on the six vertices, using the same notation as in Exercise 9.
c) Give the effect of the isometries

$$V, VG, VG^2, VG^3, VG^4, VG^5$$

on the six vertices, using the same notation as above.
d) Make up the multiplication table for the twelve elements

$$I, G, G^2, G^3, G^4, G^5, H, HG, HG^2, HG^3, HG^4, HG^5.$$

16. Using a pentagon instead of a hexagon, answer the same types of questions that were raised in the preceding exercises. Draw the picture, so that the pentagon has one vertex on the vertical axis and admits reflection through the vertical axis as a symmetry. Your picture should be similar to that of Exercise 13, but with a 5-sided figure.

17. Let G be rotation by 72°. What is the smallest positive integer k such that $G^k = I$? Express G^{-1} as a positive power of G.

§5. CHARACTERIZATION OF ISOMETRIES

The main result of this section is that an isometry can be expressed as a composite of a translation, a rotation, and possibly a reflection. We first prove an intermediate result.

Theorem 4. *Let P, Q be distinct points. Let F be an isometry which leaves P and Q fixed. Then either F is the identity, or F is a reflection through the line L_{PQ} passing through P and Q.*

Proof. Let M be a point on the perpendicular bisector of the segment $\overline{PQ}$, but not lying on this segment, as in Fig. 6–29.

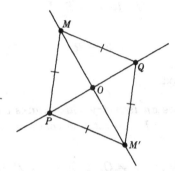

Fig. 6–29

Then

$$d(P, M) = d(Q, M).$$

If M is fixed by F, i.e. if $F(M) = M$, then we can apply Theorem 3 to conclude that F is the identity. Suppose that $F(M) \neq M$. Let $M' = F(M)$ as shown in Fig. 6–33. Since F preserves distances, we have

$$d(P, M') = d(M', Q).$$

Hence by the corollary of the Pythagoras theorem, the point M' lies on the perpendicular bisector of the segment $\overline{PQ}$.

Let O be the point of intersection of $\overline{PQ}$ and $\overline{MM'}$, i.e. the midpoint between P and Q. Again since F preserves distances, and since O is fixed under F (by Theorem 2, §2) we have

$$d(O, M) = d(O, M').$$

Hence M' is the reflection of M through the straight line L_{PQ}. Let R denote the reflection through this line, so that we have $R = R^{-1}$. We also have

$$R(M) = M' \quad \text{and} \quad R(M') = M.$$

Consider the composite isometry

$$R \circ F.$$

It leaves P and Q fixed. Furthermore,

$$R(F(M)) = R(M') = M.$$

Hence $R \circ F$ leaves M fixed. By Theorem 3, we conclude that $R \circ F = I$. Composing with $R^{-1} = R$ on the left, we find

$$R \circ R \circ F = R \circ I,$$

whence

$$F = R,$$

and our theorem is proved.

Theorem 5. *Let F be an isometry which leaves one point O fixed. Then either F is a rotation, or F is a rotation composed with a reflection through a line.*

Proof. Let P be any point $\neq O$. If $F(P) = P$, then we are in the case of Theorem 4 and we are done. Suppose that $F(P) \neq P$, and let $F(P) = P'$. Since F preserves distances, we have

$$d(O, P) = d(O, P').$$

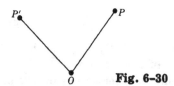

Fig. 6–30

There exists a rotation with respect to O which maps P on P'. Let us denote this rotation by G. We know that G^{-1} is a rotation, and

$$G^{-1}(P') = P.$$

Therefore

$$G^{-1}(F(P)) = G^{-1}(P') = P.$$

This means that $G^{-1} \circ F$ leaves P fixed. But $G^{-1} \circ F$ also leaves O fixed. Hence we can apply Theorem 4, and we conclude that $G^{-1} \circ F$ is either the identity or a reflection R. In the first case, we have

$$G^{-1} \circ F = I,$$

whence composing with G on the left we find

$$G \circ G^{-1} \circ F = G \circ I = G,$$

and hence

$$F = G$$

is a rotation.

In the second case, we find

$$G^{-1} \circ F = R,$$

whence

$$G \circ G^{-1} \circ F = G \circ R$$

and

$$F = G \circ R.$$

This proves our theorem.

Theorem 6. *Let F be an arbitrary isometry of the plane. If F does not leave any point fixed, then F is either a translation, or the composite of a translation and a rotation, or the composite of a translation, a rotation, and a reflection through a line.*

Proof. Suppose that F does not leave any point fixed. Let O be any point and let $P = F(O)$. Let T be the translation such that $T(O) = P$. Then T^{-1} is a translation, and

$$T^{-1}(P) = O.$$

Hence

$$T^{-1}(F(O)) = T^{-1}(P) = O.$$

This means that $T^{-1} \circ F$ leaves O fixed. But $T^{-1} \circ F$ is an isometry. We can therefore apply Theorem 5, and we see that

$$T^{-1} \circ F = G \quad \text{or} \quad T^{-1} \circ F = G \circ R,$$

where G is a rotation and R is a reflection through a line. In the first case, we find

$$F = T \circ G$$

and in the second case we find

$$F = T \circ G \circ R.$$

This proves our theorem.

EXERCISES

1. Prove that every isometry has an inverse.

2. If P is a fixed point for an isometry F, prove that P is also a fixed point for F^{-1}.

3. Let T be the translation by 1 in. to the right and let U be the translation by 1 in. upward. Draw the image of a point P under T^{-1}, T^{-2}, T^{-3}, U^{-1}, U^{-2}, U^{-3}, $T^{-1} \circ U^{-1}$.

§6. CONGRUENCES

Let S, S' be sets of points in the plane. We shall say that S is **congruent** to S' if there exists an isometry F such that the image $F(S)$ is equal to S'.

Theorem 7. *Two circles of the same radius are congruent.*

Proof. Let the first circle be $C(r, O)$, or radius r, centered at O, and let the other circle be $C(r, O')$, centered at O'. Let T be the translation which maps O on O'. We know that T preserves distances. Hence if P is at distance r from O, then $T(P)$ is at distance r from $T(O) = O'$. Hence the image of the circle $C(r, O)$ is contained in the circle $C(r, O')$. We must still show that every point on $C(r, O')$ is the image of a point on $C(r, O)$ under T. Let Q be a point at distance r from O'. Note that the point

$$P = T^{-1}(Q)$$

is at distance r from O, and that $T(P) = T(T^{-1}(Q)) = Q$. This proves our assertion.

To prove that two figures are congruent, it is often useful to use Exercise 2 at the end of this section. We can then change one figure by any number of isometries. It suffices to prove that its image under these isometries is congruent to the other figure. We shall illustrate this by an example from classical geometry, after we prove the next theorem.

Theorem 8. *Any two segments of the same length are congruent.*

Proof. Let $\overline{PQ}$ and $\overline{MN}$ be segments of the same length. Let T be the translation which maps M on P. Then $T(N)$ is at the same distance from $T(M) = P$ as Q, because T is an isometry. Hence there exists a rotation G with respect to P such that $G(T(N)) = Q$. By Theorem 1 of §2, we conclude that $G \circ T$ maps $\overline{MN}$ on $\overline{PQ}$, thus proving our theorem.

The two steps of the proof in Theorem 8 corresponding to T and G are illustrated in Fig. 6–31.

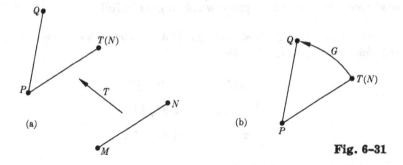

Fig. 6–31

Look at Exercise 2. In the light of this exercise, we could also have phrased the proof of Theorem 8 as follows. We let T be the translation such that $T(M) = P$. Since the image of $\overline{MN}$ under T is congruent to $\overline{MN}$, we are reduced to the case when $P = M$, which we now assume. By assumption,

$$d(P, Q) = d(P, N).$$

Hence there exists a rotation G with respect to P such that $G(N) = Q$. By Theorem 1 of §2, we conclude that $\overline{PN}$ is congruent to $\overline{PQ}$. This concludes the proof.

Using language as we did, reducing the proof to the case when $P = M$, has the slight advantage whereby we avoid having to write the composite

$G \circ T$ explicitly. We shall phrase the proof of Theorem 10 that way also. Note that Theorem 10 is a classical congruence case of elementary courses in plane geometry, finding its natural place within our present system.

Theorem 9. *Let $\triangle PQM$ and $\triangle P'Q'M'$ be right triangles whose right angles are at Q and Q', respectively. Assume that the corresponding legs have the same lengths, that is:*

$$d(P, Q) = d(P', Q')$$

and

$$d(Q, M) = d(Q', M').$$

Then the triangles are congruent.

Proof. We leave the proof as an exercise. Observe that this theorem puts our old assumption **RT** in the context of congruences, as we anticipated in Chapter 5, §3.

Actually, Theorem 9 is a special case of a more general result, stated in the next theorem, and whose proof we shall give in full.

Theorem 10. *Let $\triangle PQM$ and $\triangle P'Q'M'$ be triangles whose corresponding sides have equal lengths, that is*

$$d(P, Q) = d(P', Q'),$$
$$d(P, M) = d(P', M'),$$
$$d(Q, M) = d(Q', M').$$

These triangles are congruent.

Proof. There exists a translation which maps P on P'. Hence it suffices to prove our assertion when $P = P'$ (cf. Exercise 2). We now assume this, i.e. $P = P'$. Since $d(P, Q) = d(P, Q')$, there exists a rotation relative to P which maps Q on Q'. This rotation leaves P fixed. Again by Exercise 2, we are reduced to the case when

$$P = P'$$

and

$$Q = Q'.$$

We assume that this is the case. Now either $M = M'$, or $M \neq M'$. Suppose $M \neq M'$. We illustrate this by Fig. 6–32.

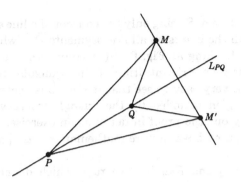

Fig. 6–32

Let L_{PQ} be the line passing through P and Q. By the Corollary of the Pythagoras theorem, and the fact that

$$d(P, M) = d(P, M')$$

and

$$d(Q, M) = d(Q, M'),$$

we conclude that L_{PQ} is the perpendicular bisector of $\overline{MM'}$. In particular, M' is the reflection of M through L_{PQ}. Hence if we reflect M' through L_{PQ}, we get M. Thus we have found a composite of isometries which map P on P', Q on Q', and M on M'. By Theorem 1 of §2 we conclude that our triangles are congruent, as was to be shown.

Remark. In Theorems 7 through 10 we have dealt with figures which consist of line segments. Of course, we may also want to deal with other types of figures, for instance discs (cf. Exercise 1), or, say, the triangular region bounded by a triangle, or the region bounded by a rectangle. Because of this, it is useful to have a description of these regions in terms of line segments. We treat the triangle as an example.

Let $\triangle PQM$ be a triangle, and let S be the region bounded by the triangle. We represent S as the shaded region in Fig. 6–33(a).

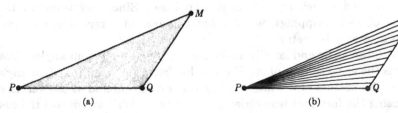

(a)   (b)

Fig. 6–33

We can give a definition of S using only the concept of a line segment by saying that S consists of all the points on all line segments $\overline{PX}$, where X ranges over all points of $\overline{QM}$. Looking at Fig. 6–33(b) convinces you that this indeed coincides with your geometric intuition of the triangular region. Using this definition, it is then very easy to see that if F is an isometry, the image $F(S)$ is the triangular region bounded by the triangle whose vertices are $F(P)$, $F(Q)$, $F(M)$. Carry out the proof in detail as an exercise. This definition is also the one that is used both in pure mathematics and applied mathematics (e.g. economics).

Similarly, let R_{PQ} and R_{PM} be two rays which define an angle whose measure is less than 180°. The angle can be described as the set of all points on all segments $\overline{XY}$, where X is a point on R_{PQ} and Y is a point on R_{PM}, as in Fig. 6–34.

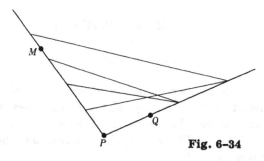

Fig. 6–34

Isometries and area. In the next chapter, we shall discuss the notion of area. We may be interested as to how the area of a region behaves under an isometry. It is natural to take the following statement as a basic axiom.

Let S be a region of the plane, whose area is equal to a. Let F be an isometry. Then the area of $F(S)$ is also equal to a.

To convince ourselves that this is a reasonable statement, we can use the characterization of isometries. If we visualize rotations, reflections, or translations, then our intuition tells us that, in each case, the area of a region is preserved under each one of these mappings. Since any isometry is a composite of such mappings, we see that the area of a region is preserved under an arbitrary isometry.

Let A be an angle and let F be an isometry. Then $F(A)$ is an angle, whose measure is the same as that of A. We see this from the definition of an angle, looking at the portion of the angle A lying in a disc centered at the vertex of A, and using the fact that isometries preserve area. We have drawn the case when the isometry is the translation $T_{PP'}$ in Fig. 6–35. Note, however, that a reflection reverses the order of the rays which are used to compute the measure of the angle in counterclockwise direction. Draw the picture.

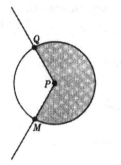 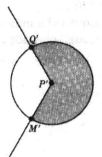

Fig. 6–35

EXERCISES

1. Prove that two discs of the same radius are congruent.

2. Let S, S', S'' be sets in the plane. Prove that if S is congruent to S', and S' is congruent to S'', then S is congruent to S''. Prove that if S is congruent to S', then S' is congruent to S.

3. Prove that two squares whose sides have the same length are congruent.

4. Prove that any two lines are congruent.

5. Let $\triangle PQM$ be a triangle whose three angles all have 60°. Prove that the sides have equal length. [*Hint:* From any vertex draw the perpendicular to the other side, and reflect through this perpendicular.]

6. Prove Theorem 9. At first you are not allowed to use Theorem 10. If you were allowed to use Theorem 10, how could you deduce Theorem 9 from it?

7. Let $\triangle PQM$ and $\triangle P'Q'M'$ be triangles having one corresponding angle of the same measure, say, $\angle PQM$ and $\angle P'Q'M'$ have the same measure, and having adjacent sides of the same length, i.e.

$$d(P, Q) = d(P', Q') \quad \text{and} \quad d(Q, M) = d(Q', M').$$

Prove that the triangles are congruent.

8. Prove that two rectangles having corresponding sides of equal lengths are congruent.

9. Give a definition of the region bounded by a square in terms of line segments. Same thing for a rectangle.

10. Let A be the angle shown on Fig. 6–36. Draw the image of A under

    a) rotation by 60°,   b) rotation by 90°,

    c) rotation by 120°,   d) reflection through O,

    e) reflection through the indicated line L,

    f) reflection through one side of the angle.

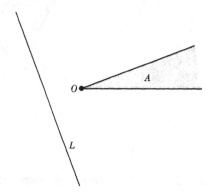

Fig. 6–36

11. Let $\triangle PQM$ and $\triangle P'Q'M'$ be triangles whose corresponding angles have the same measures (i.e. the angle with vertex at P has the same measure as the angle with vertex at P', and similarly for the angles with vertices at Q, Q' and M, M'). Assume that $d(P, Q) = d(P', Q')$. Prove that the triangles are congruent.

12. Let $\triangle PQM$ be a triangle. Let L_1, L_2, L_3 be the three lines which bisect the three angles of the triangle, respectively. Let O be the point of intersection of L_1 and L_2. Prove that O lies on L_3. [*Hint:* From O, draw the perpendicular segments to the corresponding sides. Prove that their lengths are equal.]

7 Area and Applications

§1. AREA OF A DISC OF RADIUS r

We assume that the notion of area and its basic properties, corresponding to our simple intuition of area, are known. In particular, the area of a square of side a is a^2, and the area of a rectangle whose sides have lengths a, b is ab. (Remember, a unit of length is fixed throughout, and determines a unit of area.)

Let r be a positive number, and consider dilation by r. We wish to analyze what happens to area under such a dilation. We start with the simplest case, that of a rectangle. Consider a rectangle whose sides have lengths a and b as on Fig. 7-1(a). Suppose that we multiply the lengths of the sides by 2, and obtain the rectangle illustrated on Fig. 7-1(b).

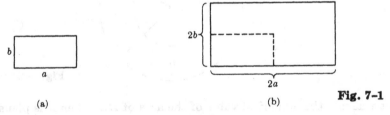

(a)

(b)

Fig. 7-1

Then the sides of this dilated rectangle have lengths $2a$ and $2b$. Hence the area of the dilated rectangle is equal to $2a2b = 4ab = 2^2ab$. Similarly, suppose that we dilate the sides by 3, as illustrated on Fig. 7-1(c).

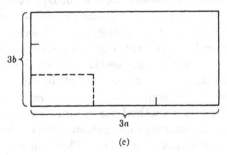

(c)

Fig. 7-1 (cont.)

173

Then the sides of the dilated rectangle have lengths $3a$ and $3b$, whence the area of the dilated rectangle is equal to $3a3b = 9ab = 3^2ab$.

In general, let S be a rectangle whose sides have lengths a, b respectively. Let rS be the dilation of S by r. Then the sides of rS have lengths ra and rb respectively, so that the area of the dilated rectangle rS is equal to

$$(ra)(rb) = r^2ab.$$

Thus the area of rectangles changes by r^2 under dilation by r.

This makes it very plausible that if S is an arbitrary region of the plane, whose area can be approximated by the area of a finite number of rectangles, then the area of S itself changes by r^2 under dilation by r. In other words,

$$\text{area of } rS = r^2(\text{area of } S).$$

For instance, let D_r be the disc of radius r, so that $D_1 = D$ is the disc of radius 1, both centered at the origin (see Fig. 7–2). Then $D_r = rD_1$.

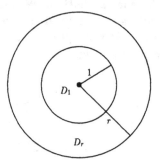

Fig. 7–2

Let π denote the numerical value of the area of D_1. Then it is plausible that

$$\text{area of } D_r = \pi r^2.$$

We used the symbol π to denote the area of D_1. It is of course a problem to determine its numerical value. Various devices allow us to do this, and we find the familiar decimal, $\pi = 3.14159\ldots$. There is the possibility that you have heard of π only as the ratio of the circumference to the diameter of a circle of radius 1. In the next section, we shall indicate how to prove that this ratio has the same value as the area of the disc of radius 1. Thus the π we are using now is the same one that you may know already. This relationship then gives us a method for computing π. For instance, get a circular pan and a soft measuring tape, measure the circumference of the pan, measure its diameter, and take the ratio. This will give you a value for π, good at least to one decimal place. There are more sophisticated ways of finding more decimal places for π, and you will learn some of these in a subsequent course in calculus.

One of the methods used to compute π is also the method which convinces us that the area of a disc of radius r is πr^2. Namely, we approximate the disc by rectangles, or even squares, as in Fig. 7–3.

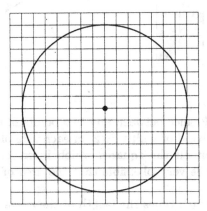

Fig. 7–3

We draw a grid consisting of vertical and horizontal lines at the same distance from each other, thus determining a decomposition of the plane into squares. If the grid is fine enough, that is, if the sides of the square are sufficiently small, then the area of the disc is approximately equal to the sum of the areas of the squares which are contained in the disc. The difference between the area of the disc and this sum will be, at most, the sum of the areas of the squares which intersect the circle. To determine the area of the disc approximately, you just count all the squares that lie inside the circle, measure their sides, add up their areas, and get the desired approximation. Using fine graph paper, you can do this yourself and arrive at your own approximation of the area of the disc.

Of course, you want to estimate how good your approximation is. The difference between the sum of the areas of all the little squares contained in the disc and the area of the disc itself is determined by all the small portions of squares which touch the boundary of the disc, i.e. which touch the circle. We have a very strong intuition that the sum of such little squares will be quite small if our grid is fine enough, and in fact, we give an estimate for this smallness in the following discussion.

Suppose that we make the grid so that the squares have sides of length c. Then the diagonal of such a square has length $c\sqrt{2}$. If a square intersects the circle, then any point on the square is at distance at most $c\sqrt{2}$ from the circle. Look at Fig. 7–4(a).

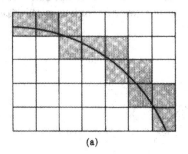

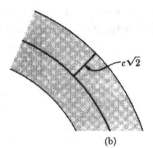

(a)   (b)

Fig. 7–4

This is because the distance between any two points of the square is at most $c\sqrt{2}$. Let us draw a band of width $c\sqrt{2}$ on each side of the circle, as shown in Fig. 7–4(b). Then all the squares which intersect the circle must lie within that band. It is very plausible that the area of the band is at most equal to

$$2c\sqrt{2} \text{ times the length of the circle.}$$

Thus if we take c to be very small, i.e. if we take the grid to be a very fine grid, then we see that the area of the disc is approximated by the area covered by the square lying entirely inside the disc.

The same type of argument also works for more general regions. For instance, we have drawn a region S inside a curve in Fig. 7–5(a), and we have drawn the dilation of S by 2 in Fig. 7–5(b). Then the area of $2S$ is equal to $4A$, where A is the area of S.

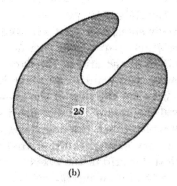

(a)   (b)

Fig. 7–5

We also illustrate the fact that these areas are approximated by squares in Fig. 7–6(a) and (b).

(a)

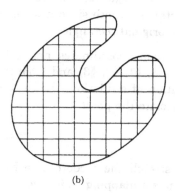

(b)

Fig. 7–6

EXERCISES

1. a) Draw a rectangle whose sides have lengths 2 in. and $\frac{1}{2}$ in., respectively. What is the area of this rectangle?
   b) Draw the rectangle whose sides have lengths equal to twice the length of the sides of the rectangle in part (a). What is the area of this rectangle?
   c) Same question for the rectangle whose sides have lengths equal to three times the lengths of the sides of the rectangle in part (a).
   d) Same question for the rectangle whose sides have lengths equal to one-half the lengths of the sides of the rectangle in part (a).

2. a) Draw a rectangle whose sides have lengths $\frac{2}{3}$ in. and 2 in. respectively. What is the area of this rectangle? Draw the rectangle whose sides have lengths equal to:
   b) twice,
   c) three times,
   d) half the lengths of the sides of the rectangle in part (a). In each case, what is the area of the rectangle?

e) Suppose that a rectangle has an area equal to 15 in². Dilate the rectangle by 2. What is the area of the dilated rectangle?

f) If a rectangle has an area equal to 25 in², what is the area of the rectangle whose sides have lengths equal to one-fifth the lengths of the original rectangle?

Read about coordinates and dilations in terms of coordinates, that is, Chapter 8, §1, §2, §3, and Chapter 9, §1. Then consider the following generalization of a dilation. Let $a > 0$, $b > 0$. To each point (x, y) of the plane, associate the point

$$(ax, by).$$

Thus we stretch the x-coordinate by a and the y-coordinate by b. This association is a mapping which we may denote by $F_{a,b}$.

3. a) Suppose that the sides of a rectangle S have lengths r and s. What are the lengths of the sides of the rectangle $F_{a,b}(S)$, i.e. of the rectangle obtained by the mixed dilation $F_{a,b}$?

b) What is the volume of $F_{a,b}(S)$?

c) If S is a bounded region in the plane with volume V, what is the volume of $F_{a,b}(S)$?

4. a) Show that the set of points (u, v) satisfying the equation

$$\left(\frac{u}{a}\right)^2 + \left(\frac{v}{b}\right)^2 = 1$$

is the image of the circle of radius 1 centered at O under the map $F_{a,b}$.

b) Let $a = 3$ and $b = 2$. Sketch this set, which is called an **ellipse**.

c) Can you guess and motivate your guess as to what the area of the region bounded by the ellipse in (a) should be?

5. What is the area of the region bounded by the following ellipses:

a) $\left(\frac{x}{7}\right)^2 + \left(\frac{y}{4}\right)^2 = 1$?   b) $\left(\frac{x}{3}\right)^2 + \left(\frac{y}{7}\right)^2 = 1$?

c) $\dfrac{x^2}{6} + \dfrac{y^2}{3} = 1$?   d) $\dfrac{x^2}{5} + \dfrac{y^2}{7} = 1$?

6. What is the area of the region bounded by the following ellipses:

a) $3x^2 + 4y^2 = 1$?   b) $2x^2 + 5y^2 = 1$?

c) $4x^2 + 9y^2 = 1$?   d) $4x^2 + 25y^2 = 1$?

7. Write up a discussion of how to give coordinates (x, y, z) to a point in 3-space. In terms of these coordinates, what would be the effect of dilation by r?

8. Generalize the discussion of this section to the 3-dimensional case. Specifically:

   a) Under dilation by r, how does the volume of a cube change?
   b) How does the volume of a rectangular box with sides a, b, c change? Draw a picture, say for $r = \frac{1}{2}$, $r = 2$, $r = 3$, arbitrary r.
   c) How would the volume of a 3-dimensional solid change under dilation by r?
   d) The volume of the solid ball of radius 1 in 3-space is equal to $\frac{4}{3}\pi$. What is the volume of the ball of radius r in 3-space?

9. Write down the equation of a sphere of radius r centered at the origin in 3-space.

10. How would you define the volume of a rectangular solid whose sides have lengths a, b, c?

11. Let a, b, c be positive numbers. Let $\mathbf{R}^3$ be 3-space, that is, the set of all triples of numbers (x, y, z). Let

$$F_{a,b,c}: \quad \mathbf{R}^3 \to \mathbf{R}^3$$

be the mapping

$$(x, y, z) \longmapsto (ax, by, cz).$$

Thus $F_{a,b,c}$ is a generalization to 3-space of our mixed dilation $F_{a,b}$.

   a) What is the image of a cube whose sides have length 1 under $F_{a,b,c}$?
   b) A rectangular box S has sides of lengths r, s, t respectively. What are the lengths of the sides of the image $F_{a,b,c}(S)$? What is the volume of $F_{a,b,c}(S)$?
   c) Let S be a solid in 3-space, and let V be its volume. In terms of V, a, b, c, what is the volume of the image of S under $F_{a,b,c}$?

12. What is the volume of the solid in 3-space consisting of all points (x, y, z) satisfying the inequality

$$\left(\frac{x}{3}\right)^2 + \left(\frac{y}{2}\right)^2 + \left(\frac{z}{7}\right)^2 \leq 1?$$

13. What is the volume of the solid in 3-space consisting of all points (x, y, z) satisfying the inequality

$$\frac{x^2}{5} + \frac{y^2}{3} + \frac{z^2}{10} \leq 1?$$

14. Let a, b, c be numbers > 0. What is the volume of the solid in 3-space consisting of all points (x, y, z) satisfying the inequality

$$\left(\frac{x}{a}\right)^2 + \left(\frac{y}{b}\right)^2 + \left(\frac{z}{c}\right)^2 \leq 1?$$

15. What about 4-space? n-space for arbitrary n?

§2. CIRCUMFERENCE OF A CIRCLE OF RADIUS r

Let C be the circle of radius 1, and let C_r be the circle of radius r. We wish to convince ourselves that

$$\boxed{\begin{aligned} \text{length of } C_1 &= 2\pi, \\ \text{length of } C_r &= 2\pi r. \end{aligned}}$$

We studied area in the preceding section by approximating a region S by means of squares. To study length, we have to approximate a curve by means of straight line segments.

On the following picture, we show how to approximate a curve by 6 segments.

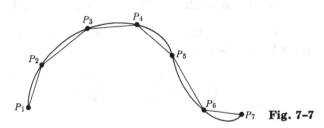

Fig. 7-7

To approximate a circle by segments, we select a special kind of segment. We decompose the disc of radius r into n sectors whose angles have

$$\frac{360}{n} \text{ degrees.}$$

Here, n is an integer. The picture is that of Fig. 7–8, drawn with $n = 7$.

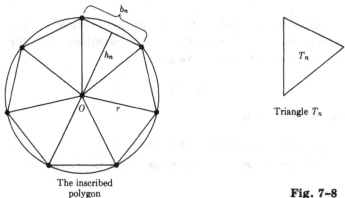

The inscribed
polygon

Triangle T_n

Fig. 7–8

You should now do Exercise 1, taking special values for n, namely $n = 4$, $n = 5$, $n = 6$, $n = 8$ (more if you wish). In each case, determine the number of degrees

$$\frac{360}{4}, \quad \frac{360}{5}, \quad \frac{360}{6}, \quad \frac{360}{8}, \quad \text{etc.,}$$

and draw the corresponding sectors. In general we then join the end points of the sectors by line segments, thus obtaining a polygon inscribed in the circle. The region bounded by this polygon consists of n triangles congruent to the same triangle T_n, lying between the segments from the origin O to the vertices of the polygon. Thus in the case of 4 sides, we call the triangle T_4. In the figure with 5 sides, we call the triangle T_5. In the figure with 6 sides, we call the triangle T_6. In the figure with 8 sides, we call the triangle T_8. And so on, to the polygon having n sides, when we call the triangle T_n.

We denote the (length of the) base of T_4 by b_4 and its height by h_4. We denote the base of T_5 by b_5 and its height by h_5. In general, we denote the base of T_n by b_n and its height by h_n, as indicated on Fig. 7–8. Since the area of a triangle whose base has length b and whose height has length h

is $\frac{1}{2}bh$, we see that the area of our triangle T_n is given by

$$\text{area of } T_n \;=\; \tfrac{1}{2}b_n h_n.$$

Let A_n be the area of the region surrounded by the polygon, and let P_n be the perimeter of the polygon. Since the polygonal region consists of n triangular regions congruent to the same triangle T_n, we find that the area of A_n is equal to n times the area of T_n, or in symbols,

$$A_n \;=\; n \cdot \text{area of } T_n \;=\; \tfrac{1}{2}n b_n h_n.$$

On the other hand, the perimeter of the polygon consists of n segments whose lengths are all equal to b_n. Hence $P_n = nb_n$. Substituting P_n for nb_n in the value for A_n which we just found, we get

$$A_n \;=\; \tfrac{1}{2}P_n h_n.$$

As n becomes arbitrarily large,

A_n approaches the area of the disc D_r,

P_n approaches the circumference of the circle C_r,

and

h_n approaches the radius r of the disc.

For instance, if we double the number of sides of the polygon successively, the picture looks like Fig. 7–9.

Fig. 7–9

Let c denote the circumference of the circle of radius r. Then A_n approaches πr^2. Since $A_n = \tfrac{1}{2}P_n h_n$, it follows that A_n also approaches $\tfrac{1}{2}cr$. Thus we obtain

$$\pi r^2 \;=\; \tfrac{1}{2}cr.$$

We cancel r from each side of this equation, and multiply both sides by 2. We conclude that

$$c = 2\pi r,$$

as was to be shown.

We have discussed above the behavior of area under dilation. We conclude this section by a discussion of the behavior of length under dilation.

Let r be a positive number. Under dilation by r, the distance between two points is multiplied by r; in other words,

$$d(rP, rQ) = r \cdot d(P, Q).$$

When we have coordinates later, this will be proved. We can already justify it to some extent by using the Pythagoras theorem. Consider a right triangle $\triangle POQ$ with sides a, b as shown in Fig. 7–10, such that the right angle is at the origin O.

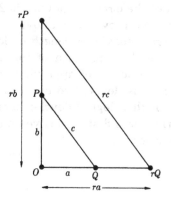

Fig. 7–10

Under dilation by r, the two sides are dilated by r, so that the three points

$$rP, \quad O, \quad rQ,$$

form a right triangle whose legs have lengths ra and rb, respectively. By the Pythagoras theorem, the hypotenuse of the dilated triangle has length

$$\sqrt{r^2 a^2 + r^2 b^2} = \sqrt{r^2(a^2 + b^2)}.$$

If we let

$$c = \sqrt{a^2 + b^2},$$

we see that the hypotenuse has length rc. Thus the length of the hypotenuse also gets multiplied by r. As we shall see later when we have coordinates, this is also true even if the vertex of the right angle is not necessarily at the origin.

To investigate what happens to the length of an arbitrary curve S under dilation, we approximate the curve by segments as on Fig. 7–7. In general, suppose that we approximate the curve by n segments where n is a positive integer. Let these segments be $S_1, \ldots, S_n$. Let $l(S_i)$ denote the length of S_i. If we dilate the curve by r, then rS is approximated by the segments $rS_1, \ldots, rS_n$. Hence the length of rS is approximated by

$$l(rS_1) + \cdots + l(rS_n) = r[l(S_1) + \cdots + l(S_n)].$$

Thus we see that whenever we can approximate S by line segments, we have the formula

$$\text{length of } rS = r(\text{length of } S).$$

In particular, the length of the circle of radius 1 is 2π, and the length of the circle of radius r is $2\pi r$. We proved this above, and we now see that it is compatible with the general behavior of length under dilation.

Our arguments are based on the idea of taking a limit as n becomes arbitrarily large, and on the notion of approximation. These are the basic ideas of the calculus, which is devoted to systematizing these notions and giving an analytic basis for their logical development. However, it is always useful to have the intuitive ideas first. Thus you may view this section as a good introduction to the calculus.

EXERCISES

1. Draw the picture of a polygon with sides of equal length, inscribed in a circle of radius 2 inches, in the cases when the polygon has:

   a) 4 sides,   b) 5 sides,   c) 6 sides,
   d) 8 sides,   e) 9 sides.

   Draw the radii from the center of the circle to the vertices of the polygon. Use a protractor for the angles of the sectors. Using a ruler, measure (approximately) the base of each triangle and measure its height. From your measurements, compute the area inside the polygon, and the circumference of the polygon. Compare these values with the area of the disc and its circumference, given as πr^2 and $2\pi r$ respectively, and $r = 2$. Use the value $\pi = 3.14$.

2. Get a tin can with as big a circular bottom as possible. Take a tape, measure the circumference of the bottom, measure the diameter, take the ratio and get a value for π, probably good to one decimal place. Do the same thing to another circular object, say a frying pan, and verify that you get the same value for π.

3. Read the definitions of coordinates at the beginning of Chapter 8. Then read the sections of Chapter 16 on induction and summation, and do the exercises at the end of these sections. The material just mentioned is logically self-contained, so you should have no trouble. It provides a direct continuation of the topics which were discussed concerning area and volumes. You will see how to compute the volume of a sphere and a cone, or other similar figures, in 3-space.

Part Three
COORDINATE GEOMETRY

As stated before, we shall see how to use coordinates to give definitions for geometric terms using only properties of numbers.

The chapters in this part are logically independent of each other to a large extent. For instance, the chapter on trigonometry could be read immediately after the introduction of coordinates in Chapter 8, §1, and the discussion of the distance formula between points. Many readers may want to do that, instead of reading first about translations, addition of points, etc. On the other hand, the chapter on segments, rays, and lines is also independent of the trigonometry and the analytic geometry. These three chapters can be read in any order. Take your pick as to what approach you like most.

Giving coordinates to points not only allows us to give analytic proofs. It allows us also to compute in a way that the "intuitive" geometry did not. For instance, given a coordinatized definition of straight lines, we can compute the point of intersection of two lines explicitly.

8 Coordinates and Geometry

§1. COORDINATE SYSTEMS

Once a unit length is selected, we can represent numbers as points on a line. We shall now extend this procedure to the plane, and to pairs of numbers.

In Fig. 8–1, we visualize a horizontal line and a vertical line intersecting at a point O, called the **origin**.

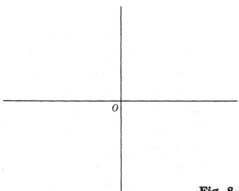

Fig. 8–1

These lines will be called **coordinate axes,** or simply **axes.**

We select a unit length and cut the horizontal line into segments of lengths 1, 2, 3, . . . to the left and to the right. We do the same to the vertical line, but up and down, as indicated on Fig. 8–2.

191

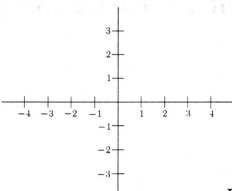

Fig. 8-2

On the vertical line, we visualize the points going below O as corresponding to the negative integers, just as we visualized points on the left of the horizontal line as corresponding to negative integers. We follow the same idea as that used in grading a thermometer, where the numbers below zero are regarded as negative.

We can now cut the plane into squares whose sides have length 1.

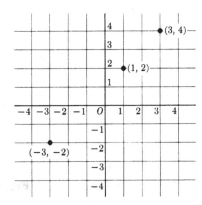

Fig. 8-3

We can describe each point where two lines intersect by a pair of integers. Suppose that we are given a pair of integers, like (1, 2). We go 1 unit to the right of the origin and up 2 units vertically to get the point (1, 2) which has been indicated in Fig. 8-3. We have also indicated the point (3, 4). The diagram is just like a map.

Furthermore, we could also use negative numbers. For instance, to describe the point $(-3, -2)$, we go 3 units to the left of the origin and 2 units vertically downward.

There is actually no reason why we should limit ourselves to points which are described by integers. For instance we can also describe the point $(\frac{1}{2}, -1)$ and the point $(-\sqrt{2}, 3)$ as on Fig. 8–4.

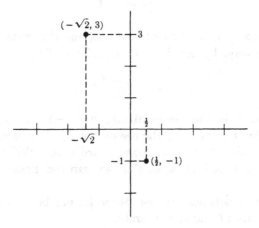

Fig. 8–4

In general, if we take any point P in the plane and draw the perpendicular lines to the horizontal axis and to the vertical axis, we obtain two numbers x, y as on Fig. 8–5.

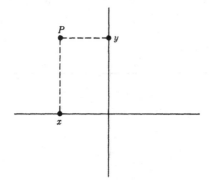

Fig. 8–5

We then say that the numbers x, y are the **coordinates** of the point P, and we write

$$P = (x, y).$$

Conversely, every pair of numbers (x, y) determines a point of the plane. If x is positive, then this point lies to the right of the vertical axis. If x is negative, then this point lies to the left of the vertical axis. If y is positive, then this point lies above the horizontal axis. If y is negative, then this point lies below the horizontal axis.

The coordinates of the origin are

$$O = (0, 0).$$

We usually call the horizontal axis the x-axis, and the vertical axis the y-axis (exceptions will always be noted explicitly). Thus if

$$P = (5, -10),$$

then we say that 5 is the x-**coordinate** and -10 is the y-**coordinate.** Of course, if we don't want to fix the use of x and y, then we say that 5 is the first coordinate, and -10 is the second coordinate. What matters here is the ordering of the coordinates, so that we can distinguish between a first and a second.

We can, and sometimes do, use other letters besides x and y for coordinates, for instance t and s, or u and v.

Our two axes separate the plane into four **quadrants,** which are numbered as indicated in Fig. 8–6.

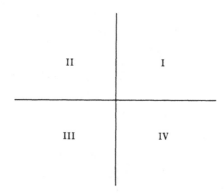

Fig. 8–6

A point (x, y) lies in the **first quadrant** if and only if both x and y are > 0. A point (x, y) lies in the **fourth quadrant** if and only if $x > 0$, but $y < 0$.

Finally, we note that we placed our coordinates horizontally and vertically for convenience. We could also place the coordinates in a slanted way, as shown on Fig. 8–7.

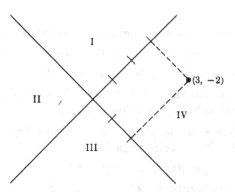

Fig. 8-7

In Fig. 8-7, we have indicated the quadrants corresponding to this coordinate system, and we have indicated the point $(3, -2)$ having coordinates 3 and -2 with respect to this coordinate system. Of course, when we change the coordinate system, we also change the coordinates of a point.

Remark. Throughout this book, when we select a coordinate system, the positive direction of the second axis will always be determined by rotating counterclockwise the positive direction of the first axis through a right angle.

We observe that the selection of a coordinate system amounts to the same procedure that is used in constructing a map. For instance, on the following (slightly distorted) map, the coordinates of Los Angeles are $(-6, -2)$, those of Chicago are $(3, 2)$, and those of New York are $(7.2, 3)$. (View the distortion in the same spirit as you view modern art.)

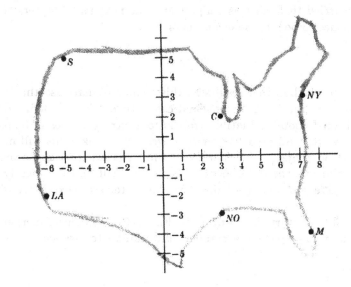

Fig. 8-8

EXERCISES

In these exercises, draw the coordinate axes horizontally and vertically.

1. Plot the following points: $(-1, 1)$, $(0, 5)$, $(-5, -2)$, $(1, 0)$.

2. Plot the following points: $(\frac{1}{2}, 3)$, $(-\frac{1}{3}, -\frac{1}{2})$, $(\frac{4}{3}, -2)$, $(-\frac{1}{4}, -\frac{1}{2})$.

3. Let (x, y) be the coordinates of a point in the second quadrant. Is x positive or negative? Is y positive or negative?

4. Let (x, y) be the coordinates of a point in the third quadrant. Is x positive or negative? Is y positive or negative?

5. Plot the following points: $(1.2, -2.3)$, $(1.7, 3)$.

6. Plot the following points: $(-2.5, \frac{1}{3})$, $(-3.5, \frac{5}{4})$.

7. Plot the following points: $(1.5, -1)$, $(-1.5, -1)$.

8. What are the coordinates of Seattle (S), Miami (M) and New Orleans (NO) on our map?

Big exercise

To be carried through this chapter and the next two chapters: Define a point in 3-space to be a triple of numbers

$$(x_1, x_2, x_3).$$

Try to formulate the same results in this 3-dimensional case that we have done in the book for the 2-dimensional case. In particular, define distance, define addition of points, dilation, translations, straight lines in 3-space, the whole lot. Write it all up as if you were writing a book. This will make you really learn the subject. Point out the similarities with the 2-dimensional case, and point out the differences if any. You will find practically no difference! To give you some guidelines, we shall often state explicitly what you should do.

So to get you started, we draw in Fig. 8–9 a system of perpendicular coordinate axes in 3-space in a manner quite similar to 2-space.

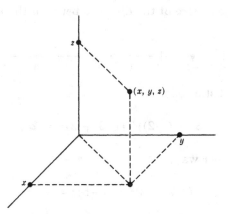

Fig. 8-9

Thus a point in 3-space is represented by three numbers, and from the analytic point of view, we define such a point to be a triple of numbers (x, y, z). For instance, $(1, -2, 7)$ is a point in 3-space. We denote 3-space by $\mathbf{R}^3$.

§2. DISTANCE BETWEEN POINTS

First let us consider the distance between two points on a line. For instance, the distance between the points 1 and 4 on the line is $4 - 1 = 3$.

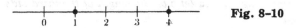

Fig. 8-10

Observe that if we take the difference between 1 and 4 in the other direction, namely $1 - 4$, then we find -3, which is negative. However, if we then take the square, we find the same number, namely

$$(-3)^2 = 3^2 = 9.$$

Thus when we take the square, it does not matter in which order we took the difference.

Example. Find the square of the distance between the points -2 and 3 on the line.

Fig. 8–11

The square of this distance is

$$(3 - (-2))^2 = (3 + 2)^2 = 25;$$

or computing the other way,

$$(-2 - 3)^2 = (-5)^2 = 25.$$

Note that again we take the difference between the coordinates of the points, and that we can deal with points having negative coordinates. If we want the distance rather than its square, then we take the square root, and we find

$$\sqrt{(-5)^2} = \sqrt{5^2} = \sqrt{25} = 5.$$

Because of our universal convention that the square root of a positive number is taken to be positive, we see that we can express the general formula for the distance between points on a line as follows.

Let x_1, x_2 be points on a line. Then the distance between x_1 and x_2 is equal to

$$\sqrt{(x_1 - x_2)^2}.$$

Next we discuss distance between points in the plane, given a coordinate system, which we draw horizontally and vertically for convenience. We recall the Pythagoras theorem from plane geometry.

In a right triangle, let a, b be the lengths of the legs (i.e. the sides forming the right angle), and let c be the length of the third side (i.e. the hypotenuse). Then

$$a^2 + b^2 = c^2.$$

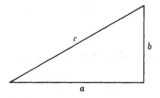

Fig. 8–12

Example. Let $(1, 2)$ and $(3, 5)$ be two points in the plane. Using the Pythagoras theorem, we wish to find the distance between them. First we draw the picture of the right triangle obtained from these two points, as on Fig. 8–13.

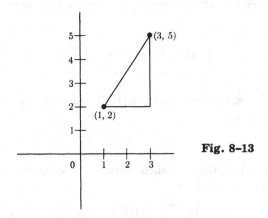

Fig. 8–13

We see that the square of the length of one side is equal to

$$(3 - 1)^2 = 4.$$

The square of the length of the other side is

$$(5 - 2)^2 = (3)^2 = 9.$$

By the Pythagoras theorem, we conclude that the square of the length between the points is $4 + 9 = 13$. Hence the distance itself is $\sqrt{13}$.

Now in general, let (x_1, y_1) and (x_2, y_2) be two points in the plane. We can again make up a right triangle, as shown in Fig. 8–14.

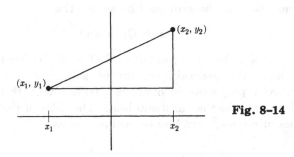

Fig. 8–14

The square of the bottom side is $(x_2 - x_1)^2$, which is also equal to $(x_1 - x_2)^2$. The square of the vertical side is $(y_2 - y_1)^2$, which is also equal to $(y_1 - y_2)^2$. If d denotes the distance between the two points, then

$$d^2 = (x_2 - x_1)^2 + (y_2 - y_1)^2,$$

and therefore we get the formula for the distance between the points, namely

$$d = \sqrt{(x_2 - x_1)^2 + (y_2 - y_1)^2}.$$

Example. Let the two points be $(1, 2)$ and $(1, 3)$. Then the distance between them is equal to

$$\sqrt{(1 - 1)^2 + (3 - 2)^2} = 1.$$

Example. Find the distance between the points $(-1, 5)$ and $(4, -3)$. This distance is equal to

$$\sqrt{(4 - (-1))^2 + (-3 - 5)^2} = \sqrt{89}.$$

Example. Find the distance between the points $(2, 4)$ and $(1, -1)$. The square of the distance is equal to

$$(1 - 2)^2 + (-1 - 4)^2 = 26.$$

Hence the distance is equal to $\sqrt{26}$.

Warning: Always be careful when you meet minus signs. Place the parentheses correctly and remember the rules of algebra.

As it should be, we can compute the distance between points in any order. In the last example, the square of the distance is equal to

$$(2 - 1)^2 + (4 - (-1))^2 = 26.$$

If we let $d(P, Q)$ denote the distance between points P and Q, then our last remark can also be expressed by saying that

$$d(P, Q) = d(Q, P).$$

Note that this is the basic property **DIST 2** of Chapter 5.

We have the general program of giving foundations for geometry assuming only properties of numbers. In line with this program, we must make clear what we take as definitions. The rules of the game are that only properties of numbers can be assumed as known.

We therefore define the **plane** to be the set of all pairs (x, y) of real numbers. We denote the plane by $\mathbf{R}^2$.

If $X = (x_1, x_2)$ and $Y = (y_1, y_2)$ are two points of the plane, then we *define* the **distance** between them to be

$$d(X, Y) = \sqrt{(x_2 - x_1)^2 + (y_2 - y_1)^2}.$$

Thus we see that we have defined geometric objects using only numbers.

EXERCISES

Find the distance between the following points P and Q. Draw these points on a sheet of graph paper.

1. $P = (2, 1)$ and $Q = (1, 5)$

2. $P = (-3, -1)$ and $Q = (-4, -6)$

3. $P = (-2, 1)$ and $Q = (3, 7)$

4. $P = (-3, -4)$ and $Q = (-2, -3)$

5. $P = (3, -2)$ and $Q = (-6, -7)$

6. $P = (-3, 2)$ and $Q = (6, 7)$

7. $P = (-3, -4)$ and $Q = (-1, -2)$

8. $P = (-1, 5)$ and $Q = (-4, -2)$

9. $P = (2, 7)$ and $Q = (-2, -7)$

10. $P = (3, 1)$ and $Q = (4, -1)$

11. Prove that if $d(P, Q) = 0$, then $P = Q$. Thus we have now proved two of the basic properties of distance.

12. Let $A = (a_1, a_2)$ and $B = (b_1. b_2)$. Let r be a positive number. Write down the formula for $d(A, B)$. Define the **dilation** rA to be

$$rA = (ra_1, ra_2).$$

For instance, if $A = (-3, 5)$ and $r = 7$, then $rA = (-21, 35)$. If $B = (4, -3)$ and $r = 8$, then $rB = (32, -24)$. Prove in general that

$$d(rA, rB) = r \cdot d(A, B).$$

We shall investigate dilations more thoroughly in the next chapter.

Exercise on 3-space

Let

$$A = (a_1, a_2, a_3) \quad \text{and} \quad B = (b_1, b_2, b_3)$$

be points in 3-space. Define the distance between them to be

$$d(A, B) = \sqrt{(b_1 - a_1)^2 + (b_2 - a_2)^2 + (b_3 - a_3)^2}.$$

This generalizes the Pythagoras theorem. Draw a picture. Draw the segment between the origin $O = (0, 0, 0)$ and a point $X = (x, y, z)$. Write down the simple formula for the distance between $(0, 0, 0)$ and (x, y, z). Draw right triangles, showing that geometrically, our formula for the distance between O and X in 3-space can be justified in terms of an iterated application of the ordinary Pythagoras theorem in a horizontal plane and a vertical plane. Look back at the exercises of Chapter 5, §3, and note their relation with the present considerations.

13. Give the value for the distance between the following.

   a) $P = (1, 2, 4)$ and $Q = (-1, 3, -2)$
   b) $P = (1, -2, 1)$ and $Q = (-1, 1, 1)$
   c) $P = (-2, -1, -3)$ and $Q = (3, 2, 1)$
   d) $P = (-4, 1, 1)$ and $Q = (1, -2, -5)$

14. Let r be a positive number. Define rA in a manner similar to the definition of Exercise 12. Prove that

$$d(rA, rB) = r \cdot d(A, B).$$

§3. EQUATION OF A CIRCLE

Let P be a given point and r a number > 0. The **circle of radius** r **centered at** P is by definition the set of all points whose distance from P is equal to r. We can now express this condition in terms of coordinates.

Example. Let $P = (1, 4)$ and let $r = 3$. A point whose coordinates are (x, y) lies on the circle of radius 3 centered at $(1, 4)$ if and only if the distance between (x, y) and $(1, 4)$ is 3. This condition can be written as

$$(1) \qquad \sqrt{(x - 1)^2 + (y - 4)^2} = 3.$$

This relationship is called the equation of the circle of center $(1, 4)$ and radius 3. Note that both sides are positive. Thus this equation holds if and only if

$$(2) \qquad (x - 1)^2 + (y - 4)^2 = 9.$$

Indeed, if (1) is true, then (2) is true because we can square each side of (1) and obtain (2). On the other hand, if (2) is true and we take the square root of each side, we obtain (1), because the numbers on each side of (1) are positive. It is often convenient to leave the equation of the circle in the form (2), to avoid writing the messy square root sign. We also call (2) the equation of the circle of radius 3 centered at $(1, 4)$.

Example. The equation

$$(x - 2)^2 + (y + 5)^2 = 16$$

is the equation of a circle of radius 4 centered at $(2, -5)$. Indeed, the square of the distance between a point (x, y) and $(2, -5)$ is

$$(x - 2)^2 + (y - (-5))^2 = (x - 2)^2 + (y + 5)^2.$$

Thus a point (x, y) lies on the prescribed circle if and only if

$$(x - 2)^2 + (y + 5)^2 = 4^2 = 16.$$

Note especially the $y + 5$ in this equation.

Example. The equation

$$(x + 2)^2 + (x + 3)^2 = 7$$

is the equation of a circle of radius $\sqrt{7}$ centered at $(-2, -3)$.

Example. The equation

$$x^2 + y^2 = 1$$

is the equation of a circle of radius 1 centered at the origin. More generally, let r be a number > 0. The equation

$$x^2 + y^2 = r^2$$

is the equation of a circle of radius r centered at the origin. We can draw this circle as in Fig. 8–15.

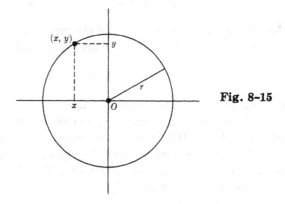

Fig. 8–15

In general, let a, b be two numbers, and r a number > 0. Then the equation of the circle of radius r, centered at (a, b), is the equation

$$(x - a)^2 + (y - b)^2 = r^2.$$

This means that the circle is the set of all points satisfying this equation.

EXERCISES

Write down the equation of a circle centered at the indicated point P, with radius r.

1. $P = (-3, 1), r = 2$ 2. $P = (1, 5), r = 3$

3. $P = (-1, -2), r = \frac{1}{3}$ 4. $P = (-1, 4), r = \frac{2}{5}$

5. $P = (3, 3), r = \sqrt{2}$ 6. $P = (0, 0), r = \sqrt{8}$

Give the coordinates of the center of the circle defined by the following equations, and also give the radius.

7. $(x - 1)^2 + (y - 2)^2 = 25$ 8. $(x + 7)^2 + (y - 3)^2 = 2$

9. $(x + 1)^2 + (y - 9)^2 = 8$ 10. $(x + 1)^2 + y^2 = \frac{5}{3}$

11. $(x - 5)^2 + y^2 = 10$ 12. $x^2 + (y - 2)^2 = \frac{3}{2}$

In each one of the following cases, we give an equation for a pair of numbers (x, y). Show that the set of all points (x, y) satisfying each equation is a circle. Give the center of the circle and its radius. [*Hint:* Complete the square to transform the equation into an equivalent one of the type studied above.]

13. $x^2 + 2x + y^2 = 5$ 14. $x^2 + y^2 - 3y - 7 = 0$

15. $x^2 + 4x + y^2 - 4y = 20$ 16. $x^2 - 4x + y^2 - 2y + 1 = 0$

17. $x^2 - 2x + y^2 + 5y = 26$ 18. $x^2 + x + y^2 - 3y - 4 = 0$

The case of 3-space.

19. a) Write down the equation for a sphere of radius 1 centered at the origin in 3-space, in terms of the coordinates (x, y, z).
   b) Same question for a sphere of radius 3.
   c) Same question for a sphere of radius r.

20. Write down the equation of a sphere centered at the given point P in 3-space, with the given radius r.
   a) $P = (1, -3, 2)$ and $r = 1$ b) $P = (-1, 5, 3)$ and $r = 2$
   c) $P = (-1, 1, 4)$ and $r = 3$ d) $P = (1, 2, -5)$ and $r = 1$
   e) $P = (-2, -1, -3)$ and $r = 2$ f) $P = (1, 3, 1)$ and $r = 7$

21. In each of the following cases, write down the center of the sphere with the given equation, and write down its radius.
   a) $(x - 2)^2 + y^2 + z^2 = 25$
   b) $x^2 + y^2 + z^2 = 1$

c) $x^2 + (y - 3)^2 + (z - 10)^2 = 3$
d) $(x + 3)^2 + y^2 + (z + 2)^2 = 8$
e) $(x - 6)^2 + (y + 4)^2 + (z + 7)^2 = 2$
f) $(x - 4)^2 + (y - 5)^2 + z^2 = 11$

§4. RATIONAL POINTS ON A CIRCLE

The result proved in this section is not essential for what follows, and may be skipped. It is, however, quite beautiful.

Let us go back to Pythagoras. We ask whether we can describe all right triangles with sides having lengths a, b, c such that

$$a^2 + b^2 = c^2$$

and a, b, c are *integers*. For instance, we have a right triangle with sides 3, 4, 5, because

$$3^2 + 4^2 = 25 = 5^2.$$

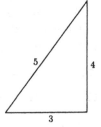

Fig. 8–16

First, it is not clear what we mean by "describe all such right triangles". Is there even another one? The answer to that is yes; for instance, a right triangle with sides (8, 15, 17). By experimenting, we might find still another, and the first question that arises is: Are there infinitely many? The answer to that is again yes, but is not immediately clear, although after we get through with our discussion, we shall see at once that there are infinitely many.

We shall transform our problem to an equivalent one. Observe that our equation

(*) $$a^2 + b^2 = c^2$$

is true if and only if

$$\left(\frac{a}{c}\right)^2 + \left(\frac{b}{c}\right)^2 = 1.$$

All we need to do is to divide by c^2 on both sides. If a, b, c are integers, then the quotients a/c and b/c are rational numbers. Thus every solution of equation (*) yields a solution of the equation

(**) $x^2 + y^2 = 1$

with rational numbers x and y.

   Conversely, suppose that x, y are rational numbers satisfying equation (**). Express x, y as fractions over a common denominator, say c. Thus write

$$x = \frac{a}{c} \quad \text{and} \quad y = \frac{b}{c}$$

with integers a, b, c such that $c \neq 0$. Then

$$\left(\frac{a}{c}\right)^2 + \left(\frac{b}{c}\right)^2 = 1.$$

If we multiply both sides of this equation by c^2, then we obtain (*), namely,

$$a^2 + b^2 = c^2.$$

Thus to solve (*) in integers, it suffices to solve (**) in rational numbers, and this is what we shall do.

   A point (x, y) satisfying the equation $x^2 + y^2 = 1$ may be viewed as a point on the circle of radius 1, centered at the origin. Thus to solve (**) in rational numbers, we may say that we want to find all rational points on the circle. (A **rational point** (x, y) is by definition a point such that its coordinates x, y are rational numbers.)

   Our next step is to give examples of such points.
   For any number t, let

$$x = \frac{1 - t^2}{1 + t^2} \quad \text{and} \quad y = \frac{2t}{1 + t^2}.$$

Simple algebraic manipulations will show you that

$$x^2 + y^2 = 1.$$

If we give t special values which are rational numbers, or integers, then both x and y will be rational numbers, and this gives us our desired examples.

Example. Let $t = 2$. Then

$$x = \frac{1 - 4}{1 + 4} = -\frac{3}{5} \quad \text{and} \quad y = \frac{2 \cdot 2}{1 + 4} = \frac{4}{5}.$$

This yields an example which we already had, namely

$$\left(\frac{-3}{5}\right)^2 + \left(\frac{4}{5}\right)^2 = 1.$$

Multiplying by 5^2 on both sides yields the relation

$$3^2 + 4^2 = 5^2,$$

and thus we recover the $(3, 4, 5)$ right triangle.

Example. Let $t = 4$. Then computing the values of x and y will show you that we recover the right triangle with sides $(8, 15, 17)$.

Example. Let $t = 5$. Then

$$x = \frac{1 - 25}{1 + 25} = \frac{-24}{26} \quad \text{and} \quad y = \frac{2 \cdot 5}{1 + 25} = \frac{10}{26}.$$

Simplifying the fractions, we find that $x = -\frac{12}{13}$ and $y = \frac{5}{13}$. This corresponds to a right triangle with sides 12, 5, and 13. Observe that

$$5^2 + 12^2 = 13^2.$$

It is clear that we have found a way of getting lots of rational points on the circle, or equivalently, lots of right triangles with integral sides. It is not difficult to show that two different values of t yield different points (x, y). (Can you prove this as an exercise?)

You can ask: How did we guess the formulas expressing x and y in terms of t in the first place? Answer: These formulas have been known for a long time. As far as I know, history does not tell us who discovered them first, but he was a good mathematician. What distinguishes someone with talent for mathematics from someone without talent is that the first person will be able to discover such beautiful formulas and the second person will not.

However, everybody is able to plug numbers in the formula once it is written down. That does not take much talent.

We still have not solved our problem of rational points on the circle completely, namely we can ask: Are the points described by our two formulas

$$x = \frac{1 - t^2}{1 + t^2} \quad \text{and} \quad y = \frac{2t}{1 + t^2}$$

with rational values for t, the only rational points on the circle? In other words, does plugging rational numbers for t in these formulas yield all points

(x, y) with rational x, y lying on the circle? The answer is yes, with only one exception. It is based on the following result.

Theorem 1. *Let* (x, y) *be a point satisfying the equation*

$$x^2 + y^2 = 1$$

and such that $x \neq -1$. (See Fig. 8–17.)
Let

$$t = \frac{y}{x + 1}.$$

Then

$$x = \frac{1 - t^2}{1 + t^2} \quad \text{and} \quad y = \frac{2t}{1 + t^2}.$$

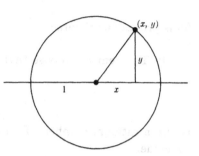

Fig. 8–17

Proof. Multiplying both sides of the equation $t = y/(x + 1)$ by $x + 1$, we find that

$$t(x + 1) = y.$$

Squaring yields

$$t^2(x + 1)^2 = y^2,$$

which gives

$$t^2(x + 1)^2 = 1 - x^2 = (1 + x)(1 - x).$$

We cancel $(x + 1)$ from both sides, and find

$$t^2(x + 1) = 1 - x.$$

Expanding the left-hand side yields

$$t^2x + t^2 = 1 - x,$$

and also

$$(t^2 + 1)x = 1 - t^2.$$

Dividing by $t^2 + 1$ gives us our expression for x. Using

$$y = t(x + 1)$$

and an easy algebraic manipulation which we leave to you, we find the expression for y, namely

$$y = \frac{2t}{1 + t^2}.$$

This proves our theorem.

In Theorem 1, suppose that x, y are rational numbers. Then

$$t = \frac{y}{x + 1}$$

is also a rational number. From the expression for x and y in terms of t we conclude:

Corollary. *Let x, y be rational numbers such that*

$$x^2 + y^2 = 1.$$

If $x \neq -1$, then there exists a rational number t such that

$$x = \frac{1 - t^2}{1 + t^2} \quad \text{and} \quad y = \frac{2t}{1 + t^2}.$$

So we have completely described the rational points on the circle, or equivalently, the right triangles with integral sides.

We could now ask further questions, like:

Determine all pairs of rational numbers (x, y) such that

$$x^3 + y^3 = 1.$$

This is harder, but it can be shown that the only solutions are

$$x = 0 \quad \text{or} \quad x = 1,$$

with the obvious corresponding value for y. In general, given a positive integer k, the problem is to find all solutions of Fermat's equation:

$$x^k + y^k = 1;$$

say, with positive x and y. It is known for many values of k that there is no solution other than $x = 0$ or $x = 1$, but a solution in general is unknown. This is the famous Fermat problem.

EXERCISES

1. Write down explicitly five examples of positive integers (a, b, c) such that

$$a^2 + b^2 = c^2,$$

which have not already been listed in the text and which are not multiples of those listed in the text.

2. Prove that if s, t are real numbers such that $0 \leq s < t$, then

$$\frac{1 - s^2}{1 + s^2} > \frac{1 - t^2}{1 + t^2}.$$

[*Hint:* Prove appropriate inequalities for the numerators and denominators, before taking the quotient.] This proves that different values for $t > 0$ already give different values for x.

3. Using the formulas of this section, give explicitly the values of x and y as quotients of integers, when t has the following values:

a) $t = \frac{1}{2}$,   b) $t = \frac{1}{3}$,   c) $t = \frac{1}{4}$,   d) $t = \frac{1}{5}$.

4. When t becomes very large positive, what happens to

$$\frac{1 - t^2}{1 + t^2}?$$

When t becomes very large negative, what happens to

$$\frac{1 - t^2}{1 + t^2}?$$

Substitute large values of t, like $10{,}000$ or $-10{,}000$, to get a feeling for what happens.

5. Analyze what happens to

$$\frac{2t}{1 + t^2}$$

when $t \leq 0$ and when t becomes very large negative. Next analyze what happens when $t \geq 0$ and t becomes very large positive.

9 Operations on Points

§1. DILATIONS AND REFLECTIONS

From now on, unless otherwise specified, we deal with a fixed coordinate system. Thus we make no distinction between a point and its associated coordinates.

We use $\mathbf{R}$ to denote the set of all real numbers, and $\mathbf{R}^2$ to denote the set of all pairs (x, y), where x, y are real numbers. Thus a point of the plane is simply an element of $\mathbf{R}^2$.

Let A be a point in the plane, with coordinates

$$A = (a_1, a_2).$$

If c is any real number, we define the product cA to be the point

$$cA = (ca_1, ca_2).$$

Thus we multiply each coordinate of A by c to get the coordinates of cA.

Example. Let $A = (2, 5)$ and $c = 6$. Then $cA = (12, 30)$.

Example. Let $A = (-3, 7)$ and $c = -4$. Then $cA = (12, -28)$.

We shall now interpret this multiplication geometrically.

Example. Suppose that c is positive, and let us draw the picture, with the point $A = (1, 2)$ and $c = 3$. Then $3A = (3, 6)$, as on Fig. 9–1(a).

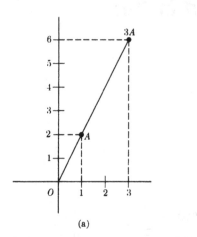
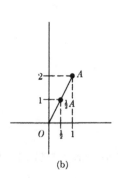

(a)   (b)

Fig. 9-1

Geometrically, we see that multiplication by 3 stretches the coordinates by 3. Similarly, if r is a positive number, then

$$rA = (r, 2r),$$

and we see that multiplication by r stretches the coordinates by r. For instance, stretching by $\frac{1}{2}$ amounts to halving, e.g. in Fig. 9–1(b),

$$\tfrac{1}{2}A = (\tfrac{1}{2}, 1).$$

Example. Let $A = (-2, 3)$ and $r = 2$. Then $rA = (-4, 6)$. Picture:

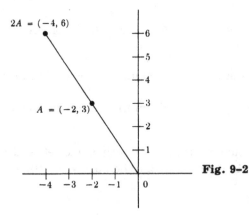

Fig. 9-2

If r is a positive number, we call rA the **dilation** of A by r. The association

$$A \mapsto rA$$

which to each point A associates rA is called **dilation** by r, and gives us an analytic definition for the concept introduced in Chapter 6. It is dilation with respect to O, and leaves O fixed.

Next, we consider the case when c is negative. If $A = (a_1, a_2)$, we define $-A$ to be $(-1)A$, so that

$$-A = (-a_1, -a_2).$$

Example. Let $A = (1, 2)$. Then $-A = (-1, -2)$. We represent $-A$ in Fig. 9–3.

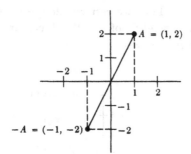

Fig. 9-3

Example. Let $A = (-2, 3)$. Then $-A = (2, -3)$. We draw A and $-A$ in Fig. 9–4.

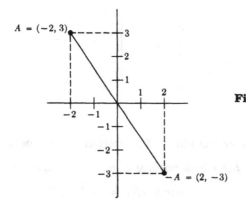

Fig. 9-4

We see that $-A$ is obtained by a certain symmetry, which justifies the next definition.

We define **reflection through our given origin** O to be the association which to each point A associates the point $-A$. As usual, this association is denoted by

$$A \mapsto -A.$$

If R denotes reflection through O, then

$$R(A) = -A.$$

Thus we have been able to give a definition of reflection using only numbers and their properties, i.e. an analytic definition.

If c is negative, we write $c = -r$, where r is positive, and we see that multiplication of A by c can be obtained by first multiplying A by r and then taking the reflection $-rA$. Thus we can say that $-rA$ points in the opposite direction from A, with a stretch of r. We have drawn an example with $A = (1, 2)$ and $c = -3$ in Fig. 9–5.

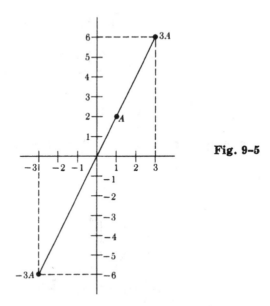

Fig. 9–5

We now consider the effect of a dilation on distances.

Theorem 1. *Let r be a positive number. If A, B are points, then*

$$d(rA, rB) = r \cdot d(A, B).$$

Proof. This was assigned as an exercise in the preceding chapter. We work it out here, so that you see how simple it is. Let $A = (a_1, a_2)$ and $B = (b_1, b_2)$ as usual. Then $rA = (ra_1, ra_2)$ and $rB = (rb_1, rb_2)$. Hence

$$\begin{aligned}
d(rA, rB)^2 &= (rb_1 - ra_1)^2 + (rb_2 - ra_2)^2 \\
&= (r(b_1 - a_1))^2 + (r(b_2 - a_2))^2 \\
&= r^2(b_1 - a_1)^2 + r^2(b_2 - a_2)^2 \\
&= r^2 \cdot d(A, B)^2.
\end{aligned}$$

Taking the square root proves our theorem.

What happens to the distance under multiplication of points by a negative number c? Recall that the absolute value of a number c is defined to be

$$|c| = \sqrt{c^2}.$$

Thus $|-3| = \sqrt{(-3)^2} = \sqrt{9} = 3$.

Theorem 2. *Let c be a number. Then*

$$d(cA, cB) = |c| \cdot d(A, B).$$

Proof. The proof follows exactly the same pattern as the proof of Theorem 1, except that, at the very end when we take $\sqrt{c^2}$ instead of r^2, we find $|c|$ instead of r. Write this proof out in full.

EXERCISES

1. Write the coordinates for cA for the following values of c and A. In each case, draw A and cA.

    a) $A = (-3, 5)$ and $c = 4$ b) $A = (4, -2)$ and $c = 3$
    c) $A = (-4, -5)$ and $c = 2$ d) $A = (2, -3)$ and $c = -2$
    e) $A = (-4, -5)$ and $c = -1$ f) $A = (2, -3)$ and $c = 2$

2. Let A be a point, $A \neq O$. If b, c are numbers such that $bA = cA$, prove that $b = c$.

3. Prove that reflection through O preserves distances. In other words, prove that

$$d(A, B) = d(-A, -B).$$

4. **The 3-dimensional case**

   a) Define the multiplication (dilation) of a point $A = (a_1, a_2, a_3)$ by a number c. Write out interpretations for this similar to those we did in the plane. Draw pictures.
   b) Define reflection of A through $O = (0, 0, 0)$.
   c) State and prove the analogs of Theorems 1 and 2.

§2. ADDITION, SUBTRACTION, AND THE PARALLELOGRAM LAW

Let A and B be points in the plane. We write their coordinates,

$$A = (a_1, a_2) \qquad \text{and} \qquad B = (b_1, b_2).$$

We define their sum $A + B$ to be

$$A + B = (a_1 + b_1, a_2 + b_2).$$

Thus we define their sum componentwise.

Example. Let $A = (1, 4)$ and $B = (-1, 5)$. Then

$$A + B = (1 - 1, 4 + 5) = (0, 9).$$

Example. Let $A = (-3, 6)$ and $B = (-2, -7)$. Then

$$A + B = (-3 - 2, 6 - 7) = (-5, -1).$$

This addition satisfies properties similar to the addition of numbers — and no wonder, since the coordinates of a point are numbers. Thus we have for any points A, B, C:

Commutativity. $A + B = B + A.$

Associativity. $A + (B + C) = (A + B) + C.$

Zero element. Let $O = (0, 0)$. Then $A + O = O + A = A$.

Additive inverse. If $A = (a_1, a_2)$ then the point

$$-A = (-a_1, -a_2)$$

is such that

$$A + (-A) = O.$$

These properties are immediately proved from the definitions. For instance, let us prove the first one. We have:

$$A + B = (a_1 + b_1, a_2 + b_2) = (b_1 + a_1, b_2 + a_2) = B + A.$$

Our proof simply reduces the property concerning points to the analogous property concerning numbers. The same principle applies to the other properties. Note especially the additive inverse. For instance,

$$\text{if } A = (2, -5), \text{ then } -A = (-2, 5).$$

As with numbers, we shall write $A - B$ instead of $A + (-B)$.

Example. If $A = (2, -5)$ and $B = (3, -4)$, then

$$A - B = (2 - 3, -5 - (-4)) = (-1, -1).$$

We shall now interpret this addition and subtraction geometrically. We consider examples.

Example. Let $A = (1, 2)$ and $B = (3, 1)$. To find $A + B$, we start at A, go 3 units to the right, and 1 unit up, as shown in Fig. 9–6(a) and (b).

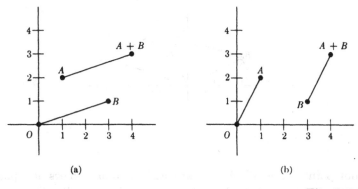

(a)   (b)

Fig. 9–6

Thus we see geometrically that $A + B$ is obtained from A by the same procedure as B is obtained from O. In this geometric representation, we also see that the line segment between A and $A + B$ is parallel to the line segment between O and B, as shown in Fig. 9–6(a). Similarly, the line segment between O and A is parallel to the line segment between B and $A + B$. Thus the four points

$$O,\ A,\ B,\ A + B$$

form the four corners of a parallelogram, which we draw in Fig. 9–6(c).

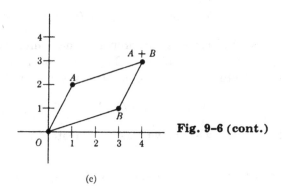

Fig. 9–6 (cont.)

(c)

This gives us a geometric interpretation of addition.

Next, we consider subtraction.

Example. Let $A = (1, 2)$ and $B = (3, 1)$. Then $A - B = (-2, 1)$. By definition, $A - B = A + (-B)$. Thus we can represent this subtraction as in Fig. 9–7.

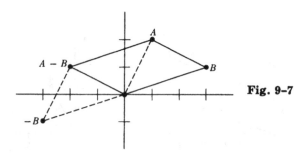

Fig. 9–7

The four points $O,\ A,\ B,\ A - B$ are still the four corners of a parallelogram, but starting from A we have to move in the opposite direction to go from A to $A - B$ than when we moved from A to $A + B$.

Observe that the point A can be written in the form

$$A = (A - B) + B.$$

Thus we also obtain a parallelogram whose corners are O, $A - B$, A, B.

Let A be a fixed element of $\mathbf{R}^2$. We define the **translation** by A to be the association which to each point P of the plane associates the point $P + A$. In Fig. 9–8, we have drawn the effect of this translation on several points.

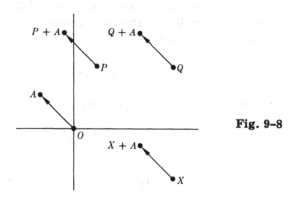

Fig. 9–8

The association

$$P \mapsto P + A$$

has been represented by arrows in Fig. 9–8.

It is useful to abbreviate "translation by A" by the symbol T_A. You should read Chapter 6, §1, which now applies to the present situation. By definition, the value of T_A at a point P is

$$T_A(P) = P + A.$$

Example. Let $A = (-2, 3)$ and $P = (1, 5)$. Then

$$T_A(P) = P + A = (-1, 8).$$

We see that we have been able to define one more of the intuitive geometric notions within our system of coordinates, based only on properties of numbers.

So far, we have given analytic definitions (i.e. definitions based only on properties of numbers) for points, distance, reflection through O, and translation. We recall that a **mapping** of the plane into itself is an association which to each point P associates another point. If the mapping is denoted by F, then this other point is denoted by $F(P)$, and is called the **value** of F at P. If F is translation by A, i.e. if $F = T_A$, then $P + A$ is the value of F at P.

Our previous definition of isometry now makes sense analytically: An **isometry** is a mapping of the plane into itself which preserves distances. In other words, F is an isometry if and only if, for every pair of points P, Q, we have

$$d(P, Q) = d(F(P), F(Q)).$$

Since we have introduced addition and subtraction for points, we shall now describe a way of expressing the distance between points by using subtraction.

We recall that the distance between two points P, Q is denoted by $d(P, Q)$. We shall use a special symbol for the distance between a point and the origin, namely the absolute value sign. We let

$$d(A, O) = |A|.$$

Thus we use two vertical bars on the sides of A. If $A = (a_1, a_2)$, then

$$|A| = \sqrt{a_1^2 + a_2^2},$$

and therefore

$$|A|^2 = a_1^2 + a_2^2.$$

We call $|A|$ the **norm** of A. The norm generalizes the absolute value of a number. We can represent the norm of A as in Fig. 9–9. Geometrically, it is the length of the line segment from O to A.

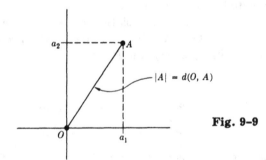

Fig. 9–9

Note that $|A| = |-A|$. Proof?

Using our addition of points, or rather subtraction, we can express the distance between two points A, B by

$$d(A, B) = |A - B| = |B - A|.$$

Indeed, if $A = (a_1, a_2)$ and $B = (b_1, b_2)$, then

$$d(A, B) = \sqrt{(b_1 - a_1)^2 + (b_2 - a_2)^2} = |B - A|$$
$$= \sqrt{(a_1 - b_1)^2 + (a_2 - b_2)^2} = |A - B|.$$

With this notation, we see that *a mapping F of the plane into itself is an isometry if and only if for every pair of points P, Q we have*

$$|F(P) - F(Q)| = |P - Q|.$$

With this notation, a special case of Theorem 2 of §1 can now be written:

$$|cA| = |c|\,|A|.$$

If r is a positive number, then

$$|rA| = r|A|.$$

We suggest that you read the section on mappings in Chapter 6 if you have not already done so. We recall some definitions. If F is a mapping of the plane into itself, and P is a point, then $F(P)$ is also called the **image** of P under F. If S is a subset of the plane, then we denote by $F(S)$ the set of all points $F(P)$ with P in S, and call $F(S)$ the **image** of S under F.

We can now prove analytically a result which was intuitively clear.

Theorem 3. *The circle of radius r and center A is the translation by A of the circle of radius r and center O.*

Proof. Let X be a point on the circle of radius r and center O. This means that

$$|X| = r.$$

The translation of X by A, which is $X + A$, satisfies the condition

$$|X + A - A| = r.$$

Thus we see that $X + A$ is at distance r from A, and hence lies on the circle of radius r centered at A. Conversely, given a point Y on this circle, so that

$$|Y - A| = r,$$

let $X = Y - A$. Then $Y = X + A$ is the translation of X by A, and $|X| = r$. Therefore every point on the circle of radius r, centered at A, is the image under T_A of a point on the circle of radius r centered at O. This proves our theorem.

Actually, you can also do Exercise 11, and then the proof given for Theorem 7 of Chapter 6, §6 is seen to be essentially the same proof as that given above.

Finally, we make a remark concerning the relation between addition and the multiplication of points by numbers as in §1. We ask whether the ordinary rules which we had for numbers also apply, and the answer is yes. Namely, we have:

Associativity. If b, c are numbers, then $b(cA) = (bc)A$.

Distributivity. If b, c are numbers, and A, B are points, then

$$(b + c)A = bA + cA \qquad and \qquad c(A + B) = cA + cB.$$

Also,

$$1A = A \qquad and \qquad 0A = O.$$

The proofs are easy, since we can reduce each statement to the analogous property for numbers. For instance, to prove one of the distributivities, we have:

$$
\begin{aligned}
(b + c)A &= ((b + c)a_1, (b + c)a_2) & \text{by definition} \\
&= (ba_1 + ca_1, ba_2 + ca_2) & \text{by distributivity for numbers} \\
&= (ba_1, ba_2) + (ca_1, ca_2) & \text{by definition} \\
&= bA + cA.
\end{aligned}
$$

We leave the proofs of the other properties to you. They are just as easy, or easier.

EXERCISES

Plot the points A, B, $A + B$, drawing appropriate parallelograms.

1. $A = (1, 4)$, $B = (3, 2)$ 2. $A = (1, 5)$, $B = (1, 1)$

3. $A = (-1, 2)$, $B = (3, 1)$ 4. $A = (-2, 1)$, $B = (1, 2)$

5. $A = (-1, 1)$, $B = (-1, 2)$ 6. $A = (-3, -2)$, $B = (-1, -1)$

7. $A = (-2, -1)$, $B = (-3, 5)$ 8. $A = (-4, -1)$, $B = (1, -3)$

9. $A = (2, -3)$, $B = (-1, -2)$ 10. $A = (2, -3)$, $B = (-1, 5)$

In each of the preceding exercises, plot the points A, $-B$, and $A - B$.

11. Let T_A be a translation by A. Prove that it is an isometry, in other words, that for any pair of points P, Q, we have

$$d(P, Q) = d(T_A(P), T_A(Q)).$$

12. Let $D(r, A)$ denote the disc of radius r centered at A. Show that $D(r, A)$ is the translation by A of the disc $D(r, O)$ of radius r centered at O. [For the definition of the disc, cf. Chapter 5, and observe that this definition is now analytic since all terms entering in it have been defined analytically.]

13. Let $S(r, P)$ denote the circle of radius r centered at P.
   a) Show that the reflection of this circle through O is again a circle. What is the center of the reflected circle?
   b) Show that the reflection of the disc $D(r, P)$ through O is a disc. What is the center of this reflected disc?

14. Let P, Q be points. Write $P = Q + A$, where $A = P - Q$. Define the **reflection of P through Q** to be the point $Q - A$. If R_Q denotes reflection through Q, then we have $R_Q(P) = 2Q - P$. (Why?) Draw the picture, showing P, Q, A, and $Q - A$ to convince yourself that this definition corresponds to our geometric intuition.

15. a) Prove that reflection through a point Q can be expressed in terms of reflection through O, followed by a translation.
   b) Let T_A be translation by A, and R_O reflection with respect to the origin. Prove that the composite $T_A \circ R_O$ is equal to R_Q for some point Q. Which one?

16. a) Let r be a positive number. Give an analytic definition of **dilation by r with respect to a point** Q, and denote this dilation by $F_{r,Q}$. To give this definition, look at Exercise 14. You may also want to look at the discussion about line segments in §4. If P is a point, draw the picture with O, P, Q, $P - Q$, and $F_{r,Q}(P)$.
   b) From your definition, it should be clear that $F_{r,Q}$ can be obtained as a composite of dilation with respect to O, and a translation. Translation by what point?

17. Let $S(r, A)$ be the circle of radius r and center A. Show that the reflection of this circle through a point Q is a circle. What is the center of this reflected circle? What is its radius? Draw a picture.

Let F be a mapping of the plane into itself. Recall that the inverse mapping G of F (if it exists) is the mapping such that $F \circ G = G \circ F = I$ (the identity mapping). This inverse mapping is denoted by F^{-1}. This is all you need to know for Exercises 18 through 21.

18. The inverse of the translation T_A is also a translation. By what? Prove your assertion.

19. Let F_r be dilation by a positive number r, with respect to O, and let T_A be translation by A.
    a) Show that F_r^{-1} is also a dilation. By what number?
    b) Show that $F_r \circ T_A \circ F_r^{-1}$ is a translation.

20. Show that the composite of two translations is a translation. If $T_A \circ T_B = T_C$, how would you express C in terms of A and B?

21. Let R be reflection through the origin.
    a) Show that R^{-1} exists.
    b) Show that $R \circ T_A \circ R^{-1}$ is a translation. By what?

22. Let $A = (a_1, a_2)$ be a point. Define its **reflection through the x-axis** to be the point $(a_1, -a_2)$. Draw A and its reflection through the x-axis in the following cases.
    a) $A = (1, 2)$ b) $A = (-1, 3)$
    c) $A = (-2, -4)$ d) $A = (5, -2)$

23. Prove that reflection through the x-axis is an isometry.

24. Define reflection through the y-axis in a similar way, and prove that it is an isometry. Draw the points of Exercise 22, and their reflections through the y-axis.

25. Recall that a **fixed point** of a mapping F is a point P such that $F(P) = P$. Using the coordinate definition, determine the fixed points of
    a) a translation,
    b) reflection through O,
    c) reflection through an arbitrary point P,
    d) reflection through the x-axis and through the y-axis.
    Prove your assertions.

26. a) Let
$$E_1 = (1, 0) \quad \text{and} \quad E_2 = (0, 1).$$

    We call E_1 and E_2 the **basic unit points** of the plane. Plot these points. If $A = (a_1, a_2)$, prove that

$$A = a_1 E_1 + a_2 E_2.$$

    b) If c is a number, what are the coordinates of cE_1, cE_2?

27. Let $A = (2, 3)$. Draw the points

$$A, \quad A + E_1, \quad A + E_2, \quad A + E_1 + E_2.$$

28. Let $A = (2, 3)$. Draw the points

$$A, \quad A + 3E_1, \quad A + 3E_2, \quad A + 3E_1 + 3E_2.$$

29. Given a number $r > 0$ and a point A, we can define the corners of a square, having sides of length r parallel to the axes, and A as its lower left-hand corner, to be the points

$$A, \quad A + rE_1, \quad A + rE_2, \quad A + rE_1 + rE_2.$$

Let s be a positive number. Show that if these four points are dilated by multiplication with s, they again form the corners of a square. What are the corners of this dilated square?

30. Let the notation be as in Exercise 29. What is the area of the dilated square? How does it compare with the area of the original square?

31. Let A be a point and r, s positive numbers. How would you define the corners of a rectangle whose sides are parallel to the axes, with A as the lower left-hand corner, and such that the vertical side has length r and the horizontal side has length s?

32. Let t be a positive number. What is the effect of dilation by t on the sides and on the area of the rectangle in Exercise 31?

33. Let A be a point and let $r = |A|$. Assume that $r \neq 0$. What is the norm of $(1/r)A$? Prove your assertion.

34. Do the exercise at the end of Chapter 7, §1.

The 3-dimensional case

Define addition of points in 3-space componentwise. Verify the basic properties of commutativity, associativity, zero element, and additive inverse. Define translations. Verify that the sphere of radius r centered at a point A is the translation of the sphere of radius r centered at the origin O. The analog of the disc in 3-space is called the **ball**, for obvious reasons.

Define reflection of a point P through a point Q. Does the definition differ from the 2-dimensional case? Note that using notation P, Q, A, B without coordinates allows us to generalize at once certain notations from 2-space to 3-space. And higher. Why not?

Define an isometry of 3-space. Prove that translations and reflections are isometries.

Show that Exercises 18, 19, 20, and 21 apply to the 3-dimensional case.

Write all this up as if you were writing a book. Part of your mathematical training should consist of making you write mathematics in full English sentences. This forces you to think clearly, and is antidote to slapping down answers to routine plugging problems. As you will notice, carrying out the theory in 3-space amounts practically to copying the theory in 2-space. There is nothing wrong or harmful in copying mathematics. Do you know one of the means Bach used to learn how to compose? He copied practically the entire works of Vivaldi. Some 300 years later rock musicians still use approximately the same technique on each other.

10 Segments, Rays and Lines

§1. SEGMENTS

Line segments

Let P, Q be points in the plane. We can write $Q = P + A$ for some A, namely $A = Q - P$. We wish to give an analytic definition of the **line segment** between P and Q, as shown in Fig. 10–1(a).

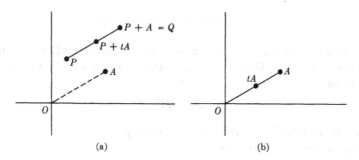

(a)   (b)

Fig. 10–1

This is easy. We define this segment to be the set of all points

$$P + tA, \qquad\qquad \text{with } 0 \leq t \leq 1.$$

Example. The point halfway between P and $P + A$ is the point

$$P + \tfrac{1}{2}A.$$

Example. The line segment between O and A consists of all dilations tA with $0 \leq t \leq 1$, as in Fig. 10–1(b). Thus we see that the line segment between P and $P + A$ is simply the translation by P of the line segment between O and A.

229

The line segment between points P and Q is denoted by $\overline{PQ}$. Let $Q = P + A$. Then in the notation of mappings, we have

$$T_P(\overline{OA}) = \overline{PQ}.$$

The line segment $\overline{PQ}$ is the image under T_P of the line segment $\overline{OA}$. The length of a line segment $\overline{PQ}$ is simply the distance between P and Q.

The line segment between P and Q consists of all points

$$P + t(Q - P), \qquad\qquad 0 \leq t \leq 1.$$

The above expression can be written in the form $P + tQ - tP$, or, in other words,

$$(1 - t)P + tQ.$$

Let $s = 1 - t$ and $t = 1 - s$. When t takes on all values from 0 to 1, we see that s takes on all values from 1 to 0. The points of the line segment between P and Q can be written in the form

$$sP + (1 - s)Q, \qquad\qquad 0 \leq s \leq 1.$$

Thus we see that the segment between P and Q consists of the same points as the segment between Q and P. Of course, we had a right to expect this. Thus we have

$$\overline{PQ} = \overline{QP}.$$

It does not matter which is written first, P or Q.

We can define another concept, that of **directed segment,** or **located vector,** in which the order *does* matter. Thus we define a **located vector** to be an ordered pair of points, which we denote by the symbols

$$\overrightarrow{PQ}.$$

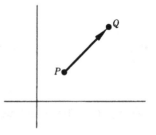

The arrow on top means that P is the first point and Q is the second. We draw a located vector as an arrow, shown in Fig. 10–2.

Fig. 10–2

We say that P is the **beginning point** and Q is the **end point** of the located vector. We say that the located vector is **located at** P. Having ordered our points, we see that

$$\overrightarrow{PQ} \neq \overrightarrow{QP}.$$

§2. RAYS

Let $\overrightarrow{OA}$ be a located vector, located at the origin, such that $A \neq O$. Let P be a point. We define the **ray with vertex P, in the direction of** $\overrightarrow{OA}$, to be the set of all points

(1) $P + tA,$ $t \geq 0.$

Observe that the set of all points tA, with $t \geq 0$ is a ray with vertex at the origin, in the direction of $\overrightarrow{OA}$, as on Fig. 10–3. Thus the ray with vertex P in the direction of $\overrightarrow{OA}$ is the translation by P of the ray consisting of all points tA with $t \geq 0$.

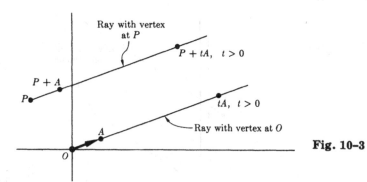

Fig. 10–3

Because the origin O has been fixed throughout our discussion, the mention of O when we speak of the ray in the direction of $\overrightarrow{OA}$ is superfluous, and we shall also say that this ray has the **direction of** A. Such a ray, with vertex P, is completely determined by the expression in (1), which involves simply P and A (and all numbers $t \geq 0$).

Example. Let

$$P = (-1, 3)$$

and

$$A = (2, 1).$$

Letting $t = 5$, we see that the point

$$(-1, 3) + 5(2, 1) = (9, 8)$$

lies on the ray with vertex P in the direction of A. Similarly, letting $t = \frac{1}{3}$, we see that the point

$$(-1, 3) + \tfrac{1}{3}(2, 1) = (-\tfrac{1}{3}, \tfrac{10}{3})$$

lies on this ray.

Given two points P, Q such that $P \neq Q$, we can define the **ray with vertex P, passing through Q** to be the ray with vertex P in the direction of $Q - P$. This ray consists therefore of all points

$$P + t(Q - P), \qquad\qquad t \geqq 0.$$

Example. Let $P = (-1, 3)$ and $Q = (2, 5)$. Then $Q - P = (3, 2)$. The ray with vertex P passing through Q is shown in Fig. 10–4. It consists of all points

$$(-1, 3) + t(2, 5) = (-1 + 2t, 3 + 5t)$$

with $t \geqq 0$.

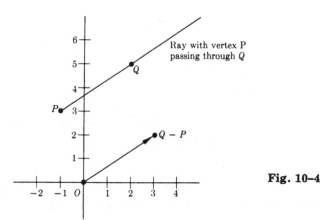

Ray with vertex P passing through Q

Fig. 10–4

Remark. *Let A be a point $\neq O$. Let c be a positive number. Then the ray having a given vertex P in the direction of A is the same as the ray having this same vertex P in the direction of cA.*

Proof. The first ray consists of all points

$$P + tA, \qquad\qquad t \geqq 0.$$

The second ray consists of all points

$$P + scA, \qquad\qquad s \geqq 0.$$

The point P is common to both of them, as we see by taking $t = s = 0$. Suppose that we have a point $P + tA$ on the first with a given value of t. Let $s = t/c$. Then this point can be written in the form $P + scA$, with $s \geqq 0$, and hence is also a point of the second ray. Conversely given a point on the second ray, we let $t = sc$, and therefore see that it is also a point on the first ray. Thus the two rays are equal.

Our remark, combined with our geometric intuition, leads us to make a definition. Let $A \neq O$ and $B \neq O$. We say that $\overrightarrow{OA}$ and $\overrightarrow{OB}$ have the **same direction** (or also that A and B have the **same direction**) if there exists a number $c > 0$ such that

$$B = cA.$$

Since we can then write $A = (1/c)B$, we see that A and B have the same direction if and only if each is a positive multiple of the other. We could also say that each is the dilation of the other by a positive number.

Similarly, let $\overrightarrow{PQ}$ and $\overrightarrow{MN}$ be located vectors. We say that they have the **same direction** if there exists a number $c > 0$ such that

$$Q - P = c(N - M).$$

We draw located vectors having the same direction in Fig. 10–5.

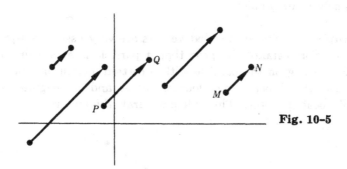

Fig. 10–5

Example. The two points $A = (3, 5)$ and $B = (9, 15)$ define the same direction. According to our convention, this means that $\overrightarrow{OA}$ and $\overrightarrow{OB}$ have the same direction. We see that B is just the dilation of A by 3.

Example. The located vectors $\overrightarrow{PQ}$ and $\overrightarrow{MN}$ have the same direction if

$$P = (1, 3), \quad Q = (-4, 2), \quad M = (7, 1), \quad N = (-3, -1)$$

because

$$Q - P = (-5, -1)$$

and

$$N - M = (-10, -2)$$

so that $N - M = 2(Q - P)$.

Example. Let

$$P = (-3, 5), \quad Q = (1, 2), \quad M = (4, 1).$$

Find the point N such that $\overrightarrow{MN}$ has the same direction as $\overrightarrow{PQ}$, and such that the length of $\overrightarrow{PQ}$ is the same as the length of $\overrightarrow{MN}$.

We need to find N such that

$$N - M = Q - P,$$

and hence

$$N = M + Q - P.$$

We can solve this easily by adding and subtracting components, and we get

$$N = (4, 1) + (1, 2) - (-3, 5)$$
$$= (8, -2).$$

This solves our problem.

Example. In physics, located vectors are very useful to represent physical forces. For instance, suppose that a particle is at a point P, and that a force is acting on the particle with a certain magnitude and direction. We represent this direction by a located vector, and the magnitude by the length of this located vector. Thus when we draw the picture

Fig. 10-6

we can interpret it as a force acting on the particle at P.

Similarly, suppose that an airplane is located at O as in Fig. 10–7. We can interpret $\overrightarrow{OA}$ as the force of the wind acting on the plane. If the pilot runs his engines with a certain force, and gives direction to the airplane by placing his rudders a certain way, we can also represent this force and direction by a located vector $\overrightarrow{OB}$.

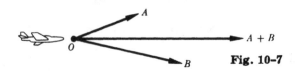

Fig. 10-7

It can be verified experimentally that under these two forces, the airplane moves in a manner described by addition as we defined it; namely, in the direction of $A + B$. Furthermore, the resultant of the two forces acting on the airplane has a magnitude equal to the norm of $A + B$, i.e. $|A + B|$. Thus our concepts of distance, located vectors, and vector sums are used constantly in the sciences.

EXERCISES

Let P, Q be the indicated points. Give the coordinates of the point
a) halfway,
b) one-third of the way,
c) two-thirds of the way
between P and Q.

1. $P = (1, 5)$, $Q = (3, -1)$ 2. $P = (2, 4)$, $Q = (3, 7)$
3. $P = (-3, -2)$, $Q = (-4, 5)$ 4. $P = (-5, -1)$, $Q = (4, 6)$

5. Prove that the image of a line segment $\overline{PQ}$ under translation T_A is also a line segment. What are the end points of this image?

Let P, Q, M be the indicated points. In Exercises 6 through 9, find the point N such that $\overrightarrow{PQ}$ has the same direction as $\overrightarrow{MN}$ and such that the length of $\overrightarrow{MN}$ is
a) 3 times the length of $\overrightarrow{PQ}$,
b) one-third the length of $\overrightarrow{PQ}$.

6. $P = (1, 4)$, $Q = (1, -5)$, $M = (-2, 3)$
7. $P = (-1, -1)$, $Q = (3, -2)$, $M = (4, 4)$
8. $P = (1, -2)$, $Q = (5, 2)$, $M = (-4, 3)$
9. $P = (-1, 3)$, $Q = (\frac{1}{2}, 4)$, $M = (\frac{1}{3}, -1)$

10. Let F be
a) translation T_A,
b) reflection through O,
c) reflection through the x-axis,
d) reflection through the y-axis,
e) dilation by a number $r > 0$.
In each one of these cases, prove that the image under F of (i) a segment, (ii) a ray, is again (i) a segment, (ii) a ray, respectively. Thus you really have 10 cases to consider ($10 = 5 \times 2$), but they are all easy.

11. After you have read the definition of a straight line in the next section, prove that the image under F of a straight line is again a straight line. [Here F is any one of the mappings of Exercise 10.]

12. Give a definition for two located vectors to have **opposite direction**. Similarly, if $A \neq O$ and $B \neq O$, give a definition for A and B to have opposite direction. Draw the corresponding pictures.

The 3-dimensional case

Give the definition for a segment, a ray in 3-space. Does the discussion about $\overline{PQ}$ being equal to $\overline{QP}$ apply? What about the definitions for having the same direction, opposite direction, etc?

13. Give the coordinates of the point
   a) one-third of the distance,
   b) two-thirds of the distance,
   c) one-half of the distance
   between the points

$$P = (3, 1, 5) \qquad \text{and} \qquad Q = (-1, 4, 3)$$

14. Same question for $P = (6, -1, -2)$ and $Q = (-4, 2, -3)$.

Let P, Q, M be the indicated points. In Exercises 15 through 17, find the point N such that $\overrightarrow{PQ}$ has the same direction as $\overrightarrow{MN}$ and such that the length of $\overrightarrow{MN}$ is
   a) 3 times the length of $\overrightarrow{PQ}$,
   b) one-third the length of PQ.

15. $P = (1, 2, 3)$, $Q = (-1, 4, 5)$, $M = (-1, 5, 4)$

16. $P = (-2, 4, 1)$, $Q = (-3, 5, -1)$, $M = (2, 3, -1)$

17. $P = (3, -2, -2)$, $Q = (-1, -3, -4)$, $M = (3, 1, 1)$

§3. LINES

We first discuss the notion of parallelism, and give an analytic definition for it.

Let $\overrightarrow{PQ}$ and $\overrightarrow{MN}$ be located vectors such that $P \neq Q$ and $M \neq N$. We shall say that they are **parallel** if there exists a number c such that

$$Q - P = c(M - N).$$

Observe that this time we may take c to be positive or negative. Our definition applies equally well to segments $\overline{PQ}$ or $\overline{MN}$ instead of located vectors, i.e. we can take P, Q in any order, and we can take M, N in any order.

As an exercise, prove:

If $\overrightarrow{P_1Q_1}$ is parallel to $\overrightarrow{P_2Q_2}$, and if $\overrightarrow{P_2Q_2}$, is parallel to $\overrightarrow{P_3Q_3}$, then $\overrightarrow{P_1Q_1}$ is parallel to $\overrightarrow{P_3Q_3}$.

If $\overline{PQ}$ is parallel to $\overline{MN}$, then $\overline{MN}$ is parallel to $\overline{PQ}$.

Similarly, let $A \neq O$ and $B \neq O$. We define A to be **parallel** to B if there exists a number $c \neq 0$ such that $A = cB$. This amounts to saying that $\overrightarrow{OA}$ is parallel to $\overrightarrow{OB}$. Having fixed our origin O, it suffices to give the end point of the segment $\overrightarrow{OA}$ to determine parallelism with respect to $\overrightarrow{OA}$.

In Fig. 10–8(a) we illustrate parallel located vectors. In Fig. 10–8(b), we see that the end points of located vectors $\overrightarrow{OA}$ and $\overrightarrow{OB}$ lie on the same line, when A, B are parallel.

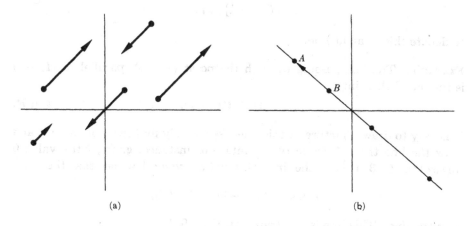

(a)   (b)

Fig. 10–8

Figure 10–8(b) suggests to us how to give an analytic definition for a line. Let A be a point $\neq O$. We define the **straight line** (or simply the line) parallel to $\overrightarrow{OA}$ (or to A) passing through the origin to be the set of all points tA, with all real numbers t.

For an arbitrary line, not necessarily passing through the origin, we just take a translation.

We define the **straight line** (or simply the **line**) passing through a given point P, parallel to $\overrightarrow{OA}$ (or more simply, parallel to A) to be the set of all points

$$P + tA,$$

with all real numbers t, positive, negative, or 0. Picture:

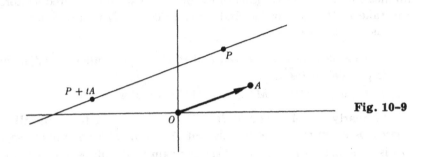

Fig. 10-9

In Fig. 10–9, we have drawn the point $P + tA$ corresponding to a negative value of t.

We shall also use the symbols

$$\{P + tA\}_{t \text{ in } \mathbf{R}}$$

to denote this straight line.

Example. The line passing through the point $(-3, 4)$, parallel to $(1, -5)$ is the set of all points

$$(-3, 4) + t(1, -1), \qquad\qquad t \text{ in } \mathbf{R}.$$

It is easy to draw a picture of this line. We merely find two points on it and draw the line through these two points. For instance, giving t the value 0 we see that $(-3, 4)$ is on the line. Giving t the value 1 we see that the point

$$(-3, 4) + (1, -1) = (-2, 3)$$

is on the line. This line is illustrated on Fig. 10–10.

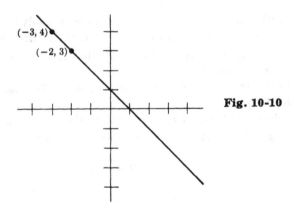

Fig. 10-10

The representation of a line in the form $P + tA$ is called a **parametric representation,** and we call t the **parameter.** One sometimes interprets the point $P + tA$ as describing the position of a bug, or a particle, moving along the line with uniform speed, and we interpret t as the time. Thus at time $t = 0$, the bug is at the point P. The coordinates (x, y) of a point on the line then depend on t, and it is customary to write them in the form

$$(x(t), y(t))$$

to indicate this dependence on t. This parametric representation of a straight line is advantageous for at least two reasons. First, it generalizes easily to 3-space. Second, it represents our physical intuition of the moving bug, and allows us to give a simple coordinate representation for the position of the bug at a given time.

Example. Let

$$P = (-1, 4) \qquad \text{and} \qquad A = (2, 3).$$

Then the coordinates for an arbitrary point on the line passing through P parallel to A are given by

$$x(t) = -1 + 2t$$

and

$$y(t) = 4 + 3t.$$

Example. Find a parametric representation of a line passing through two points $P = (1, 5)$ and $Q = (-2, 3)$; see Fig. 10–11.

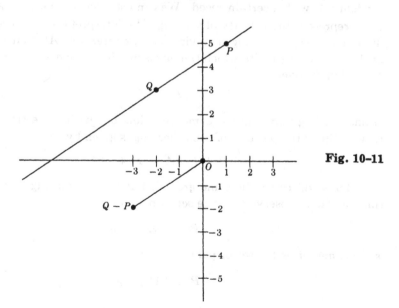

Fig. 10–11

Let $A = Q - P$. Then the parametric representation is

$$P + tA = P + t(Q - P).$$

In terms of the individual coordinates, $Q - P = (-3, -2)$, and hence

$$x(t) = 1 - 3t, \qquad y(t) = 5 - 2t.$$

Observe that when $t = 0$, we obtain the point P, and when $t = 1$, we obtain the point Q. Thus we have found a parametric representation of a line passing through P and Q. We can also write this representation in the form

$$\{(1, 5) - t(3, 2)\}_{t\,in\,\mathbf{R}}.$$

Example. Find the point at which the line of the preceding example crosses the x-axis.

The second coordinate of a point on this line is

$$5 - 2t.$$

Thus we must find the value for t such that $5 - 2t = 0$. This is easily solved, and gives $t = \frac{5}{2}$. Hence the point at which the line of the preceding example crosses the x-axis is

$$(1, 5) - \frac{5}{2}(3, 2) = \left(-\frac{13}{2}, 0\right).$$

Remark. Suppose that a bug starts from a point P and moves along a straight line with a certain speed. We can use a located vector $\overrightarrow{OA}$, or simply A, to represent the velocity of the bug. We interpret $\overrightarrow{OA}$ as representing the direction in which the bug is moving, and we interpret $|A|$ (that is, the length of $\overrightarrow{OA}$) as the speed with which the bug is moving. Then at time t the position of the bug is given by

$$P + tA.$$

If another bug moves in the same direction, but with three times the speed, then at time t the position of the other bug is given by

$$P + t \cdot 3A = P + 3tA.$$

Both bugs will cover the same ground, but the second bug will do so three times as fast. Observe that the set of points

$$\{P + 3tA\}_{t\,in\,\mathbf{R}}$$

is the same set as the set of points

$$\{P + tA\}_{t\,in\,\mathbf{R}}.$$

In intuitive geometry, we assume that two lines which are not parallel have exactly one point in common. We now have a means of determining this point.

Example. Let two lines be represented parametrically by

$$(1, 2) + t(3, 4) \qquad \text{and} \qquad (-1, 1) + s(2, -1),$$

where t, s are the respective parameters. Find the point of intersection of these two lines.

We must find the values of s and t such that

$$1 + 3t = -1 + 2s,$$
$$2 + 4t = 1 - s,$$

or, in other words,

$$3t - 2s = -2,$$
$$4t + s = -1.$$

This is a system of two equations in two unknowns which we know how to solve from algebra. The solutions are

$$t = -\frac{4}{11} \qquad \text{and} \qquad s = \frac{5}{11}.$$

Hence the common point of the two lines is the point

$$(1, 2) - \frac{4}{11}(3, 4) = \left(\frac{-1}{11}, \frac{6}{11}\right).$$

We can also find the point of intersection of a line and other geometric figures given by equations.

Example. Find the points of intersection of the line given parametrically by

$$(-1, 2) + t(3, -4), \qquad\qquad\qquad t \text{ in } \mathbf{R},$$

and the circle

$$x^2 + y^2 = 4.$$

The first and second coordinates of points on the line are given by

$$x(t) = -1 + 3t \qquad \text{and} \qquad y(t) = 2 - 4t.$$

We must find those values of t which are such that $x(t)$ and $y(t)$ satisfy the equation of the circle. This means that we must find those values of t such that

$$(-1 + 3t)^2 + (2 - 4t)^2 = 4.$$

Expanding out, this amounts to solving for t the equation

$$1 - 6t + 9t^2 + 4 - 16t + 16t^2 = 4,$$

or in other words

$$25t^2 - 22t + 1 = 0.$$

This is a quadratic equation, which we know how to solve. We obtain:

$$t = \frac{22 \pm \sqrt{(22)^2 - 4 \cdot 25 \cdot 1}}{50}$$

$$= \frac{14 \pm \sqrt{384}}{50}.$$

Thus we obtain two values for t. We can simplify slightly, writing $384 = 4 \cdot 96$. The two values for t are then

$$t = \frac{7 + \sqrt{96}}{25}$$

and

$$t = \frac{7 - \sqrt{96}}{25}.$$

The points of intersection of the line and the circle are then given by (x_1, y_1) and (x_2, y_2) where x_1, y_1, x_2, y_2 have the following values:

$$x_1 = -1 + 3 \cdot \frac{7 + \sqrt{96}}{25},$$

$$y_1 = 2 - 4 \cdot \frac{7 + \sqrt{96}}{25},$$

$$x_2 = -1 + 3 \cdot \frac{7 - \sqrt{96}}{25},$$

$$y_2 = 2 - 4 \cdot \frac{7 - \sqrt{96}}{25}.$$

The intersection of the line and the circle can be illustrated as follows. Giving t the special values $t = 0$ and $t = 1$, we find that the two points

$$(-1, 2) \qquad \text{and} \qquad (2, -2)$$

lie on the line. Thus we draw the line passing through these two points, and we see it intersect the circle in Fig. 10–12.

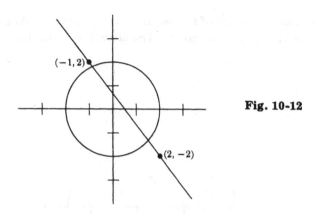

Fig. 10-12

Example. On the other hand, there may be cases when the line and the circle do not intersect. For instance, consider the line given by

$$(-5, 0) + t(1, 1), \qquad\qquad t \text{ in } \mathbf{R}.$$

We wish to determine the points of intersection of this line, and the circle having the equation

$$x^2 + y^2 = 4.$$

Proceeding as before, we note that the coordinates of the line are given by

$$x(t) = -5 + t \qquad \text{and} \qquad y(t) = t.$$

Substituting these in the equation for the circle, we get an equation for t, namely

$$(-5 + t)^2 + t^2 = 4,$$

which expanded out yields

$$25 - 10t + t^2 + t^2 = 4,$$

or in other words,

$$2t^2 - 10t + 21 = 0.$$

Using the quadratic formula to solve for t gives us

$$t = \frac{10 \pm \sqrt{100 - 168}}{4} = \frac{10 \pm \sqrt{-68}}{4}.$$

We see that the expression under the square root sign is negative, and hence there is no real value of t satisfying our equation. This means that the circle and the line do not intersect. The situation is illustrated in Fig. 10–13.

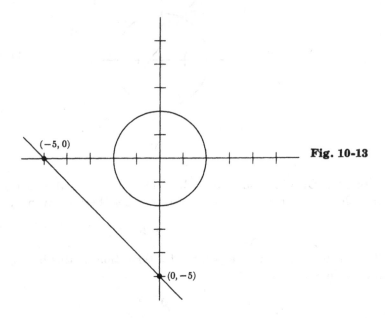

Fig. 10-13

To draw the line in this figure, all we need are two points on it. Using the special values $t = 0$, $t = 5$ shows that the points

$$(-5, 0) \qquad \text{and} \qquad (0, 5)$$

are on the line. These points have also been indicated on the figure.

For Exercises 1 through 6: (a) write down parametric representations of the lines passing through the indicated points P and Q, (b) find the point of intersection of the line and the x-axis, (c) find the point of intersection of the line and the y-axis.

1. $P = (1, -1), Q = (3, 5)$ 2. $P = (2, 1), Q = (4, -1)$

3. $P = (-4, -2)$, $Q = (-1, 1)$ 4. $P = (3, -1)$, $Q = (1, 1)$
5. $P = (3, -5)$, $Q = (-2, 6)$ 6. $P = (-1, 4)$, $Q = (-1, -3)$

One airplane moves along a straight line in the plane, starting at a point P in the direction of A. Another plane also moves along a straight line, starting at a point Q in the direction of B. Find the point at which they may collide if P, Q, A, B are given by the following values. Draw the two lines.

7. $P = (1, -1)$, $Q = (3, 5)$, $A = (-3, 1)$, $B = (2, -1)$
8. $P = (-4, -2)$, $Q = (-1, 1)$, $A = (5, 1)$, $B = (2, -5)$
9. $P = (1, -1)$, $Q = (4, 1)$, $A = (-1, -2)$, $B = (-3, -1)$
10. $P = (1, 1)$, $Q = (2, -1)$, $A = (3, 3)$, $B = (-4, 3)$
11. Find the point of intersection of the lines of Exercises

a) 1 and 2,   b) 2 and 3,   c) 1 and 3,   d) 5 and 6.

12. Let $A = (a_1, a_2)$ and $B = (b_1, b_2)$. Assume that $A \neq O$ and $B \neq O$. Prove that A is parallel to B if and only if

$$a_1 b_2 - a_2 b_1 = 0.$$

13. Prove: If two lines are not parallel, then they have exactly one point in common. [*Hint:* Let the two lines be represented parametrically by

$$\{P + tA\}_{t \, in \, \mathbf{R}} = \{(p_1, p_2) + t(a_1, a_2)\}_{t \, in \, \mathbf{R}}$$
$$\{Q + sB\}_{s \, in \, \mathbf{R}} = \{(q_1, q_2) + s(b_1, b_2)\}_{s \, in \, \mathbf{R}}.$$

Write down the general system of two equations for s and t and show that it can be solved.]

14. Find the points of intersection of the given line $\{P + tA\}_{t \, in \, \mathbf{R}}$ and the circle of radius 8 centered at the origin, when P, A are as indicated. If there are no such points, say so, and why. Draw the line and circle.

a) $P = (1, -1)$, $A = (-3, 1)$ b) $P = (2, -3)$, $A = (1, 1)$
c) $P = (-2, 1)$, $A = (3, 5)$ d) $P = (3, 1)$, $A = (-2, -4)$

[*Hint:* Substitute the values for $x(t)$, $y(t)$ in the equation of the circle, and solve for t.]

15. Given (a), (b), (c), (d) as in Exercise 14. Find the points of intersection of the lines and the circle of radius 2 centered at the origin.

16. Given (a), (b), (c), (d) as in Exercise 14. Find the points of intersection of the lines and the circle of radius 1 centered at $(1, 1)$.

17. Given (a), (b), (c), (d) as in Exercise 14. Find the points of intersection of the lines and the circle of radius 2 centered at $(-1, -1)$.

18. (Slightly harder.) Let S be the circle of radius $r > 0$ centered at the origin. Let $P = (p, q)$ be a point such that

$$p^2 + q^2 \leq r^2.$$

In other words, P is a point in the disc of radius r centered at O. Show that any line passing through P must intersect the circle, and find the points of intersection. [*Hint*: Write the line in the form

$$P + tA,$$

where $A = (a, b)$, substitute in the equation of the circle, and find the coordinates of the points of intersection in terms of p, q, a, b. Show that the quantity you get under the square root sign is ≥ 0.]

§4. ORDINARY EQUATION FOR A LINE

So far we have described a line in terms of a parameter t. We can eliminate this parameter t and get another type of equation for the line. We show this by an example.

Example. Let $L = \{P + tA\}_{t \text{ in } \mathbf{R}}$, where $P = (3, 5)$ and $A = (-2, 7)$. Thus a point $(x(t), y(t))$ on the line is given by

$$x = 3 - 2t, \qquad y = 5 + 7t.$$

Multiply the expression for x by 7, multiply the expression for y by 2, and add. We get

$$7x + 2y = 3 \cdot 7 - 14t + 2 \cdot 5 + 14t$$
$$= 31.$$

Thus any point (x, y) on the line L satisfies the equation

$$7x + 2y = 31.$$

Conversely, let (x, y) be a point satisfying this equation. We want to solve for t such that

$$x = 3 - 2t \qquad \text{and} \qquad y = 5 + 7t.$$

Let

$$t = -\frac{x-3}{2}.$$

Then certainly $x = 3 - 2t$. But $y = (31 - 7x)/2$. We must verify that this is equal to

$$5 - 7\frac{(x-3)}{2}.$$

This is obvious by a simple algebraic manipulation, and we have solved for the desired t.

We shall call

$$7x + 2y = 31$$

the **ordinary equation** of the line.

Similarly, any equation of the form

$$ax + by = c$$

where a, b, c are numbers, is the equation of a straight line.

EXERCISES

Find the ordinary equation of the line $\{P + tA\}_{t\,in\,\mathbf{R}}$ in each one of the following cases.

1. $P = (3, 1)$, $A = (7, -2)$

2. $P = (-2, 5)$, $A = (5, 3)$

3. $P = (-4, -2)$, $A = (7, 1)$

4. $P = (-2, -5)$, $A = (5, 4)$

5. $P = (-1, 5)$, $A = (2, 4)$

6. $P = (-3, 2)$, $A = (-3, -2)$

7. $P = (1, 1)$, $A = (1, 1)$

8. $P = (-5, -6)$, $A = (-4, 3)$

11 Trigonometry

This chapter can be read immediately after the definition of coordinates and distance. We don't need anything about segments, lines, etc. covered in the other chapters of this part.

§1. RADIAN MEASURE

In a sense, the measurement of angles by degrees is not a natural measurement. It is much more reasonable to take another measure which we now describe.

Let π be the (numerical value of the) area of a disc of radius 1. The approximate value for π is 3.14159 .... (Look at our comments about π in Chapter 7, §1.) Let A be an angle with vertex at P and let S be the sector determined by A in the disc D of radius 1 centered at P. Let x be a number between 0 and 2π. We shall say that

$$\textbf{A has } x \textbf{ radians}$$

to mean that

$$\frac{\text{area of } S}{\text{area of } D} = \frac{x}{2\pi}.$$

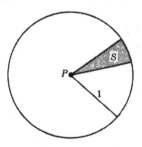

Fig. 11-1

Thus if D is the disc of radius 1 centered at P, our definition is adjusted so that the area of the sector S determined by an angle of x radians is $x/2$; see Fig. 11–1. We draw various angles and indicate their radian measure, in Fig. 11–2.

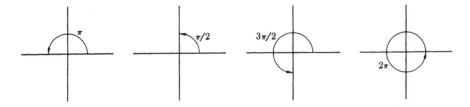

Fig. 11-2

Of course, degrees are related to radians. For instance:

$$360 \text{ degrees} = 2\pi \text{ radians}$$
$$180 \text{ degrees} = \pi \text{ radians}$$
$$60 \text{ degrees} = \pi/3 \text{ radians}$$
$$45 \text{ degrees} = \pi/4 \text{ radians}$$
$$30 \text{ degrees} = \pi/6 \text{ radians}.$$

In general,

$$x \text{ degrees} = \frac{\pi}{180} x \text{ radians}.$$

However, from now on, unless otherwise specified, we *always* deal with radian measure.

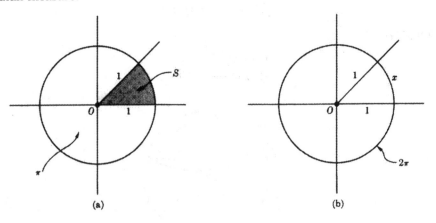

(a)   (b)

Fig. 11-3

Observe that the length of the circle of radius 1 is 2π. (Cf. Chapter 7.) The angle A determines an arc on this circle, and radian measure is so adjusted that

$$\frac{\text{length of this arc}}{\text{total length of circle}} = \frac{x}{2\pi}.$$

Thus x is the length of arc determined by A; this is illustrated in Fig. 11–3.

The Greeks realized that they had to choose a constant which appears very frequently in mathematics, relating the radius and circumference of a circle or its diameter and circumference. The way they chose π, however, is somewhat inconvenient, because it introduces a factor of 2 in front of π in most of mathematics. It would have been more useful to use the constant c such that

$$\frac{\text{area of } S}{\text{area of } D} = \frac{x}{c}.$$

Too late to change, however.

For convenience of language, we shall sometimes speak, incorrectly but usefully, of an angle of x radians even if x does not lie between 0 and 2π. To do this, we write

$$x = 2n\pi + w,$$

with a number w such that

$$0 \leq w < 2\pi.$$

Then, by an angle of x radians, we mean the angle of w radians.

Also for convenience of language, if x is negative, say

$$x = -z$$

where z is positive, we shall speak of an angle of x radians to mean an angle of z radians in the clockwise direction. We draw such an angle in Fig. 11–4.

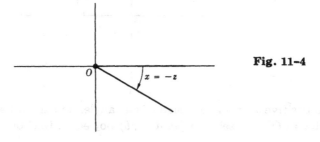

Fig. 11–4

EXERCISES

1. Give the following values of angle in radians, as a fractional multiple of π.

   a) 15°   b) 75°   c) 105°   d) 120°
   e) 135°   f) 150°   g) 165°

2. Same question in the following cases.

   a) 20°   b) 40°   c) 140°   d) 310°

3. Find the measure in degrees (between 0° and 360°) for the following angles given in radians.

   a) $-\dfrac{\pi}{4}$ b) $\dfrac{8\pi}{9}$ c) $\dfrac{5\pi}{9}$ d) $\dfrac{7\pi}{4}$

   e) $\dfrac{14\pi}{3}$ f) $\dfrac{22\pi}{3}$ g) $-\dfrac{\pi}{3}$

§2. SINE AND COSINE

Suppose that we have given a coordinate system and an angle A with vertex at the origin O as shown in Fig. 11–5.

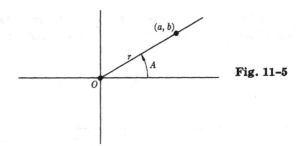

Fig. 11–5

The positive x-axis is one side of our angle, and the other side is a ray with vertex at O. We select a point (a, b) not equal to O on this ray, and we let

$$r = \sqrt{a^2 + b^2}.$$

Then r is the distance from $(0, 0)$ to the point (a, b). We define

$$\text{sine } A = \frac{b}{r} = \frac{b}{\sqrt{a^2 + b^2}}$$

$$\text{cosine } A = \frac{a}{r} = \frac{a}{\sqrt{a^2 + b^2}} \cdot$$

If we select another point (a_1, b_1) on the ray determining our angle A, and use its coordinates to get the sine and cosine, then we obtain the same values as with (a, b). Indeed, there is a positive number c such that

$$a_1 = ca \qquad \text{and} \qquad b_1 = cb.$$

Hence

$$\frac{b_1}{\sqrt{a_1^2 + b_1^2}} = \frac{cb}{\sqrt{c^2 a^2 + c^2 b^2}} \cdot$$

We can factor c from the denominator, and then cancel c in both the numerator and denominator on the right-hand side, to get

$$\frac{b}{\sqrt{a^2 + b^2}} \cdot$$

This proves that our definition of sine A does not depend on the choice of coordinates (a, b) on the ray. The proof for the cosine is similar.

The geometric interpretation of the above argument simply states that the triangles in Fig. 11–6 are similar, i.e. are obtained by a dilation of each other.

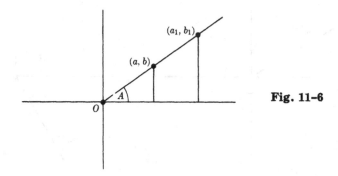

Fig. 11–6

In particular, we can select the point (a, b) at any distance from the origin that we find convenient. For many purposes, it is convenient to select (a, b)

on the circle of radius 1, so that $r = 1$. *In that case,*

$$\sin A = b \quad \text{and} \quad \cos A = a.$$

Consequently, by definition, the coordinates of a point on the circle of radius 1 are

$$(\cos \theta, \sin \theta)$$

if θ is the angle in radians; see Fig. 11–7.

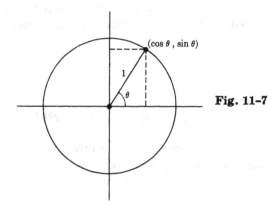

Fig. 11–7

An angle A can be determined by a ray in any one of the four quadrants. Fig. 11–8 depicts both the case when the ray is in the second quadrant and the case when it is in the third.

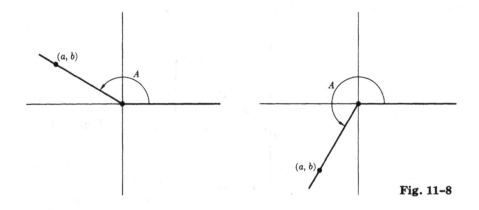

Fig. 11–8

When the ray is in the first quadrant, then both the sine and cosine are positive because both a and b are positive. When the ray is in the second quadrant, then the sine is positive because b is positive. The cosine is negative because a is negative. When the ray is in the third quadrant, then sine A is negative and cosine A is also negative.

It is also convenient to remember the sine in the context of a right triangle. Let A be one of the angles in a right triangle, other than the right angle, as shown in Fig. 11–9. Let a be the length of the opposite side of A, let b be the length of the adjacent side, and let c be the length of the hypotenuse. Then we have

$$\sin A = \frac{a}{c} = \frac{\text{opposite side}}{\text{hypotenuse}}.$$

Similarly,

$$\cos A = \frac{b}{c} = \frac{\text{adjacent side}}{\text{hypotenuse}}.$$

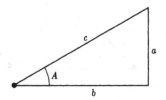

Fig. 11–9

We want to define the sine and cosine of a number. Let x be a number. We write

$$x = 2n\pi + w$$

with some integer n and some number w such that

$$0 \leq w < 2\pi.$$

We then *define*:

$$\sin x = \text{sine of } w \text{ radians}$$
$$\cos x = \text{cosine of } w \text{ radians}.$$

Thus we have for all numbers x:

$$\sin (x + 2\pi) = \sin x$$
$$\cos (x + 2\pi) = \cos x.$$

Instead of adding 2π, we could also add any integral multiple of 2π, and find the similar property, namely

$$\sin (x + 2n\pi) = \sin x$$
$$\cos (x + 2n\pi) = \cos x.$$

We can compute a few values of the sine and cosine, as shown in the following table.

Number	Sine	Cosine
$\pi/6$	$\frac{1}{2}$	$\sqrt{3}/2$
$\pi/4$	$1/\sqrt{2}$	$1/\sqrt{2}$
$\pi/3$	$\sqrt{3}/2$	$\frac{1}{2}$
$\pi/2$	1	0
π	0	-1
2π	0	1

These can be determined by plane geometry and the Pythagoras theorem. For instance, we get the sine of the angle $\pi/4$ radians from a right triangle with two equal legs:

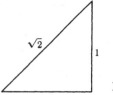

Fig. 11–10

We see that the sine is equal to $1/\sqrt{2}$.

Similarly, consider a right triangle whose angles other than the right angle have $\pi/6$ and $\pi/3$ radians (in other words 30° and 60°, respectively).

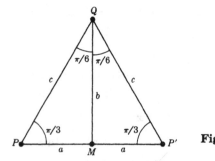

Fig. 11–11

In Fig. 11–11, we reflect our triangle $\triangle PQM$ through the side $\overline{QM}$ and let P' be the reflection of P through M. Then $\triangle PQP'$ is a triangle all of whose angles have the same measure, so that the three sides have equal length. Hence $c = 2a$ and

$$a^2 + b^2 = c^2 = 4a^2.$$

Therefore

$$b^2 = 3a^2$$

and

$$b = \sqrt{3}a.$$

Hence

$$\sin \frac{\pi}{6} = \frac{1}{2} \quad \text{and} \quad \cos \frac{\pi}{6} = \frac{\sqrt{3}}{2}.$$

Of course we could have taken our triangle to be normalized so that $b = 1$. In this case, we have a triangle with sides of lengths 1, 2, $\sqrt{3}$.

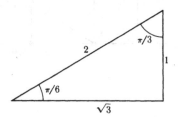

Fig. 11–12

The sine and cosine satisfy a basic relation, namely for all numbers x, we have

$$\sin^2 x + \cos^2 x = 1.$$

[Notation: $\sin^2 x$ means $(\sin x)^2$, and similarly for the cosine.] This is immediate because

$$\left(\frac{a}{r}\right)^2 + \left(\frac{b}{r}\right)^2 = \frac{a^2 + b^2}{r^2} = \frac{r^2}{r^2} = 1.$$

Figure 11–13 illustrates this argument.

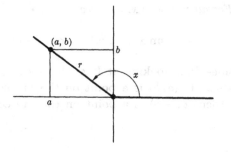

Fig. 11–13

Theorem 1. *For any number x, we have*

$$\cos x = \sin\left(x + \frac{\pi}{2}\right) \quad \text{and} \quad \sin x = \cos\left(x - \frac{\pi}{2}\right).$$

Proof. We may assume that $0 \leq x < 2\pi$. Let A be an angle of x radians, let $P = (a, b)$ be a point on the ray which forms one side of A as shown in Fig. 11–14 and such that P lies on the circle of radius 1.

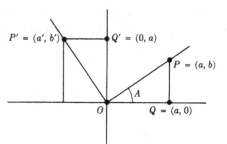

Fig. 11–14

Let $Q = (a, 0)$. Let G be rotation by an angle of $\pi/2$. Let $P' = G(P)$ and $Q' = G(Q)$. Then $Q' = (0, a)$. If $P = Q$, then the first formula is clear. If $P \neq Q$, then the segment $\overline{PQ}$ is perpendicular to the first axis, and hence $\overline{P'Q'}$ is perpendicular to the second axis. Hence the second coordinate of Q' is a, in other words, we have $b' = a$. Since we took P on the circle of radius 1, it follows that

$$\cos x = a \quad \text{and} \quad \sin\left(x + \frac{\pi}{2}\right) = b'.$$

This proves the first formula. The second is proved similarly, and we leave it to you.

Theorem 2. *For all numbers x, we have*

$$\sin(-x) = -\sin x \quad \text{and} \quad \cos(-x) = \cos x.$$

Proof. This comes from looking at Fig. 11–15 and using the definition of the sine and cosine. If (a, b) is a point on the ray corresponding to an angle of x radians, then $(a, -b)$ is a point on the ray corresponding to an angle of $-x$ radians.

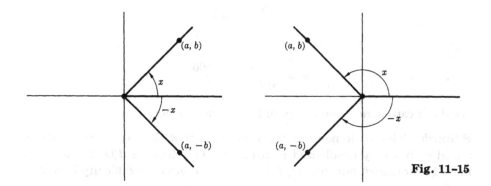

Fig. 11-15

We can take (a, b) on the circle of radius 1, in which case

$$\sin x = b \quad \text{and} \quad \cos x = a.$$

Our assertion is then clear.

The sine and cosine have various applications. We give one example.

Example. A boat B starts from a point P on a straight river and moves down the river. An observer O stands at a distance of 1,000 ft from P, on the line perpendicular to the river passing through P. After 10 min, the observer finds that the angle θ formed by P, himself, and the boat, is such that $\cos \theta = 0.7$. What is the distance between the observer and the boat at that time?

Picture first:

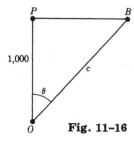

Fig. 11-16

We want to find c. We have

$$\frac{1,000}{c} = \cos \theta.$$

Hence

$$\frac{1,000}{c} = 0.7$$

and

$$c = \frac{1,000}{0.7} = \frac{10,000}{7}.$$

We don't care here if you compute the decimal or not.

Remark. The whole point in this example is that the angle θ can be determined with a very small-sized instrument. The distance $d(O, P)$ of course must be measured, but both O, P are fixed so there is no difficulty in that.

Polar coordinates

Referring to Fig. 11–17, let (x, y) be a point in the plane. We can describe this point by using other coordinates. Let

$$r = \sqrt{x^2 + y^2}.$$

Thus (x, y) lies at distance r from the origin. Then by definition we have

$$\frac{x}{r} = \cos \theta \qquad \text{and} \qquad \frac{y}{r} = \sin \theta$$

for some number θ, provided that $r \neq 0$. We can rewrite these in the form

$$x = r \cos \theta \qquad \text{and} \qquad y = r \sin \theta,$$

and since we don't divide by r, these are valid whether $r = 0$ or not.

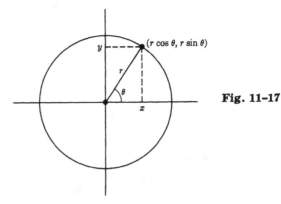

Fig. 11–17

We call (r, θ) **polar coordinates** for the point (x, y). When we deal simultaneously with polar coordinates and the other coordinates (x, y), then we call (x, y) the **Cartesian,** or **rectangular coordinates.**

Example. Find polar coordinates for the point whose rectangular coordinates are $(1, \sqrt{3})$.

We have $x = 1$ and $y = \sqrt{3}$, so that

$$r = \sqrt{1 + 3} = 2.$$

Also

$$\cos \theta = \frac{1}{2}$$

and

$$\sin \theta = \frac{\sqrt{3}}{2}.$$

We see that $\theta = \pi/3$, and that the polar coordinates for the point are $(2, \pi/3)$. According to our convention, we note that if (r, θ) are polar coordinates for a point, then $(r, \theta + 2\pi)$ are also polar coordinates. Thus in our example, our given point also has polar coordinates given by

$$\left(2, \frac{\pi}{3} + 2\pi \right).$$

In practice, we usually select the value of θ such that $0 \leq \theta < 2\pi$.

Example. Given numbers a, b such that $a^2 + b^2 = 1$, we can always find a number θ such that $a = \cos \theta$ and $b = \sin \theta$. Therefore the point whose rectangular coordinates are (a, b) has polar coordinates $(1, \theta)$ for such θ. If a is positive, then $-\pi/2 < \theta < \pi/2$, and our choice of θ is restricted to two possibilities as shown in Fig. 11–18.

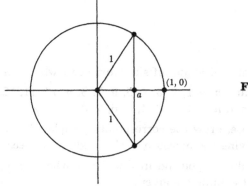

Fig. 11–18

The number a must lie between 0 and 1 because $a^2 \leq 1$. The angle of θ radians is determined by the point of intersection of the circle of radius 1 centered at the origin, and the line through the point $(a, 0)$, perpendicular to the x-axis. Now depending on whether b is positive or negative, we eliminate one of the two possibilities for θ; namely, we take θ positive if b is positive, and θ negative if b is negative, such that $\sin \theta = b$.

When a is negative, we argue in a similar way.

Example. Let $a = \frac{1}{2}$ and $b = -\sqrt{3}/2$. Then $a^2 + b^2 = 1$. In this case, illustrated in Fig. 11–19, we take $\theta = -\pi/3$.

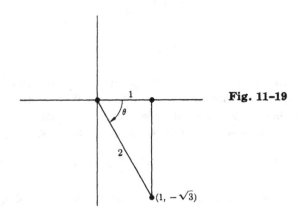

Fig. 11–19

EXERCISES

1. Make a table of values of $\sin x$ and $\cos x$ when x is equal to:

   a) $n\pi/6$, and $n = 1, 2, 3, 4, 5, 6, 7, 8, 9, 10, 11, 12$;
   b) $n\pi/4$, and $n = 1, 2, 3, 4, 5, 6, 7, 8$.

   In each case, draw the corresponding angle. In your table, also make a column giving the measure of the angle in degrees.

2. Make a table as you did in Exercise 1 when n ranges over the negative values of the numbers given.

3. When the angle A has its defining ray in the fourth quadrant, determine whether the sine is positive or negative. Repeat for the cosine.

4. A boat B starts from a point P and moves along a straight river. An observer O stands at a distance of 600 ft from P, on the line perpendicular to the river passing through P. Find the distance from the boat to the observer when the angle θ formed by B, O, and P has

   a) $\pi/6$ radians,   b) $\pi/4$ radians,   c) $\pi/3$ radians.

5. A balloon B starts from a point P on earth and goes straight up. A man M is at a distance of $\frac{1}{2}$ mi from P. After 2 min, the angle θ formed by P, M, B has a cosine equal to

   a) 0.3,   b) 0.4,   c) 0.2.

   Find the distance between the man and the balloon at that time.

6. Repeat Exercise 5 if, after 10 min, θ itself has

   a) $\pi/3$,   b) $\pi/4$,   c) $\pi/6$ radians.

7. Plot the following points with polar coordinates (r, θ).

   a) $(2, \pi/4)$ b) $(3, \pi/6)$ c) $(1, -\pi/4)$ d) $(2, -5\pi/6)$

8. Find polar coordinates for the following points given in the usual rectangular coordinates.

   a) $(1, 1)$ b) $(-1, -1)$ c) $(3, 3\sqrt{3})$ d) $(-1, 0)$

9. Let a, b be the lengths of the legs of a right triangle. Let θ be the angle between these legs, as shown in Fig. 11–20.

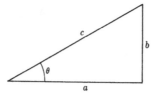

Fig. 11–20

Find the length of the hypotenuse c under the following conditions.

   a) $a = 4$ and $\cos \theta = 0.3$ b) $a = 5$ and $\cos \theta = 0.5$
   c) $a = 2$ and $\cos \theta = 0.8$ d) $b = 1$ and $\sin \theta = 0.2$
   e) $b = 4$ and $\sin \theta = 0.3$ f) $b = 3$ and $\sin \theta = 0.2$

10. Find the value of b in the preceding exercise under the following conditions.

   a) $c = 4$ and $\sin \theta = 0.2$ b) $c = 6$ and $\sin \theta = 0.3$
   c) $c = 5$ and $\sin \theta = 0.25$ d) $c = 3$ and $\sin \theta = 0.6$

11. Find the value of a in the preceding exercise under the following conditions.

   a) $c = 3$ and $\cos \theta = 0.1$
   b) $c = 6$ and $\cos \theta = 0.2$
   c) $c = 4$ and $\cos \theta = 0.6$
   d) $c = 5$ and $\cos \theta = 0.25$

12. Find the number θ such that $0 \leq \theta \leq \pi/2$ and satisfying the following conditions:

   a) $\sin \theta = \dfrac{1}{2}$,   b) $\cos \theta = \dfrac{1}{\sqrt{2}}$,   c) $\sin \theta = \dfrac{\sqrt{3}}{2}$,

   d) $\sin \theta = \dfrac{1}{\sqrt{2}}$,   e) $\cos \theta = \dfrac{1}{2}$.

13. Find a number θ satisfying the following conditions. In case of the sine, your number should satisfy $-\pi/2 \leq \theta \leq \pi/2$. In case of the cosine, it should satisfy $0 \leq \theta \leq \pi/2$.

   a) $\sin \theta = -\dfrac{1}{2}$,   b) $\cos \theta = -\dfrac{1}{2}$,   c) $\sin \theta = -\dfrac{\sqrt{3}}{2}$,

   d) $\cos \theta = -\dfrac{\sqrt{3}}{2}$,   e) $\sin \theta = \dfrac{-1}{\sqrt{2}}$,   f) $\cos \theta = \dfrac{-1}{\sqrt{2}}$,

   g) $\sin \theta = \dfrac{\sqrt{3}}{2}$,   h) $\cos \theta = \dfrac{\sqrt{3}}{2}$,   i) $\sin \theta = \dfrac{1}{2}$,

   j) $\cos \theta = \dfrac{1}{2}$.

§3. THE GRAPHS

We consider the values of $\sin x$ when x goes from 0 to 2π. We take a circle of radius 1 centered at the origin, and determine $\sin x$ from a point (a, b) on this circle, as in Fig. 11–21.

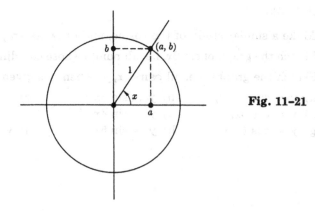

Fig. 11-21

Then

$$\sin x = b.$$

We start with $\sin 0 = 0$. As x goes from 0 to $\pi/2$, the sine of x increases (namely b increases) until x reaches $\pi/2$, at which point the sine is equal to 1.

As x ranges from $\pi/2$ to π, the sine decreases until it becomes $\sin \pi = 0$.

As x ranges from π to $3\pi/2$, the sine becomes negative, but otherwise behaves in a way similar to the sine in the first quadrant. The sine in this range decreases until it reaches

$$\sin 3\pi/2 = -1.$$

Finally, as x goes from $3\pi/2$ to 2π, the sine of x goes from -1 to 0 and increases.

At this point we are ready to start all over again.

It is interesting to plot the points whose coordinates are

$$(x, \sin x).$$

The set of points in the plane having these coordinates is called the **graph** of the sine. According to the preceding remarks, we see that the graph of the sine looks approximately as in Fig. 11-22.

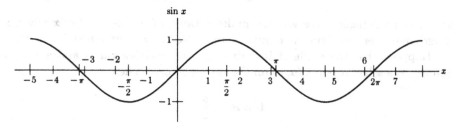

Fig. 11-22

EXERCISES

1. Make a similar study of the cosine, and draw its graph.

2. Sketch the graph of sin $2x$, i.e. all points whose coordinates are $(x, \sin 2x)$.

3. Sketch the graphs, i.e. all pairs (x, y) when y is given by:

   a) $y = \sin(-x)$,   b) $y = \sin 3x$,   c) $y = \cos 2x$,
   d) $y = \cos 3x$,   e) $y = \sin 4x$,   f) $y = \cos 4x$,
   g) $y = \cos(-x)$,   h) $y = \sin 5x$,   i) $y = \cos 5x$.

§4. THE TANGENT

We define the **tangent** of an angle A to be

$$\tan A = \frac{\sin A}{\cos A}.$$

This is defined only when $\cos A \neq 0$, and therefore is defined for angles other than $\pi/2$ radians or $3\pi/2$ radians.

Similarly, we define

$$\tan x = \frac{\sin x}{\cos x},$$

whenever x is a number which is not of the form

$$\frac{\pi}{2} + n\pi,$$

and n is an integer. We let you make a table of values for $\tan x$ when x ranges over the usual simple numbers, multiples of $\pi/4$ or multiples of $\pi/6$.

Suppose that the angle A is defined by the positive x-axis and a ray as before. We select a point (a, b) on the ray as before. Then

$$\tan A = \frac{b}{a}.$$

This is seen at once by taking the quotient of b/r and a/r: the r cancels.

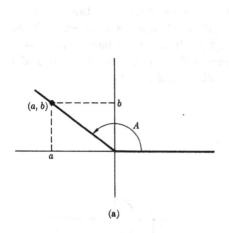

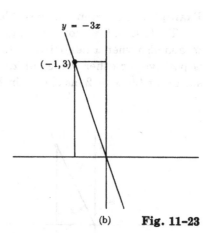

(a)   (b)   **Fig. 11-23**

Remark. If you read about the slope of a straight line in the next chapter, you will observe that the tangent of the angle which the line makes with the x-axis is precisely its slope. For instance, if the line is given by the equation

$$y = -3x,$$

then the tangent of this angle is -3, as shown in Fig. 11-23(b). We see this by picking a point on the line, say $(1, -3)$ and then using the definition of the tangent.

Example. If A is the angle as indicated in Fig. 11-24(a), in a 3, 4, 5 right triangle, then

$$\tan A = \frac{4}{3}.$$

We also have

$$\tan \frac{\pi}{3} = \sqrt{3}.$$

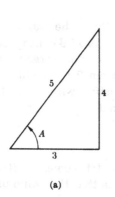

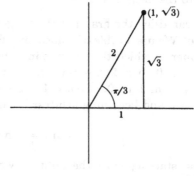

(a)   (b)   **Fig. 11-24**

Example. Determine all possible values of cos *x* given that tan *x* = 2.

To do this, first note that tan *x* is negative when *x* lies between $\pi/2$ and π, and also when *x* lies between $3\pi/2$ and 2π. On the other hand, the tangent is positive for other values of *x*, and there will be two possible values of *x* such that tan *x* = 2, as shown in Fig. 11–25(a) and (b).

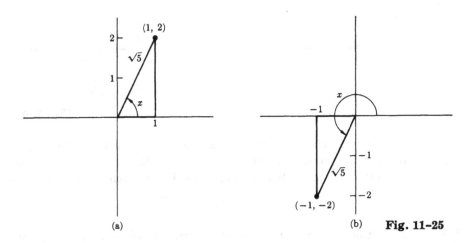

Fig. 11–25

These values correspond to right triangles whose legs have lengths 1 and 2, respectively. Therefore the hypotenuse has length $\sqrt{5}$. In the case of Fig. 11–25(a), it follows that

$$\cos x = \frac{1}{\sqrt{5}},$$

and in the case of Fig. 11–25(b), it follows that

$$\cos x = -\frac{1}{\sqrt{5}}.$$

These are the desired values.

We can draw the graph of the tangent just as easily as the graph of sine or cosine. We use a table of values for a few points, and also take into account the manner in which the tangent increases or decreases. For instance, when *x* goes from 0 to $\pi/2$, we note that sin *x* increases from 0 to 1, while cos *x* decreases from 1 to 0. Hence 1/cos *x* increases, starting with the value 1 when *x* = 0 and becoming arbitrarily large. Thus finally

$$\tan x = \sin x \cdot \frac{1}{\cos x}$$

increases, starting with the value 0 when *x* = 0, and becomes arbitrarily large. A similar discussion for other intervals shows us that the graph of the tangent looks like this.

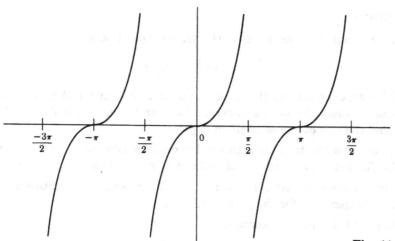

Fig. 11-26

The tangent is more practical than the sine or cosine for many purposes.

Example. Suppose we want to determine the height of a tower without climbing the tower. We go a distance a from the tower, as shown on Fig. 11–27.

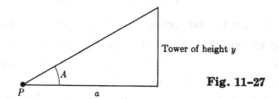

Tower of height y

Fig. 11-27

The distance a is known, and we wish to determine the height y. We can determine the angle A easily with any mechanical device available for that purpose. We can then look up the tangent, $\tan A$, in tables. Since

$$\frac{y}{a} = \tan A,$$

we can then solve for y, namely $y = a \cdot \tan A$.

For instance, suppose that the distance a is equal to 100 ft and that the angle A has $\pi/3$ radians. Then

$$\tan \frac{\pi}{3} = \sqrt{3},$$

and hence the height of the tower is equal to

$$100\sqrt{3} \text{ ft.}$$

EXERCISES

1. Make a table for the values of the tangent at all points

$$\frac{n\pi}{4} \quad \text{and} \quad \frac{n\pi}{6},$$

with integers n having the same values as in Exercise 1 of §1, except for those n where the cosine is zero. In your table, also make a column giving the values of the angle in degrees.

2. Discuss how the tangent is increasing or decreasing for $-\pi/2 < x \leq 0$. Also for $\pi/2 < x < 3\pi/2$ and $-3\pi/2 < x < -\pi/2$.

3. Define the **cotangent** $\cot x = 1/\tan x$. Draw an approximate graph for the cotangent, i.e. for the points $(x, \cot x)$.

4. Define the secant and cosecant by

$$\sec x = \frac{1}{\cos x} \quad \text{and} \quad \operatorname{cosec} x = \frac{1}{\sin x}$$

for values of x where $\cos x \neq 0$ and $\sin x \neq 0$, respectively. Find enough values of the secant and cosecant until you feel that you have the hang of things. Draw their graphs.

5. Prove that

$$1 + \tan^2 x = \sec^2 x.$$

6. State and prove a similar formula relating the cotangent and the cosecant.

7. Determine all possible values of $\cos x$ if $\tan x$ has the following values.
   a) $\tan x = 1$,   b) $\tan x = -1$,   c) $\tan x = \sqrt{3}$,
   d) $\tan x = 1/\sqrt{3}$,   e) $\tan x = 0$.

8. Determine all possible values of $\sin x$ in each one of the cases of Exercise 7.

9. You are looking at a tall building from a distance of 500 ft. The angle formed by the base of the building, your eyes, and the top of the building has
   a) $\pi/4$ radians,   b) $\pi/3$ radians,   c) $\pi/6$ radians.
   Find the height of the building.

10. A balloon B starts from a point P on earth and goes straight up. An observer O stands at a distance of 500 ft from P. After 20 min, the angle θ formed by P, O, B has
    a) $\pi/3$ radians,   b) $\pi/4$ radians,   c) $\pi/6$ radians.
    Find the height of the balloon at that time.

11. A boat B starting from a point P moves along a straight river. A man M stands $\frac{1}{2}$ mi from P on the line through P perpendicular to the river.

After 5 hr, the boat has traveled 10 mi. Let θ be the angle $\angle BMP$. Find

a) $\cos \theta$,   b) $\sin \theta$,   c) $\tan \theta$

at that time.

A billiard ball table is rectangular, and its sides have 10 ft and 7 ft, respectively. A billiard ball is hit starting at a point P on one side as drawn on the picture. It hits the next side at Q, bounces off, hits the third side at M, bounces off, and hits the fourth side at N. Each time it bounces off a side, the angle of approach to this side has the same measure as the angle of departure. Let θ be the first angle of departure as drawn in Fig. 11–28.

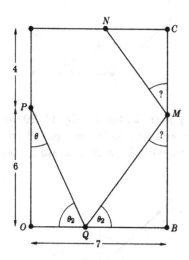

Fig. 11-28

Assume that P is at a distance of 6 ft from the corner O.

12. Assume that θ has 30°. Find:

a) $d(O, Q)$,   b) $d(P, Q)$,   c) $d(Q, B)$,   d) $d(B, M)$,
e) $d(Q, M)$,   f) $d(M, C)$,   g) $d(M, N)$.

13. Repeat the problem for (a) through (g) of Exercise 12 if $\tan \theta = \frac{1}{2}$.

14. Find the general formula for

a) $d(O, Q)$,   b) $d(P, Q)$,   c) $d(Q, M)$,
d) $d(Q, B)$,   e) $d(B, M)$

in terms of $\tan \theta$.

15. Find the general formula for the distances of Exercise 12 (a) through (e) in terms of $\sin \theta$.

§5. ADDITION FORMULAS

Our main results are the addition formulas for sine and cosine.

Theorem 3. *For any angles A, B, we have*

$$\sin(A + B) = \sin A \cos B + \cos A \sin B$$
$$\cos(A + B) = \cos A \cos B - \sin A \sin B.$$

Proof. We shall prove the second formula first.
We consider two angles A, B and their sum:

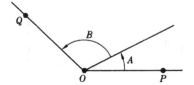

Fig. 11–29

We take two points P, Q as indicated in Fig. 11–29, at a distance 1 from the origin O. We shall compute the distance from P to Q, using two different coordinate systems. First we take a coordinate system as usual, illustrated in Fig. 11–30.

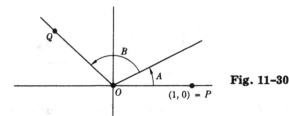

Fig. 11–30

Then the coordinates of P are $(1, 0)$ and those of Q are

$$(\cos(A + B), \sin(A + B)).$$

The square of the distance between P and Q is

$$\sin^2(A + B) + (\cos(A + B) - 1)^2,$$

which is equal to $\sin^2(A + B) + \cos^2(A + B) - 2\cos(A + B) + 1$, and hence equal to

$$-2\cos(A + B) + 2.$$

Next we place the coordinate system as shown in Fig. 11–31.

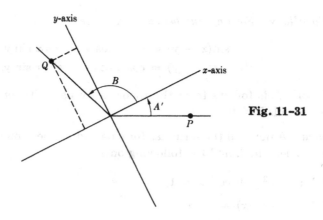

Fig. 11-31

Then the coordinates of P become

$$(\cos A, \sin(-A)) = (\cos A, -\sin A).$$

Those of Q are simply $(\cos B, \sin B)$. The square of the distance between P and Q is equal to

$$(\sin B + \sin A)^2 + (\cos B - \cos A)^2,$$

which is equal to

$$\sin^2 B + 2 \sin B \sin A + \sin^2 A + \cos^2 B - 2 \cos B \cos A + \cos^2 A$$
$$= 2 + 2 \sin A \sin B - 2 \cos A \cos B.$$

If we set the squares of the two distances equal to each other, then we get the desired addition formula for the cosine.

The addition formula for the sine can be obtained by the following device, using Theorem 2:

$$\sin (A + B) = \cos\left(A + B - \frac{\pi}{2}\right)$$
$$= \cos A \cos\left(B - \frac{\pi}{2}\right) - \sin A \sin\left(B - \frac{\pi}{2}\right)$$
$$= \cos A \sin B + \sin A \sin\left(\frac{\pi}{2} - B\right)$$
$$= \cos A \sin B + \sin A \cos B.$$

This proves the addition formula for the sine.

Corollary. *For any numbers x, y, we have:*

$$\sin(x - y) = \sin x \cos y - \cos x \sin y$$
$$\cos(x - y) = \cos x \cos y + \sin x \sin y.$$

Proof. This follows from the theorem by using Theorem 2, §2. Write out the details in full.

Remark. Among all the formulas for sine and cosine, you should remember those of Theorem 3, and the following ones:

SC 1. $\sin^2 x + \cos^2 x = 1$,

SC 2. $\cos(-x) = \cos x$,

SC 3. $\sin(-x) = -\sin x$,

SC 4. $\sin\left(x + \dfrac{\pi}{2}\right) = \cos x$,

SC 5. $\cos\left(x - \dfrac{\pi}{2}\right) = \sin x$.

Read these out loud and get an aural memory of them. All other formulas are immediate consequences of these, and can be derived each time you need them. Experience shows that this is the most economical use of your brain in dealing with these formulas.

Example. Find $\sin(\pi/12)$.
We write

$$\frac{\pi}{12} = \frac{\pi}{3} - \frac{\pi}{4}.$$

Then

$$\sin\left(\frac{\pi}{12}\right) = \sin\left(\frac{\pi}{3}\right)\cos\left(\frac{\pi}{4}\right) - \cos\left(\frac{\pi}{3}\right)\sin\left(\frac{\pi}{4}\right).$$

We know all the values on the right-hand side, and a simple computation shows that

$$\sin\left(\frac{\pi}{12}\right) = \frac{\sqrt{3} - 1}{2\sqrt{2}}.$$

From Theorem 3 we can also easily prove four other formulas which are used quite frequently and which should be memorized. They are:

SC 6. $\sin 2x = 2 \sin x \cos x$,

SC 7. $\cos 2x = \cos^2 x - \sin^2 x$,

SC 8. $\cos^2 x = \dfrac{1 + \cos 2x}{2}$,

SC 9. $\sin^2 x = \dfrac{1 - \cos 2x}{2}$.

You should have fun carrying out the easy proofs, and we leave them to you as Exercise 4. We shall now see how to use these formulas in examples to compute new values for the sine and cosine.

Example. Suppose that x is a number such that

$$\sin x = 0.8 \quad \text{and} \quad 0 < x < \frac{\pi}{2}.$$

To find $\sin 2x$ we use formula **SC 6**. Note that

$$\cos x = \sqrt{1 - \sin^2 x} = \sqrt{1 - 0.64} = \sqrt{0.36} = 0.6.$$

We took the positive square root because we prescribed that x should lie between 0 and $\pi/2$, so that $\cos x$ must be positive. Applying **SC 6** now yields

$$\sin 2x = 2 \sin x \cos x = 2(0.8)(0.6) = 0.96.$$

Example. We wish to compute $\cos \pi/8$. Let us use **SC 8**, with $x = \pi/8$. We know $\cos 2x$, namely

$$\cos \frac{\pi}{4} = \frac{1}{\sqrt{2}}.$$

Hence

$$\cos^2 \frac{\pi}{8} = \frac{1 + \cos \pi/4}{2} = \frac{1}{2} + \frac{1}{2\sqrt{2}}.$$

Taking the square root yields the desired answer, namely

$$\cos \frac{\pi}{8} = \sqrt{\frac{1}{2} + \frac{1}{2\sqrt{2}}}.$$

To put this into decimal form, it is easier to use a computing machine than your brain, and we leave the answer in the correct form above.

Example. To find $\sin \pi/8$ we use **SC 1**, and the value for $\cos \pi/8$ which we have just determined. Thus we obtain

$$\sin \frac{\pi}{8} = \sqrt{1 - \cos^2 \frac{\pi}{8}} = \sqrt{\frac{1}{2} - \frac{1}{2\sqrt{2}}}.$$

EXERCISES

1. Find $\sin 7\pi/12$. [*Hint:* Write $7\pi/12 = 4\pi/12 + 3\pi/12$.]

2. Find $\cos 7\pi/12$.

3. Find the following values:
   a) $\sin \pi/12$,   b) $\cos \pi/12$,
   c) $\sin 5\pi/12$,   d) $\cos 5\pi/12$,
   e) $\sin 11\pi/12$,   f) $\cos 11\pi/12$.

4. Prove the following formulas. They should be memorized.

   a) $\sin 2x = 2 \sin x \cos x$ b) $\cos 2x = \cos^2 x - \sin^2 x$.

   c) $\cos^2 x = \dfrac{1 + \cos 2x}{2}$ d) $\sin^2 x = \dfrac{1 - \cos 2x}{2}$

   Of course, you may assume Theorem 3 in proving these formulas. For formula (c), start with (b) and substitute $1 - \cos^2 x$ for $\sin^2 x$. Use a similar idea for (d).

5. In each one of the following cases give a numerical value for $\sin 2x$, when $\sin x$ has the indicated value.
   a) $\sin x = 0.7$ b) $\sin x = 0.6$
   c) $\sin x = 0.4$ d) $\sin x = 0.3$
   e) $\sin x = 0.2$

6. In each one of the following cases give a numerical value for $\cos 2x$ when $\sin x$ has the indicated value.
   a) $\sin x = 0.7$ b) $\sin x = 0.6$
   c) $\sin x = 0.4$ d) $\sin x = 0.3$
   e) $\sin x = 0.2$

7. In each case give a numerical value for $\cos x/2$ when $\cos x$ has the following value, and $0 \leq x \leq \pi/2$.
   a) $\cos x = 0.7$ b) $\cos x = 0.6$
   c) $\cos x = 0.4$ d) $\cos x = 0.3$
   e) $\cos x = 0.2$

8. In each of the cases of Exercise 7, what is the value for $\cos x/2$ if we assume that $-\pi/2 \leq x \leq 0$?

9. In each case give a numerical value for $\sin x/2$ when $\cos x$ has the following value, and $0 \leq x \leq \pi/2$.
   a) $\cos x = 0.7$ b) $\cos x = 0.6$
   c) $\cos x = 0.4$ d) $\cos x = 0.3$
   e) $\cos x = 0.2$

10. Find a formula for $\sin 3x$ in terms of $\sin x$ and $\cos x$. Similarly, for $\sin 4x$ and $\sin 5x$.

11. Find a formula for sin $x/2$ if $0 \leq x \leq \pi/2$ in terms of cos x and possible square root signs.

12. a) A person throws a heavy ball at an angle θ from the ground. Let d be the distance from the person to the point where the ball strikes the ground. Then d is given by

$$d = \frac{2v^2}{g} \sin \theta \cos \theta,$$

where v, g are constants. For what value of θ is the distance a maximum? [*Hint:* Give another expression for 2 sin θ cos θ.]

b) You are watering the lawn, and point the watering hose at an angle of θ degrees from the ground. The distance from the nozzle at which the water strikes the ground is given by

$$d = 2c \sin \theta \cos \theta,$$

where c is a constant. For what value of θ is the distance a maximum?

13. Prove the following formulas for any integers m, n:

$$\sin mx \sin nx = \tfrac{1}{2}[\cos(m - n)x - \cos(m + n)x],$$

$$\sin mx \cos nx = \tfrac{1}{2}[\sin(m + n)x + \sin(m - n)x],$$

$$\cos mx \cos nx = \tfrac{1}{2}[\cos(m + n)x + \cos(m - n)x].$$

[*Hint:* Expand the right-hand side using the addition formulas.]

§6. ROTATIONS

We have not yet investigated rotations from the point of view of co-ordinates, and we now fill this gap. We ask the basic question: Given a point P with coordinates (x, y), let G_φ be the rotation with respect to the origin $O = (0, 0)$ by an angle of φ radians. Let

$$G_\varphi(P) = P' = (x', y').$$

How do we describe the coordinates (x', y') of P' in terms of those of P? The answer is quite simple, and we shall use polar coordinates, as well as the addition formula, to get this answer.

Let (r, θ) be the polar coordinates of P. Then the polar coordinates of the point P' are simple $(r, \theta + \varphi)$.

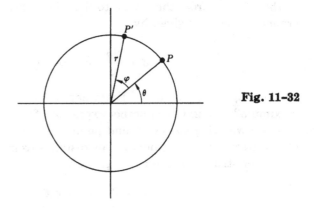

Fig. 11-32

On the other hand, we know that

$$x = r \cos \theta, \qquad x' = r \cos(\theta + \varphi),$$

$$y = r \sin \theta, \qquad y' = r \sin(\theta + \varphi).$$

We use the addition formula for the sine and cosine on the expressions on the right, and find

$$x' = r[\cos \theta \cos \varphi - \sin \theta \sin \varphi],$$

$$y' = r[\sin \theta \cos \varphi + \sin \varphi \cos \theta].$$

Expanding this out, and using the expression for x, y in terms of r, $\cos \theta$ and $\sin \theta$, we find:

$$x' = (\cos \varphi)x - (\sin \varphi)y,$$

$$y' = (\sin \varphi)x + (\cos \varphi)y.$$

Thus rotation by an angle of φ radians is described by the four numbers

$$\begin{pmatrix} \cos \varphi & -\sin \varphi \\ \sin \varphi & \cos \varphi \end{pmatrix}.$$

An array of numbers like

$$\begin{pmatrix} a & b \\ c & d \end{pmatrix}$$

has a technical name: It is called a **matrix** (in fact a 2 × 2 **matrix**).

Example. For instance,

$$\begin{pmatrix} 2 & -1 \\ 3 & 7 \end{pmatrix}$$

is a 2 × 2 matrix.

Example. The matrix associated with the rotation $G_{\pi/2}$ is the matrix

$$\begin{pmatrix} \cos \pi/2 & -\sin \pi/2 \\ \sin \pi/2 & \cos \pi/2 \end{pmatrix} = \begin{pmatrix} 0 & -1 \\ 1 & 0 \end{pmatrix}.$$

If (x, y) are the coordinates of a point P, then the coordinates of the point P' obtained by rotating P by an angle of $\pi/2$ are

$$x' = -y,$$
$$y' = x.$$

The boxed formula for the coordinates (x', y') in terms of (x, y) is sometimes written in the form of a "product",

$$\begin{pmatrix} \cos \varphi & -\sin \varphi \\ \sin \varphi & \cos \varphi \end{pmatrix} \begin{pmatrix} x \\ y \end{pmatrix} = \begin{pmatrix} x' \\ y' \end{pmatrix}.$$

This is something like a multiplication.

In general, if

$$\begin{pmatrix} a & b \\ c & d \end{pmatrix}$$

is a 2 × 2 matrix, we write the product

$$\begin{pmatrix} a & b \\ c & d \end{pmatrix} \begin{pmatrix} x \\ y \end{pmatrix} = \begin{pmatrix} ax + by \\ cx + dy \end{pmatrix}.$$

For example,

$$\begin{pmatrix} 3 & 2 \\ -1 & 5 \end{pmatrix} \begin{pmatrix} x \\ y \end{pmatrix} = \begin{pmatrix} 3x + 2y \\ -x + 5y \end{pmatrix}.$$

The theory of matrices can be considerably generalized, and we refer you to texts in linear algebra for this.

EXERCISES

In each one of the following cases, write the matrix associated with the rotation G_φ when φ has the indicated value. Write explicitly the coordinates (x', y') of $G_\varphi(P)$ if P has coordinates (x, y).

1. $\varphi = \pi$
2. $\varphi = \pi/4$
3. $\varphi = \pi/3$
4. $\varphi = \pi/6$
5. $\varphi = 3\pi/4$
6. $\varphi = 3\pi/2$
7. $\varphi = -\pi/2$
8. $\varphi = -\pi/3$
9. $\varphi = -\pi/4$
10. $\varphi = 5\pi/4$

When (x, y) have the following numerical values, give the numerical values for (x', y') in each one of the rotations of Exercises 1 through 10. Draw the picture in each case.

11. $P = (3, 1)$
12. $P = (5, -2)$
13. $P = (-2, 4)$
14. $P = (2, -1)$
15. $P = (0, 5)$
16. $P = (3, 0)$
17. $P = (-1, 1)$
18. $P = (-2, -1)$
19. $P = (2, 1)$
20. $P = (-2, -2)$

21. Associate a matrix with a dilation by r. Interpret dilation by r in terms of the matrix multiplication. Do the same for mixed dilations of type which we have written $F_{a,b}$.

22. Write down the matrices for the rotations G_φ, G_ψ, $G_{\varphi+\psi}$.

12 Some Analytic Geometry

§1. THE STRAIGHT LINE AGAIN

Let $F(x, y)$ be an expression involving a pair of numbers (x, y). Let c be a number. We consider the equation

$$F(x, y) = c.$$

The set of points (x, y) for which this equation holds is called the **graph** of the equation. In this section we study the simplest case of equations like

$$3x - 2y = 5.$$

If you have read Chapter 10, you already know this equation, but we shall not assume any knowledge from this chapter. We develop everything from scratch.

Consider first a simple example, namely

$$y = 3x.$$

The set of points $(x, 3x)$ is the graph of this equation, or equivalently of the equation

$$y - 3x = 0.$$

We can give x an arbitrary value, and thus we see that the graph looks like Fig. 12–1(a).

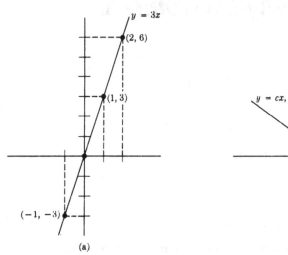

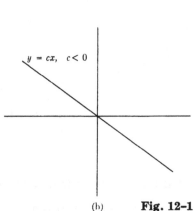

(a)   (b)   **Fig. 12-1**

If we consider the graph of

$$y = 4x,$$

we see that it is a line which slants more steeply. In general, the graph of the equation

$$y = ax,$$

where a is a number, represents a straight line. An arbitrary point on this line is of type

$$(x, ax) = x(1, a).$$

Thus the (x, y) coordinates of a point on the line are obtained by making the dilation of $(1, a)$ by x.

If a is positive, then the line slants to the right. If a is negative, then the line slants to the left, as shown on Fig. 12–1(b). For instance, the graph of

$$y = -x$$

consists of all points $(x, -x)$, and looks like this (Fig. 12–2).

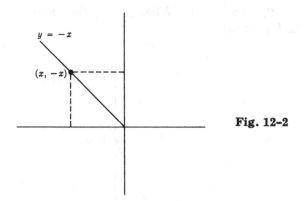

Fig. 12-2

If we drop the perpendiculars from the point $(x, -x)$ to the axes, we obtain a right triangle, which in this case has legs of equal length.

Let a, b be numbers. The graph of the equation

(1) $$y = ax + b$$

is also a straight line, which is parallel to the graph of the equation

(2) $$y = ax.$$

To convince ourselves of this, we observe the following. Let

$$y' = y - b.$$

The equation

(3) $$y' = ax$$

is of the type just discussed. If we have a point (x, y') on the graph of (3), then we get a point $(x, y' + b)$ on the graph of (1), simply by adding b to the second coordinate. This means that the graph of the equation

$$y = ax + b$$

is the straight line parallel to the line determined by the equation

$$y = ax,$$

and passing through the point $(0, b)$.

Example. We want to draw the graph of the equation

$$y = 2x + 1.$$

When $x = 0$, then $y = 1$. When $y = 0$, then $x = -\frac{1}{2}$. Hence the graph
looks like Fig. 12–3(a).

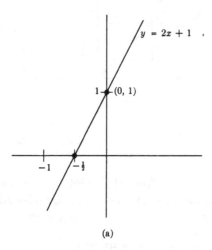

(a)   **Fig. 12–3**

Example. We want to draw the graph of the equation

$$y = -2x - 5.$$

When $x = 0$, then $y = -5$. When $y = 0$, then $x = -\frac{5}{2}$. Hence the graph
looks like Fig. 12–3(b).

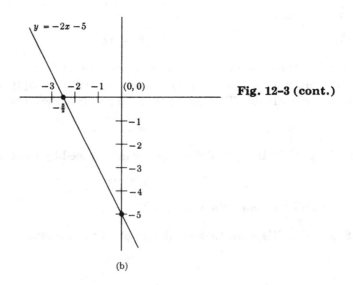

(b)

Fig. 12–3 (cont.)

If a line L is the graph of the equation

$$y = ax + b,$$

then we call the number a the **slope** of the line. For instance, the slope of the line whose equation is

$$y = -4x + 7$$

is -4. The slope determines how the line is slanted.

Let $y = ax + b$ be the equation of a line.

The slope of this line can be obtained from two distinct points on the line. Let (x_1, y_1) and (x_2, y_2) be the two points. By definition, we know that

$$y_1 = ax_1 + b,$$
$$y_2 = ax_2 + b.$$

Subtracting, we find that

$$y_2 - y_1 = a(x_2 - x_1).$$

Consequently, if the two points are distinct, $x_2 \neq x_1$, and we can divide by $(x_2 - x_1)$ to find

$$a = \frac{y_2 - y_1}{x_2 - x_1}.$$

This formula gives us the slope in terms of the coordinates of two distinct points.

Example. Consider the line defined by the equation

$$y = 2x + 5.$$

The two points $(1, 7)$ and $(-1, 3)$ lie on the line. The slope is equal to 2, and, in fact,

$$2 = \frac{7 - 3}{1 - (-1)}.$$

as it should be.

Geometrically, our quotient

$$\frac{y_2 - y_1}{x_2 - x_1}$$

is the ratio of the vertical side and horizontal side of the triangle in Fig. 12–4.

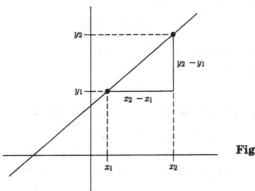

Fig. 12-4

Observe that it does not matter which point we call (x_1, y_1) and which one we call (x_2, y_2). We would get the same value for the slope. This is because if we invert their order, then in the quotient expressing the slope both the numerator and denominator will change by a sign, so that their quotient does not change.

Conversely, given two points, it is easy to determine the equation of the line passing through the two points.

Example. Find the equation of the line passing through the points $(1, 2)$ and $(2, -1)$.

We first find the slope. It must be the quotient

$$\frac{y_2 - y_1}{x_2 - x_1}$$

which in this case is equal to

$$\frac{-1 - 2}{2 - 1} = -3.$$

Thus we know that our line is the graph of the equation

$$y = -3x + b$$

for some number b, which we must determine. We know that the line passes through the point $(1, 2)$. Hence

$$2 = -3 \cdot 1 + b,$$

and we get the value of b, namely

$$b = 5.$$

Thus the equation of the line is

$$y = -3x + 5.$$

We can also determine the equation of a line given the slope and one point.

Example. Find the equation of the line having slope -7 and passing through the point $(-1, 2)$.

The equation must be of the form

$$y = -7x + b$$

with some number b. Furthermore, when $x = -1$ the corresponding value of y must be 2. Hence

$$2 = (-7)(-1) + b,$$

and

$$b = -5.$$

Hence the equation of the line is

$$y = -7x - 5.$$

In general, the equation of the line passing through the point (x_1, y_1) and having slope a is

$$y - y_1 = a(x - x_1).$$

For points such that $x \neq x_1$, we can also write this in the form

$$\frac{y - y_1}{x - x_1} = a.$$

The equation of the line passing through two points (x_1, y_1) and (x_2, y_2) is

$$\frac{y - y_1}{x - x_1} = \frac{y_2 - y_1}{x_2 - x_1}.$$

for all points such that $x \neq x_1$.

We should also mention vertical lines. These cannot be represented by equations of type

$$y = ax + b.$$

Suppose that we have a vertical line intersecting the x-axis at the point $(2, 0)$. The y-coordinate of any point on the line can be an arbitrary number, while the x-coordinate is always 2. Hence the equation of this line is simply

$$x = 2.$$

Similarly, the equation of a vertical line intersecting the x-axis at the point $(c, 0)$ is

$$x = c.$$

When a line is given in the form $y = ax + b$, it is easy to find its intersection with a circle. We give an example.

Example. Find the points of intersection of the line and the circle given by the following equations:

$$y = 3x + 2 \qquad \text{and} \qquad x^2 + y^2 = 1.$$

To do this, note that a point (x, y) lies on the intersection if and only if $y = 3x + 2$ and

$$x^2 + (3x + 2)^2 = 1.$$

Thus we must solve for x in this last equation, which is equivalent to

$$x^2 + 9x^2 + 12x + 4 = 1.$$

Again, this equation is equivalent to

$$10x^2 + 12x + 3 = 0,$$

which we can solve by the quadratic formula. We find

$$x = \frac{-12 \pm \sqrt{144 - 4 \cdot 30}}{20},$$

or in other words, we find the two values

$$x = \frac{-6 \pm \sqrt{6}}{10}.$$

These two possible values for x give us the two points of intersection, as illustrated in Fig. 12–5(a). The y-coordinates can be found by the expression $y = 3x + 2$, and hence the points of intersection are

$$\left(\frac{-6 + \sqrt{6}}{10}, \frac{2 + 3\sqrt{6}}{10} \right) \qquad \text{and} \qquad \left(\frac{-6 - \sqrt{6}}{2}, \frac{2 - 3\sqrt{6}}{10} \right).$$

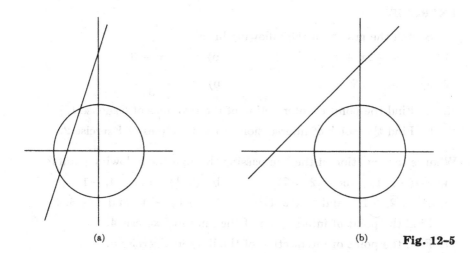

(a)   (b)   **Fig. 12–5**

Of course, it may happen that a circle and a line do not intersect, as shown in Fig. 12–5(b). If we followed the same procedure to solve for x as before, we would find in this case that the quadratic equation has no real solution because the number under the square root sign is negative.

Example. Find the points of intersection (if any) of the line and the circle given by the following equations:

$$y = x + 5 \qquad \text{and} \qquad x^2 + y^2 = 1.$$

We follow the same procedure as before, and we see that the x-coordinate of a point of intersection must satisfy the equation

$$x^2 + (x + 5)^2 = 1,$$

or equivalently

$$x^2 + x^2 + 10x + 25 = 1.$$

This amounts to solving

$$2x^2 + 10x + 24 = 0.$$

The quadratic formula gives

$$x = \frac{-10 \pm \sqrt{100 - 192}}{4} = \frac{-10 \pm \sqrt{-92}}{4}.$$

In this case, the number under the square root sign is negative, and hence there is no real solution. Hence the circle and the line do not intersect.

EXERCISES

Sketch the graphs of the following lines.

1. a) $y = -2x + 5$ b) $y = 5x - 3$

2. a) $y = \dfrac{x}{2} + 7$ b) $y = -\dfrac{x}{3} + 1$

3. a) Find the point of intersection of the two lines of Exercise 1.

   b) Find the point of intersection of the two lines of Exercise 2.

What is the equation of the line passing through the following points?

4. a) $(-1, 1)$ and $(2, -7)$ b) $(3, \frac{1}{2})$ and $(4, -1)$

5. a) $(\sqrt{2}, -1)$ and $(\sqrt{2}, 1)$ b) $(-3, -5)$ and $(\sqrt{3}, 4)$

6. Find the point of intersection of the lines in Exercise 4.

7. Find the point of intersection of the lines in Exercise 5.

8. Find the point of intersection of the following pairs of lines.

   a) $3x - 2y = 1$ and $4x - y = 2$

   b) $x - 5y = 1$ and $2x + y = 0$

   c) $3x + 2y = 2$ and $4x - y = 1$

   d) $-2x + 3y = 4$ and $-5x - 3y = -2$

What is the equation of the line having the given slope and passing through the given point?

9. slope 4 and point $(1, 1)$ 10. slope -2 and point $(\frac{1}{2}, 1)$

11. slope $-\frac{1}{2}$ and point $(\sqrt{2}, 3)$ 12. slope 3 and point $(-1, 5)$

Sketch the graphs of the following lines:

13. $x = 5$ 14. $x = -1$ 15. $x = -3$

16. $y = -4$ 17. $y = 2$ 18. $y = 0$

What is the slope of the line passing through the following points?

19. $(1, \frac{1}{2})$ and $(-1, 1)$ 20. $(\frac{1}{4}, 1)$ and $(\frac{1}{2}, -1)$

21. $(2, 1)$ and $(\sqrt{2}, 1)$ 22. $(\sqrt{3}, 1)$ and $(3, 2)$

What is the equation of the line passing through the following points?

23. $(\pi, 1)$ and $(\sqrt{2}, 3)$ 24. $(\sqrt{2}, 2)$ and $(1, \pi)$

25. $(-1, 2)$ and $(\sqrt{2}, -1)$ 26. $(-1, \sqrt{2})$ and $(-2, -3)$

27. Sketch the graphs of the following lines.

   a) $y = 2x$ b) $y = 2x + 1$ c) $y = 2x + 5$

   d) $y = 2x - 1$ e) $y = 2x - 5$

28. For our purposes here, define two straight lines to be parallel if they have the same slope. Let

$$y = ax + b \quad \text{and} \quad y = cx + d$$

be the equations of two lines with $b \neq d$.

a) If they are parallel, show that they have no point in common.

b) If they are not parallel, show that they have exactly one point in common.

29. Find the common point of the following pairs of lines.

a) $y = 3x + 5$ and $y = 2x + 1$

b) $y = 3x - 2$ and $y = -x + 4$

c) $y = 2x + 3$ and $y = -x + 2$

d) $y = x + 1$ and $y = 2x + 7$

30. If a straight line is expressed in parametric form,

$$\{P + tA\}_{t \text{ in } \mathbf{R}}$$

and $A = (a_1, a_2)$, what is the slope of the line in terms of the coordinates of A? Does this slope depend on the coordinates of P?

31. Find the points of intersection, if any, of the indicated line and circle. Draw the pictures.

a) $y = 2x - 1$ and $x^2 + y^2 = 1$

b) $y = 3x$ and $x^2 + y^2 = 4$

c) $y = 3x - 2$ and $x^2 + y^2 = 2$

d) $y = x - 1$ and $(x - 1)^2 + (y - 2)^2 = 4$

e) $y = 4x + 1$ and $(x - 3)^2 + (y - 4)^2 = 1$

f) $y = -2x - 1$ and $(x - 3)^2 + y^2 = 1$

g) $y = -2x + 3$ and $(x - 3)^2 + y^2 = 1$

§2. THE PARABOLA

Next we consider the graph of the equation

$$y = x^2.$$

In this case, we see that for any value of x, the corresponding value of y is positive. We make a small table of values.

x	$y = x^2$
1	1
2	4
3	9
4	16

We see that as x increases, so does x^2. Furthermore $(-x)^2 = x^2$. Thus our graph (x, x^2) is symmetric with respect to the y-axis. It looks like this.

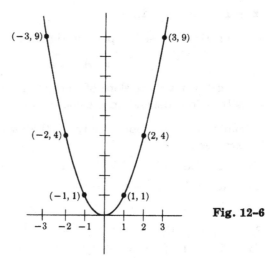

Fig. 12–6

Suppose that we want to draw the graph of the equation

(1) $$y = (x - 1)^2.$$

We shall find that it looks exactly the same, but as if the origin were placed at the point $(1, 0)$. (See Fig. 12–7.) Namely, let

$$x' = x - 1.$$

Then our equation (1) is equivalent to

(2) $$y = x'^2.$$

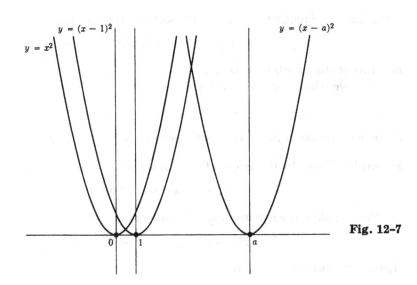

Fig. 12-7

Thus if we place our coordinate system with origin at the point $(1, 0)$, we see that equation (1) becomes equation (2) in terms of the new coordinates (x', y). Similarly, the equation

$$y = (x - a)^2$$

has a graph which looks like the others, but translated to $(a, 0)$.

We can also perform a translation on y. Let (a, b) be a given point. We let

$$x' = x - a \qquad \text{and} \qquad y' = y - b.$$

Thus when $x = a$, we have $x' = 0$ and when $y = b$, we have $y' = 0$. As we can see in Fig. 12–8, the graph of the equation

$$y' = x'^2$$

looks the same as the graph of the equation $y = x^2$, but translated to the point (a, b).

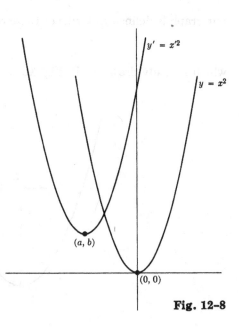

Fig. 12-8

Note that the equation $y' = x'^2$ is the same as the equation

$$(y - b) = (x - a)^2$$

in terms of the coordinates (x, y).

A curve which has an equation

$$(y - b) = c(x - a)^2$$

in some coordinate system is called a **parabola**.

Example. Describe the graph of the equation

$$2y - x^2 - 4x + 6 = 0.$$

We complete the square in x. We can write

$$x^2 + 4x = (x + 2)^2 - 4.$$

Hence our equation can be rewritten

$$2y = (x + 2)^2 - 10,$$

or

$$2(y + 5) = (x + 2)^2.$$

Choosing the new coordinates

$$x' = x + 2 \qquad \text{and} \qquad y' = y + 5,$$

our graph is defined in terms of these coordinates by the equation

$$y' = \tfrac{1}{2}x'^2,$$

which is easily drawn as in Fig. 12–9.

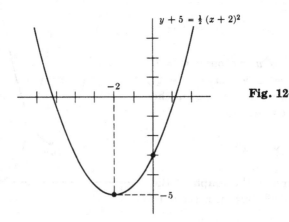

Fig. 12–9

Finally, we remark that if we have an equation

$$x - y^2 = 0$$

or

$$x = y^2,$$

then we get the graph of a curve which is also called a parabola, and is tilted horizontally as in Fig. 12–10.

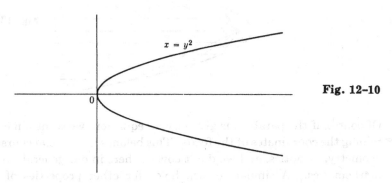

Fig. 12–10

In this case, all values of x are positive when $y \neq 0$ because then $y^2 > 0$.

We can apply the same technique of changing the coordinate system to see what the graph of a more general equation looks like.

Example. Sketch the graph of the equation

$$x - y^2 + 2y + 5 = 0.$$

After completing the square in y, we can write this equation in the form

$$(x + 6) = (y - 1)^2,$$

and hence its graph looks like this (Fig. 12–11).

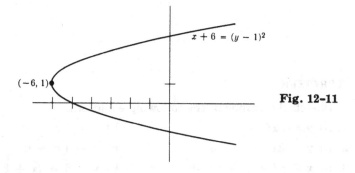

Fig. 12–11

We don't go into further properties of the parabola. However, we mention one of them here which shows how parabolas arise in optics. If one makes a mirror in the shape of a parabola, then any horizontal ray coming into the mirror gets reflected, and all these reflections meet at one point, called the **focus** F of the parabola. We have drawn this in Fig. 12–12.

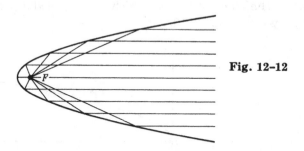

Fig. 12–12

Of course, if the parabola is given by an equation, we want a means of determining the coordinates of the focus. This belongs in a special course in analytic geometry, or optics, and we don't cover it here in our general survey of basic mathematics. A similar remark holds for other properties of the various curves discussed in this chapter (ellipse and hyperbola, in addition to the parabola). They have a number of interesting properties which are traditionally covered at this elementary level. However, I feel that it is best to postpone the discussion of such properties to the time when they are needed. All we are trying to do here is to get a quick acquaintance with the graphs of simple curves defined by simple equations. This is more than sufficient background to do calculus and many other applications (in physics, engineering, or economics).

EXERCISES

Sketch the graphs of the following equations.

1. a) $y = 3x^2$ b) $y = \frac{1}{2}x^2$

2. a) $y = 2x^2 + x - 3$ b) $x - 4y^2 = 0$

3. a) $x - y^2 + y + 1 = 0$ b) $y - 5 = (x + 3)^2$

4. a) $(y + 4) = (x + 2)^2$ b) $(y - 1)^2 = x + 7$

Complete the square in the following equations and change the coordinate system to put them into the form

$$x'^2 + y'^2 = r^2 \quad \text{or} \quad y' = cx'^2, \quad \text{or} \quad x' = cy'^2.$$

with a suitable constant c. Sketch the graphs.

5. $x^2 + y^2 - 4x + 2y - 20 = 0$ 6. $x^2 + y^2 - 2y - 8 = 0$

7. $x^2 + y^2 + 2x - 2 = 0$ 8. $y - 2x^2 - x + 3 = 0$

9. $y - x^2 - 4x - 5 = 0$ 10. $y - x^2 + 2x + 3 = 0$

11. $x^2 + y^2 + 2x - 4y = -3$ 12. $x^2 + y^2 - 4x - 2y = -3$

13. $x - 2y^2 - y + 3 = 0$ 14. $x - y^2 - 4y = 5$

15. $y = -x^2$ 16. $y = -2x^2$

17. $y = -3x^2$ 18. $y = -(x - 1)^2$

19. $y = -(x + 2)^2$ 20. $y = -(x + 3)^2$

21. $y = -(x - 3)^2$ 22. $y - 1 = -(x + 5)^2$

23. $y + 2 = -(x + 2)^2$ 24. $y + 2 = -(x - 3)^2$

25. Find the point of intersection of the parabola and the straight line given by the following equations.

a) $3x + y = 1$ and $y = 3x^2$

b) $4x - 2y = 3$ and $y = -2x^2$

c) $y = 4x - 1$ and $y = -(x - 2)^2$

d) $y = -3x + 5$ and $y = (x + 3)^2$

e) $y = 4x - 1$ and $(y - 2) = (x - 3)^2$

§3. THE ELLIPSE

Let a be a positive number. If (x, y) is a point in the plane, then we call the point

$$(ax, ay)$$

the dilation of (x, y) by a. Multiplication of each coordinate by a amounts to stretching by a.

We can generalize this slightly. Let a, b be positive numbers. To each point (x, y) we associate the point (ax, by). Thus we stretch the first co-ordinate by a and the second coordinate by b. The association

$$(x, y) \mapsto (ax, by)$$

will be denoted by $F_{a,b}$.

Consider the circle consisting of all points (x, y) such that

(1) $$x^2 + y^2 = 1.$$

It is the unit circle centered at the origin. What happens to this circle when we impose on it this mixed type of dilation? Let

$$u = ax$$

and

$$v = by.$$

Then u, v satisfy the equation

(2) $$\frac{u^2}{a^2} + \frac{v^2}{b^2} = 1.$$

Conversely, if (u, v) is a point satisfying this equation, letting

$$x = \frac{u}{a}$$

and

$$y = \frac{u}{b}$$

shows that (u, v) is the image of (x, y) under $F_{a,b}$. Hence the image of the circle defined by equation (1) is the set of points (u, v) satisfying equation (2). This image is called an **ellipse**. In general, an **ellipse** is a curve in the plane which is the graph of an equation of type (2) in some coordinate system.

Example. The set of points (u, v) satisfying the equation

$$\frac{u^2}{3^2} + \frac{v^2}{2^2} = 1$$

is an ellipse. We wish to draw its graph. Call the horizontal axis the u-axis, and the vertical axis the v-axis. When $u = 0$, we see that $v^2 = 2^2$ so that $v = \pm 2$. When $v = 0$, we see that $u^2 = 3^2$ so that $u = \pm 3$. If we visualize the circle of radius 1 undergoing the mixed stretching $F_{3,2}$, then we see that the graph of the ellipse looks like this (Fig. 12–13).

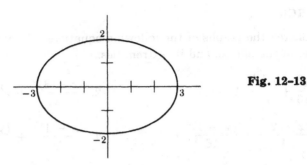

Fig. 12–13

We can also perform a translation on an ellipse just as we did for circles and parabolas.

Example. Draw the graph of the ellipse defined by the equation

$$\frac{(x-1)^2}{25} + \frac{(y-4)^2}{16} = 1.$$

Let $x' = x - 1$ and $y' = y - 4$. Then (x', y') satisfy the equation

$$\frac{x'^2}{5^2} + \frac{y'^2}{4^2} = 1.$$

Thus we have the equation of an ellipse centered at the point $(1, 4)$, as shown in Fig. 12–14.

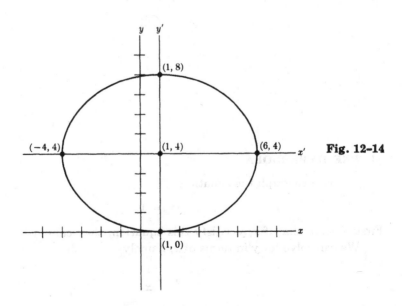

Fig. 12–14

EXERCISES

Sketch the graphs of the following equations. In each case, indicate the center of the ellipse, and its extremities.

1. $\dfrac{x^2}{16} + \dfrac{y^2}{4} = 1$

2. $\dfrac{(x-3)^2}{4} + \dfrac{(y-1)^2}{9} = 1$

3. $\dfrac{(x+2)^2}{4} + \dfrac{(y-1)^2}{16} = 1$

4. $\dfrac{(x-1)^2}{25} + \dfrac{(y+3)^2}{9} = 1$

5. $\dfrac{x^2}{16} + \dfrac{(y+1)^2}{9} = 1$

6. $\dfrac{(x+3)^2}{25} + \dfrac{y^2}{4} = 1$

7. $4x^2 + 9y^2 = 1$

8. $25x^2 + 16y^2 = 1$

9. $9x^2 + 16y^2 = 1$

10. $4x^2 + 25y^2 = 1$

11. $4(x-1)^2 + 16(y+2)^2 = 1$

12. $16(x+3)^2 + 9(y+1)^2 = 1$

13. Find the points of intersection of the ellipses given in Exercises 1 through 12 with the straight line given by the equation $y = 2x - 1$.

§4. THE HYPERBOLA

We wish to graph the equation

$$xy = 1.$$

First observe that if x, y satisfy this equation, they cannot be equal to 0.

We can solve for y in terms of x, namely

$$y = \frac{1}{x}.$$

We make a small table of values.

x	y		x	y
1	1		$\frac{1}{2}$	2
2	$\frac{1}{2}$		$\frac{1}{3}$	3
3	$\frac{1}{3}$		$\frac{1}{4}$	4
100	$\frac{1}{100}$			

We see that as x grows larger, y grows smaller. Also we have an obvious symmetry when x takes on negative values. Thus the graph looks like this (Fig. 12–15).

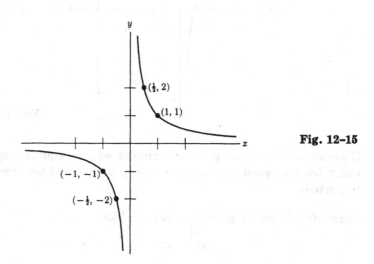

Fig. 12–15

Similarly, we can perform a translation as we did before.

Example. Sketch the graph of the equation

$$y - 2 = \frac{1}{x + 3}.$$

Let $x' = x + 3$ and $y' = y - 2$. Then the coordinates x', y' satisfy the equation

$$x'y' = 1.$$

Thus our graph looks like this (Fig. 12–16).

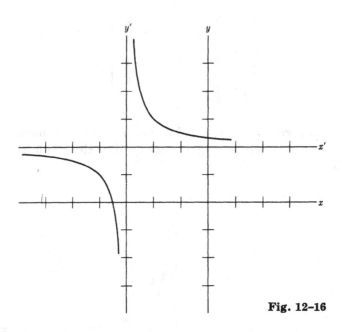

Fig. 12–16

If c is a number, we can graph in a similar way the equation $xy = c$. A curve which has an equation of type $xy = c$ in some coordinate system is called a **hyperbola.**

Example. If we are given an equation like

$$xy - 2x + 3y + 4 = 5,$$

then we can factor the left-hand side and rewrite the equation as

$$(x + 3)(y - 2) + 6 + 4 = 5,$$

or

$$(x + 3)(y - 2) = -5.$$

In terms of the new coordinates $x' = x + 3$ and $y' = y - 2$, we can rewrite our equation in the form

$$x'y' = -5.$$

Using a table of values, and a similar analysis of what happens to y' when x' increases or decreases, we see that the graph looks like this (Fig. 12–17).

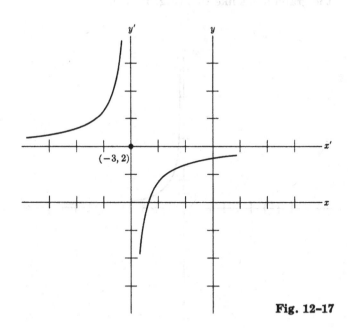

Fig. 12-17

Example. In physics we encounter cases where an object is repelled from the origin along a straight line, with a force whose magnitude is inversely proportional to the distance from the origin. Suppose this straight line is the x-axis. Let y be the magnitude of the force acting on the particle. Then we can write

$$y = \frac{k}{x},$$

for some constant k. For instance, we can take $k = 2$. We can then draw the graph of the force, which is similar to a hyperbola. We make a small table of values to get an idea of the behavior of the curve.

x	y	x	y
1	2	$\frac{1}{2}$	4
2	1	$\frac{1}{3}$	6
3	$\frac{2}{3}$	$\frac{1}{4}$	8
4	$\frac{1}{2}$	$\frac{1}{5}$	10
5	$\frac{2}{5}$		

We see that the graph looks like this (Fig. 12–18).

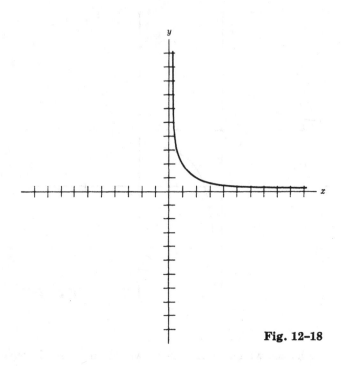

Fig. 12–18

EXERCISES

Sketch the graphs of the following curves, defined by the given equations. If need be, make small tables of values. Try to rewrite the equation in a form that makes it obvious what the graph looks like.

1. a) $y = \dfrac{3}{x}$ b) $y = \dfrac{1}{2x}$ c) $y = \dfrac{1}{3x}$

2. a) $y = \dfrac{1}{x-1}$ b) $y - 4 = \dfrac{2}{x-2}$ c) $y = \dfrac{1}{2(x-3)}$

3. a) $(x-1)(y-2) = 1$ b) $x(y+1) = 1$

4. a) $(x-1)(y-2) = 2$ b) $x(y+1) = 3$

5. $xy - 4 = 0$

6. $y = \dfrac{2}{1 - x}$

7. $y = \dfrac{3}{x + 1}$

8. $(x + 2)(y - 1) = 2$

9. $(x - 1)(y - 1) = 2$

10. $(x - 1)(y - 1) = 1$

11. $y = \dfrac{1}{x - 2} + 4$

12. $y = \dfrac{1}{x + 1} - 2$

13. $y = \dfrac{4x - 7}{x - 2}$

14. $y = \dfrac{-2x - 1}{x + 1}$

15. $y = \dfrac{x + 1}{x - 1}$

16. $y = \dfrac{x - 1}{x + 1}$

17. $xy = -1$

18. $xy = -2$

19. $xy = -3$

20. $xy = -4$

21. $(x - 1)y = -2$

22. $(x + 1)y = -2$

23. $(x - 1)(y + 2) = -1$

24. $(x - 1)(y - 2) = -1$

25. $(x + 1)(y - 3) = -4$

26. $(x - 1)(y + 3) = -4$

27. Find the point of intersection of the hyperbolas given in Exercises 1 through 26 with the straight line given by the equation $y = x - 3$.

The next section is slightly less important than the preceding one, and may be omitted.

§5. ROTATION OF HYPERBOLAS

There is another standard equation for a hyperbola which occurs frequently, and which it is useful to recognize, namely, the equation

(1) $$y^2 - x^2 = 1,$$

or more generally, the equation

$$y^2 - x^2 = c,$$

with some constant c. We shall now see that these equations represent hyperbolas, obtained by rotating the hyperbola defined by our standard equation

$$xy = 1,$$

and then perhaps performing a dilation.

Theorem 1. *Let H be the set of points* (x, y) *satisfying the equation*

$$xy = 1.$$

Let G be rotation by $\pi/4$, *counterclockwise, as usual. Then the image of H under G is the curve H′ consisting of all points* (u, v) *satisfying the equation*

(2) $$v^2 - u^2 = 2.$$

We may illustrate Theorem 1 in Fig. 12–19.

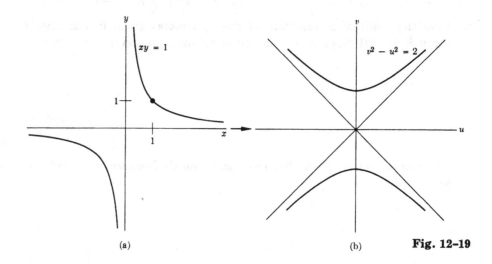

(a)   (b)   **Fig. 12–19**

Fig. 12–19(b) represents the hyperbola of Fig. 12–19(a) rotated counterclockwise by $\pi/4$. We have also drawn the axes of Fig. 12–19(a), rotated by $\pi/4$.

Proof of Theorem 1. We recall that under rotation by an angle, θ, the coordinates of a point (x, y) change by the matrix

$$\begin{pmatrix} \cos\theta & -\sin\theta \\ \sin\theta & \cos\theta \end{pmatrix}.$$

(Cf. Chapter 11, §6.) This means that if (u, v) are the coordinates of the rotated point, then

$$u = (\cos \theta)x - (\sin \theta)y$$
$$v = (\sin \theta)x + (\cos \theta)y.$$

We apply this when $\theta = \pi/4$, in which case we have

$$\cos \theta = \sin \theta = \frac{1}{\sqrt{2}}.$$

We then see that the coordinates of the rotated point (u, v) are given by

$$\text{(3)} \qquad u = \frac{1}{\sqrt{2}}x - \frac{1}{\sqrt{2}}y = \frac{1}{\sqrt{2}}(x - y)$$

$$v = \frac{1}{\sqrt{2}}x + \frac{1}{\sqrt{2}}y = \frac{1}{\sqrt{2}}(x + y).$$

Furthermore, we get:

$$v^2 - u^2 = \frac{1}{2}(x + y)^2 - \frac{1}{2}(x - y)^2$$

$$= \frac{1}{2}[x^2 + 2xy + y^2 - (x^2 - 2xy + y^2)]$$

$$= \frac{1}{2} \cdot 4xy$$

$$= 2xy.$$

If a point (x, y) satisfies the condition $xy = 1$, we conclude that its image (u, v) satisfies the equation

$$v^2 - u^2 = 2.$$

Hence we have proved that the image of H is contained in the curve H'. In other words, we have proved that if (x, y) satisfies (1), then its rotation by $\pi/4$ satisfies (2).

Conversely, let $P' = (u, v)$ be a point satisfying (2), i.e. such that

$$v^2 - u^2 = 2.$$

We must show that $P' = G(P)$ for some point $P = (x, y)$ satisfying (1). Our intuition immediately tells us how to find (x, y): these coordinates are obtained by rotating (u, v) through an angle of $-\pi/4$. Thus we let

$$x = \frac{1}{\sqrt{2}}(u + v)$$

$$y = \frac{1}{\sqrt{2}}(-u + v).$$

Taking the product, we find that

$$xy = \tfrac{1}{2}(u + v)(v - u) = \tfrac{1}{2}(v^2 - u^2) = 1.$$

If you now use formulas (3) applied to this point (x, y), you will find precisely that they yield (u, v). This proves what we wanted, and concludes the proof of Theorem 1.

Remark. Instead of rotating the hyperbola defined by the equation $xy = 1$, suppose that we rotate the hyperbola represented by the equation

$$xy = 6.$$

Denote this hyperbola by $H_{\sqrt{6}}$. It is obtained from H by a dilation, namely, we can rewrite our equation in the form

$$\frac{x}{\sqrt{6}} \cdot \frac{y}{\sqrt{6}} = 1.$$

Let $x' = x/\sqrt{6}$ and $y' = y/\sqrt{6}$. Then (x', y') satisfy the equation

$$x'y' = 1.$$

We also have

$$x = \sqrt{6} \cdot x' \quad \text{and} \quad y = \sqrt{6} \cdot y'.$$

Thus $H_{\sqrt{6}}$ is the dilation by $\sqrt{6}$ of the hyperbola H.

In general, let r be a positive number. The same argument shows that the hyperbola H_r represented by the equation

$$xy = r^2$$

is the dilation by r of the hyperbola $H = H_1$.

Let G be any rotation. For any positive number r, and any point P, we note that

$$G(rP) = rG(P).$$

In other words, dilating by r followed by the rotation G is the same as first taking the rotation G and then dilating by r. We could say that a rotation commutes with a dilation. We then see that rotating the hyperbola H_r by $\pi/4$ yields the hyperbola defined by the equation

$$v^2 - u^2 = 2r^2.$$

Example. The hyperbola

$$v^2 - u^2 = 18$$

is the rotation by $\pi/4$ of the hyperbola defined by the equation

$$xy = 9.$$

Example. The hyperbola

$$v^2 - u^2 = 10$$

is the rotation by $\pi/4$ of the hyperbola represented by the equation

$$xy = 5.$$

(Watch out: We have $10 = 2 \cdot 5$, so that we use $r = \sqrt{5}$.)

Remark. When sketching the graph of a hyperbola with an equation like

$$v^2 - u^2 = 10,$$

it is useful to note the point at which $u = 0$, i.e. the points where the hyperbola meets the vertical axis. The v-coordinates of these points satisfy

$$v^2 = 10,$$

or in other words,

$$v = \sqrt{10} \quad \text{or} \quad v = -\sqrt{10}.$$

Thus the graph of this equation can be sketched as in Fig. 12–20.

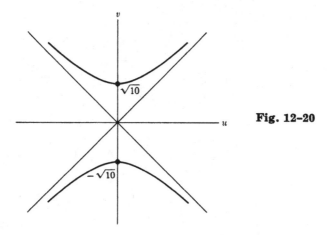

Fig. 12-20

EXERCISES

1. Sketch the graphs of the following hyperbolas.

   a) $v^2 - u^2 = 8$ b) $v^2 - u^2 = 1$ c) $v^2 - u^2 = 4$

2. Rotate the hyperbola H defined by the equation $xy = 1$ by $-\pi/4$ (i.e. clockwise by an angle of $\pi/4$). What is the equation satisfied by the image of H?

3. Rotate the hyperbola H defined by the equation $xy = 1$ by

   a) $3\pi/4$, and   b) $-3\pi/4$.

   In each case give the equation satisfied by the image of H under the prescribed rotation, and sketch this image.

4. Sketch the graph of the hyperbolas:

   a) $xy = -1$,   b) $xy = -2$,   c) $xy = -3$,   d) $xy = -4$.

5. In each one of the cases of Exercise 4, rotate the given hyperbola by $\pi/4$. Give the equation of the image.

6. In each one of the cases of Exercise 4, rotate the given hyperbola by $-\pi/4$. Give the equation of the image.

7. Sketch the graph of the following hyperbolas.

   a) $u^2 - v^2 = 2$ b) $u^2 - v^2 = 1$
   c) $u^2 - v^2 = 18$ d) $u^2 - v^2 = 4$

8. Prove the statement made in the text: If G is a rotation and F_r is dilation by r, then $G \circ F_r = F_r \circ G$.

Part Four
MISCELLANEOUS

13 Functions

§1. DEFINITION OF A FUNCTION

A **function,** defined for all numbers, is an association which to each number associates another number. If we denote a function by f, then this association is denoted by

$$x \mapsto f(x).$$

We call $f(x)$ the **value** of the function at x, or the **image** of x under f.

Example. The association

$$f: x \mapsto x^2$$

is a function called the square. Similarly, the association

$$g: x \mapsto x + 1$$

is a function. We have $f(2) = 4$, $f(3) = 9$, $f(-1) = 1$ and $f(-2) = 4$. We have $g(1) = 2$, $g(2) = 3$, $g(50) = 51$.

Example. If you have read the chapter on trigonometry, then you now see that the association

$$x \mapsto \sin x$$

is a function, which we already called the sine.

Example. The association which to each number x associates the number 4 is called the **constant function, with constant value 4.** Similarly, if c is a given number, the association

$$x \mapsto c, \qquad\qquad \text{for all numbers } x$$

is called the **constant function with value c.**

Remark on terminology. We have adopted the convention that the values of a function are numbers. Although this is not a universal convention, I find it useful. For an analogous notion when other types of values are allowed, see mappings in the next chapter.

We would like to say that the square root is also a function, but we know that negative numbers do not have real square roots. Thus we extend our definition of function as follows.

Let S be a set of numbers. A **function defined on** S is an association which to each element of S associates a number.

Example. Let S be the set of numbers ≥ 0. Then the square root is a function defined on S. Its value at 4 is 2. Its value at 10 is $\sqrt{10}$.

Example. Let S be the set of numbers $\neq 0$. Let f be the function defined on S such that

$$f(x) = \frac{1}{x}$$

for all x in S. Then $f(1) = 1$, $f(2) = \frac{1}{2}$, $f(100) = \frac{1}{100}$, and $f(\frac{3}{2}) = \frac{2}{3}$.

Example. Let n be a positive integer. Then the association

$$x \longmapsto x^n$$

is a function, which we called the n-th power.

Warning. It is very convenient sometimes to use slightly incorrect language. Often you will find written a sentence like:

"Let $f(x)$ be such and such a function."

The trouble here is that there is no quantification of x. What one means in that case, is the function whose value at a number x is such and such. We shall try to avoid incorrect language at least at the beginning of our discussion.

The preceding examples of functions have been given by formulas. Functions can, of course, be defined quite arbitrarily.

Example. Let G be the function such that

$$G(x) = 0 \text{ if } x \text{ is a rational number,}$$
$$G(x) = 1 \text{ if } x \text{ is not a rational number.}$$

Then in particular, $G(2) = G(\frac{2}{3}) = G(-\frac{3}{4}) = 0$, but

$$G(\sqrt{2}) = 1.$$

To describe a function amounts to giving its values at all numbers for which it is defined.

Let f, g be functions defined on the same set S. We can define the **sum** $f + g$ of the two functions to be the function whose value at an element x of S is

$$f(x) + g(x).$$

Then associativity for addition of numbers gives us associativity for addition of functions; namely, for any functions f, g, h defined on S, we have

$$(f + g) + h = f + (g + h).$$

Similarly, for commutativity, we have

$$f + g = g + f.$$

Example. Let $f(x) = x^2$ and $g(x) = \sin x$. Then

$$(f + g)(x) = x^2 + \sin x.$$

Also,

$$(f + g)(\pi) = \pi^2 + \sin \pi = \pi^2,$$

because $\sin \pi = 0$.

We have the **zero function**, whose value at every element x of S is 0. We denote this function also by 0. Thus for any function f defined on S, we have

$$f + 0 = 0 + f = f.$$

If f is a function defined on S, then we can define minus f, written $-f$, to be the function whose value at an element x of S is

$$-f(x).$$

Example. If $f(x) = x^2$, then $(-f)(x) = -x^2$, and $(-f)(5) = -25$. We see that

$$f + (-f) = 0 \qquad \text{(the zero function)}.$$

Thus we see that functions satisfy the same basic rules for addition as numbers.

The same is true for multiplication. If f, g are functions defined on the same set S, we define their **product** fg to be the function whose value at an element x of S is the product

$$f(x)g(x).$$

Thus we have by definition

$$(fg)(x) = f(x)g(x).$$

This product is commutative and associative. Furthermore, if 1 denotes the constant function having the value 1 for all x in S, then we have the usual rules

$$1f = f \quad \text{and} \quad 0f = 0.$$

Finally, our multiplication is distributive with respect to addition, because

$$
\begin{aligned}
((f + g)h)(x) &= (f + g)(x) \cdot h(x) \\
&= (f(x) + g(x))h(x) \\
&= f(x)h(x) + g(x)h(x) \\
&= (fh)(x) + (gh)(x) \\
&= (fh + fg)(x).
\end{aligned}
$$

Thus

$$(f + g)h = fh + gh.$$

In physical life, functions of numbers occur when we describe one quantity in terms of another.

Example. To each year we associate the population of the United States. Then the function is defined only for those years which are ≥ 1776. If P denotes this function (P for population), then

$$
\begin{aligned}
P(1800) &= 7.2 \cdot 10^6, \\
P(1900) &= 76.0 \cdot 10^6, \\
P(1940) &= 140 \cdot 10^6, \\
P(1970) &= 200 \cdot 10^6.
\end{aligned}
$$

Example. To each year we associate the price (in cents) of subway fare in New York City. Let S denote this function (S for subway). Then

$$
\begin{aligned}
S(1950) &= 15, \\
S(1969) &= 20, \\
S(1970) &= 30.
\end{aligned}
$$

We suggest that you look at the section on arbitrary mappings for a more general notion.

EXERCISES

1. Let $f(x) = 1/x$. What is $f(\frac{3}{4})$, $f(-\frac{2}{3})$?

2. For what numbers could you define a function f by the formula

$$f(x) = \frac{1}{x^2 - 2}?$$

   What is the value of this function for $x = 5$?

3. For what numbers could you define a function f by the formula

$$f(x) = \sqrt[3]{x} \qquad \text{(cube root of } x)?$$

   What is $f(27)$?

4. Let $x \mapsto |x|$ be the function which we already met earlier, namely

$$|x| = \sqrt{x^2}.$$

   What is a) $f(1)$?   b) $f(-3)$?   c) $f(-\frac{4}{3})$?

5. Let $f(x) = x + |x|$. What is

   a) $f(\frac{1}{2})$?   b) $f(2)$?

   c) $f(-4)$?   d) $f(-5)$?

6. Let $f(x) = 2x + x^2 - 5$. What is

   a) $f(1)$?   b) $f(-1)$?

7. For what numbers could you define a function f by the formula $f(x) = \sqrt[4]{x}$ (fourth root of x)? What is $f(16)$?

8. A function (defined for all numbers) is said to be an **even** function if $f(x) = f(-x)$ for all numbers x. It is said to be an **odd** function if $f(x) = -f(-x)$ for all x. Determine which of the following functions are odd or even.

   a) $f(x) = x$ b) $f(x) = x^2$ c) $f(x) = x^3$

   d) $f(x) = 1/x$ if $x \neq 0$ and $f(0) = 0$

9. Show that any function defined for all numbers can be written as a sum of an even function and an odd function. [*Hint:* The term

$$\frac{f(x) + f(-x)}{2}.$$

   will be the even function.]

10. Which of the following functions is odd or even, or neither?

   a) $\sin x$ b) $\cos x$

   c) $\tan x$ d) $\cot x$

   e) $\sin^2 x$ f) $\cos^2 x$

   g) $\tan^2 x$

11. a) Show that the sum of odd functions is odd.

   b) Show that the sum of even functions is even.

12. Determine whether the product of the following types of functions is odd, even, or neither. Prove your assertions.

   a) Product of odd function with odd function.

   b) Product of even function with odd function.

   c) Product of even function with even function.

§2. POLYNOMIAL FUNCTIONS

A function f defined for all numbers is called a **polynomial** if there exists numbers $a_0, a_1, \ldots, a_n$ such that for all numbers x we have

$$(1) \qquad f(x) = a_n x^n + a_{n-1} x^{n-1} + \cdots + a_1 x + a_0.$$

Example. The function f such that

$$f(x) = 3x^3 - 2x + 1$$

is a polynomial function. We have $f(1) = 3 - 2 + 1 = 2$.

Example. The function

$$g : x \mapsto \tfrac{1}{2}x^4 + 3x^2 - x + 5$$

is a polynomial. We have

$$g(2) = \tfrac{1}{2}2^4 + 3 \cdot 2^2 - 2 + 5 = 23.$$

When a polynomial can be written as in (1) above, we shall say that it is of **degree** $\leq n$. If $a_n \neq 0$, then we want to say that the polynomial has degree n. However, we must be careful. Is it possible that there are other numbers $b_0, \ldots, b_m$ such that for all x we have

$$f(x) = b_m x^m + \cdots + b_0?$$

For instance, can we have

$$7x^5 - 5x^4 + 2x + 1 = x^6 - 17x^3 + x + 1$$

for all numbers x? The answer is not immediately clear just by looking. Suppose that the coefficients were big complicated numbers. We would have no simple test for the equality of values. If the answer were YES, then it would be hopeless to define the degree, because in the example just written down, for instance, we would not know whether the degree is 5 or 6. That the answer is NO will be proved in the corollary to Theorem 2 below.

Let f be a polynomial. If c is a number such that $f(c) = 0$, then we call c a **root** of f. We shall see in a moment that a non-zero polynomial can have only a finite number of roots, and we shall give a bound for the number of these roots.

Example. Let $f(x) = x^2 - 3x + 2$. Then $f(1) = 0$. Hence 1 is a root of f. Also, $f(2) = 0$. Hence 2 is also a root of f.

Example. Let $f(x) = ax^2 + bx + c$. If $b^2 - 4ac = 0$, then the polynomial has one real root, which is

$$-\frac{b}{2a}.$$

If $b^2 - 4ac > 0$, then the polynomial has two distinct real roots which are

$$\frac{-b + \sqrt{b^2 - 4ac}}{2a} \quad \text{and} \quad \frac{-b - \sqrt{b^2 - 4ac}}{2a}.$$

These assertions are merely reformulations of the theory of quadratic equations in Chapter 4.

Theorem 1. *Let f be a polynomial of degree $\leq n$ and let c be a root. Then there exists a polynomial g of degree $\leq n - 1$ such that for all numbers x we have*

$$f(x) = (x - c)g(x).$$

Proof. Write

$$f(x) = a_0 + a_1 x + a_2 x^2 + \cdots + a_n x^n.$$

Substitute the value

$$x = (x - c) + c$$

for x. Each k-th power, for $k = 0, \ldots, n$, of the form

$$((x - c) + c)^k$$

can be expanded out as a sum of powers of $(x - c)$ times a number. Hence there exist numbers $b_0, b_1, \ldots, b_n$ such that

$$f(x) = b_0 + b_1(x - c) + b_2(x - c)^2 + \cdots + b_n(x - c)^n,$$

for all x. But $f(c) = 0$. Hence

$$0 = f(c) = b_0,$$

and all the other terms on the right have the value 0 for $x = c$. This proves that $b_0 = 0$. But then we can factor

$$f(x) = (x - c)(b_1 + b_2(x - c) + \cdots + b_n(x - c)^{n-1}).$$

We let

$$g(x) = b_1 + b_2(x - c) + \cdots + b_n(x - c)^{n-1},$$

and we see that our theorem is proved.

Remark. Let us look more carefully at the polynomial g which we obtained in the proof. When we expand out a power

$$((x - c) + c)^k,$$

we get a term $(x - c)^k$, and all other terms in the sum involve a lower power of $(x - c)$. Hence the highest power of $(x - c)$ which we get is the n-th power, and this power comes from the expansion of

$$((x - c) + c)^n.$$

Thus the term with the highest power of $(x - c)$ can be determined explicitly, and it is precisely

$$a_n(x - c)^n.$$

In other words, we have

$$b_n = a_n.$$

Hence the polynomial g which we obtain has an expansion of the form

$$g(x) = a_n x^{n-1} + \text{lower terms}.$$

This remark will be useful later.

Theorem 2. *Let f be a polynomial. Let $a_0, \ldots, a_n$ be numbers such that $a_n \neq 0$, and such that we have*

$$f(x) = a_n x^n + a_{n-1} x^{n-1} + \cdots + a_0$$

for all x. Then f has at most n roots.

Proof. Let $c_1, c_2, \ldots, c_r$ be distinct roots of f. Suppose that $r \geq n$. Write

$$f(x) = (x - c_1)g_1(x)$$

for some polynomial g_1 of degree $\leq n - 1$. Then

$$0 = f(c_2) = (c_2 - c_1)g_1(c_2).$$

Since $c_2 \neq c_1$, it follows that $g_1(c_2) = 0$; in other words, c_2 is a root of g_1, which has degree $\leq n - 1$. We can now give the same argument, factoring

$$g_1(x) = (x - c_2)g_2(x).$$

We see that

$$g_2(c_3) = 0,$$

and that g_2 has degree $\leq n - 2$. We keep on going, until we find g_n to be constant. Thus we can write

$$f(x) = (x - c_1)(x - c_2) \cdots (x - c_n)c$$

for all x. In fact, the remark following Theorem 1 shows that

$$c = a_n \neq 0.$$

Thus if $x \neq c_i$ for any $i = 1, \ldots, n$, we see that $f(x) \neq 0$. This means that f has at most n roots, and our theorem is proved.

Corollary. *Let f be a polynomial, which can be written in the form*

$$f(x) = a_n x^n + a_{n-1}x^{n-1} + \cdots + a_0$$

and also in the form

$$f(x) = b_n x^n + b_{n-1}x^{n-1} + \cdots + b_0.$$

Then

$$a_i = b_i \qquad \text{for every } i = 0, \ldots, n.$$

Proof. Consider the polynomial

$$0 = f(x) - f(x) = (a_n - b_n)x^n + (a_{n-1} - b_{n-1})x^{n-1} + (a_0 - b_0).$$

We have to show that

$$a_i - b_i = 0 \qquad \text{for all } i = 1, \ldots, n.$$

Let $d_i = a_i - b_i$. Suppose that there exists some index i such that $d_i \neq 0$. Let m be the largest of these indices, so that we can write

$$0 = d_m x^m + \cdots + d_0$$

for all x and $d_m \neq 0$. This contradicts Theorem 2. Therefore we conclude that $d_i = 0$ for all $i = 1, \ldots, n$, thus proving our corollary.

The corollary shows that if f is a polynomial, then there is a unique way of writing f in the form

$$f(x) = a_n x^n + \cdots + a_0$$

for all x. In other words, the numbers $a_n, \ldots, a_0$ are uniquely determined. They will be called the **coefficients** of f, and we call a_n the **leading coefficient** if $a_n \neq 0$. We call a_0 the **constant term**.

Example. Let f be the polynomial such that for all x we have

$$f(x) = 4x^5 - 7x^3 + x - 20.$$

Then the coefficients of f are 4, 0, 7, 0, 1, -20. (We have included the coefficients of all powers of x up to the fifth. Note that the fourth and second power have coefficients equal to 0.) The leading coefficient is 4. The constant term is -20.

If a polynomial f is such that

$$f(x) = a_n x^n + \cdots + a_0,$$

and $a_n \neq 0$, then we say that f has **degree** n. The polynomial in the preceding example has degree 5.

Remark. It often happens that if f is a polynomial, then

$$f(x) = 0$$

for some x. In other words, there may exist a number c such that

$$f(c) = 0.$$

This does not mean that f is the zero polynomial. By definition, we call f the **zero polynomial** if and only if

$$f(x) = 0 \qquad\qquad \text{for \textbf{all} numbers } x.$$

Thus the zero polynomial is the polynomial all of whose coefficients are 0. If some coefficient of a polynomial is not equal to 0, then f is not the zero polynomial. If f is the zero polynomial, we also sometimes say that f is **identically zero** (to distinguish this case from the one in which f may take on the value 0 at some number).

In Chapter 4 we found a way of determining all roots of a polynomial of degree 2. For polynomials of higher degree, it is much more difficult to determine the roots, except in very special cases. For polynomials of degree 3 and 4, one can give formulas involving radicals, but it is a classical result that such formulas cannot be given in general for polynomials of degree at least 5.

For polynomials whose coefficients are integers, one may ask for the rational roots, and much time is often spent in elementary classes finding such roots by factoring the polynomial. It is very unusual that a polynomial can be factored with integral or rational roots, and I think much too much emphasis is placed on this kind of accident. One result of this kind of emphasis (which I have found among students) is that they try to factor even in the quadratic case, when the systematic answer is available from the formula. My experience indicates that such training in factoring is not worth the time spent, and therefore we do not emphasize it here. However, we give two important examples of factoring.

Example. Let $f(x) = 3x^2 - 5x + 1$. Then the roots of f are given by the quadratic formula of Chapter 4, namely

$$c_1 = \frac{5 + \sqrt{25 - 12}}{6} = \frac{5 + \sqrt{13}}{6}$$

$$c_2 = \frac{5 - \sqrt{25 - 12}}{6} = \frac{5 - \sqrt{13}}{6}.$$

If we look back either at Theorem 1 or at the beginning of the proof of Theorem 2, we see that

$$f(x) = 3\left(x - \frac{5 + \sqrt{13}}{6}\right)\left(x - \frac{5 - \sqrt{13}}{6}\right).$$

We have therefore factored f into factors of degree 1.

Example. The general quadratic case follows the same pattern. Let

$$f(x) = ax^2 + bx + c,$$

and $a \neq 0$. Assume that $b^2 - 4ac > 0$. Then there are two distinct roots c_1 and c_2 of f, and therefore we have the factorization

$$f(x) = a(x - c_1)(x - c_2).$$

Example. The other important case of factoring occurs for the polynomial

$$f(x) = x^n - 1.$$

We see that 1 is a root, because $f(1) = 1^n - 1 = 1 - 1 = 0$. Hence we know that f must have $(x - 1)$ as a factor, i.e. we have

$$f(x) = (x - 1)g(x)$$

for some polynomial $g(x)$. What is $g(x)$? We suggest that you look at Exercise 2 of Chapter 1, §6, and do Exercise 6 of this chapter.

We shall now discuss what is sometimes called "long division". For polynomials it is the analog of the division of two positive integers, with a remainder. Therefore we recall this division process for positive integers.

First, we carry out an example.

Example. We want to divide 327 by 17, with a possible remainder. As you know from elementary school, this division can be represented schematically as follows.

$$
\begin{array}{r}
19 \\
17\overline{)327} \\
17 \\
\hline
157 \\
153 \\
\hline
4
\end{array}
$$

This procedure tells us that

$$327 = 19 \cdot 17 + 4.$$

We call 4 the remainder. We can describe the preceding steps as follows. We determined the first digit 1 of 19 as being the largest positive integer whose product with 17 would still be ≤ 32. We then multiply 1 by 17, write it under 32, subtract, get 15, and bring down the 7. We then repeat the process. We determine 9 as the largest positive integer whose product with 17 is ≤ 157. We multiply 17 by 9, get 153, write it under 157, subtract, and get 4. This number 4 is now less than 17, so we stop.

We can summarize what we have done in general by the following statement.

Let n, d be positive integers. Then there exists an integer r such that $0 \leq r < d$, and an integer $q \geq 0$ such that

$$n = qd + r.$$

Note that even though the standard procedure of the example, which gives us q, r, is called "long division", in fact our procedure uses only multiplication and subtraction.

We shall now describe an analogous procedure for polynomials, which gives us the

Euclidean algorithm. *Let f and g be non-zero polynomials. Then there exist polynomials q, r such that deg r < deg g and such that*

$$f(x) = q(x)g(x) + r(x).$$

Example. Let

$$f(x) = 4x^3 - 3x^2 + x + 2 \quad \text{and} \quad g(x) = x^2 + 1.$$

We want to find $q(x)$ and $r(x)$. We first lay out what we do in a diagram similar to that of long division for integers, and then explain how each step is obtained.

$$
\begin{array}{r}
4x - 3 \\
x^2 + 1 \overline{)\, 4x^3 - 3x^2 + x + 2\,} \\
4x^3 + 4x \\
\hline
-3x^2 - 3x + 2 \\
-3x^2 - 3 \\
\hline
-3x + 5
\end{array}
$$

We have

$$q(x) = 4x - 3 \quad \text{and} \quad r(x) = -3x + 5,$$

so that

(1) $\qquad 4x^3 - 3x^2 + x + 2 = (4x - 3)(x^2 + 1) + (3x + 5).$

Now we describe each step in the computation. We first determine $4x$ because

$$4x \cdot x^2$$

is equal to the term of highest degree in $f(x)$; that is, equal to $4x^3$. We then multiply $4x$ by $x^2 + 1$, we obtain $4x^3 + 4x$, which we write under $f(x)$, placing corresponding powers of x under each other. We then subtract $4x^3 + 4x$ from $4x^3 - 3x^2 + x + 2$, and obtain $-3x^2 - 3x + 2$. We then repeat our procedure, and determine -3 because

$$(-3) \cdot x^2$$

is equal to the term of highest degree in $-3x^2 - 3x + 2$. We multiply -3 by $x^2 + 1$, obtain $-3x^2 - 3$, which we write under $-3x^2 - 3x + 2$. We subtract, and obtain $-3x + 5$. We note that the polynomial $-3x + 5$ has degree 1, which is smaller than the degree of $g(x) = x^2 + 2$. Our computation is therefore finished.

Why does the above procedure actually provide us with polynomials $q(x)$ and $r(x)$ satisfying relation (1)? This is easily seen. Write $f(x) = f_3(x)$ to indicate the fact that f has degree 3. We determined $4x$ in such a way that

$$f_3(x) - 4x(x^2 + 1) = f_3(x) - 4x^3 - 4x$$

has degree 2; i.e. in such a way that the term $4x^3$ would cancel. Write

$$f_2(x) = f_3(x) - 4x^3 - 4x = -3x^2 - 3x + 2.$$

Then f_2 has degree 2. We determine -3 so that

$$f_2(x) - (-3)(x^2 + 1)$$

has degree 1; i.e. in such a way that the term $-3x^2$ cancels. Then

$$f_2(x) - (-3)(x^2 + 1) = f_1(x)$$

has degree 1. Now we see that

$$f_3(x) - 4x \cdot g(x) - (-3) \cdot g(x) = -3x + 5$$

has degree 1. Thus

$$f_3(x) - (4x - 3)g(x) = -3x + 5.$$

This shows why

$$f_3(x) = q(x)g(x) + r(x).$$

Example. Let

$$f(x) = 2x^4 - 3x^2 + 1 \quad \text{and} \quad g(x) = x^2 - x + 3.$$

We wish to find $q(x)$ and $r(x)$ as in the Euclidean algorithm. We write down our computation in the standard pattern.

$$
\begin{array}{r}
2x^2 + 2x - 7 \\
x^2 - x + 3 \overline{)2x^4 \qquad\quad - 3x^2 \qquad\quad + 1} \\
2x^4 - 2x^3 + 6x^2 \\
\hline
+2x^3 - 9x^2 \qquad\quad + 1 \\
2x^3 - 2x^2 + 6x \\
\hline
-7x^2 - 6x + 1 \\
-7x^2 + 7x - 21 \\
\hline
-13x + 22
\end{array}
$$

Hence we get

$$q(x) = 2x^2 + 2x - 7 \quad \text{and} \quad r(x) = -13x + 22.$$

As a matter of terminology, we call $r(x)$ the **remainder** in the Euclidean algorithm.

Remark. The Euclidean algorithm allows us once more to prove that if $f(x)$ has a root c, then we can write

$$f(x) = (x - c)q(x)$$

for some polynomial q. Indeed, in the Euclidean algorithm, we have

$$f(x) = q(x)(x - c) + r(x),$$

where $\deg r < 1$. Hence r must be constant, say equal to a number a. Thus

$$f(x) = q(x)(x - c) + a.$$

Now evaluate the left-hand side and the right-hand side at $x = c$. We get

$$0 = 0 + a,$$

whence $a = 0$. The remainder is equal to 0, and this gives what we wanted.

We do not prove the Euclidean algorithm. The proof would consist of carrying out the procedure of the example with general coefficients.

EXERCISES

1. What is the degree of the following polynomials?

    a) $3x^2 - 4x + 5$ b) $-5x^5 + x$

    c) $-38x^4 + x^3 - x - 1$

    d) $(3x^2 - 4x + 5)(-5x^5 + x)$ e) $(-5x^5 + x)(-7x + 3)$

    f) $(-4x^2 + 5x - 4)(3x^3 + x - 1)$

    g) $(6x^7 - x^3 + 5)(7x^4 - 3x^2 + x - 1)$

    h) Let f, g be polynomials which are not the zero polynomials. Show that

$$\deg(fg) = \deg f + \deg g.$$

2. Factor the quadratic polynomials of the exercises in Chapter 4 into factors of degree 1.

3. Let f be a polynomial of degree 3. If there exist polynomials g, h of degree ≥ 1 such that $f = gh$, show that f has a root.

4. a) Give an example of a polynomial of degree 2 which has no root in the real numbers.

    b) Give an example of a polynomial of degree 3 which has only one root in the real numbers.

    c) Give an example of a polynomial of degree 4 which has no root in the real numbers.

5. Let
$$f(x) = x^n + a_{n-1}x^{n-1} + \cdots + a_0$$
be a polynomial whose coefficients are integers and whose leading coefficient is 1. If c is an integer and is a root of f, show that c divides a_0.

6. What are all the roots in the real numbers of the following polynomials?

   a) $x^3 - 1$ b) $x^4 - 1$ c) $x^5 - 1$ d) $x^n - 1$

   e) $x^3 + 1$ f) $x^4 + 1$ g) $x^5 + 1$ h) $x^n + 1$

7. Find the polynomials $q(x)$ and $r(x)$ of the Euclidean algorithm when
$f(x) = 4x^3 - x + 2$, and:

   a) $g(x) = x - 2$,   b) $g(x) = x^2 - 1$,

   c) $g(x) = x^2 + 1$,   d) $g(x) = x^2 - x$,

   e) $g(x) = x^2 - x + 1$,   f) $g(x) = x^2 + x - 1$,

   g) $g(x) = x^3 + 2$,   h) $g(x) = x^3 - x + 1$.

8. Repeat Exercise 7 for the case in which $f(x) = 6x^4 - x^3 + x^2 - 2x + 5$
and:

   a) $g(x) = x^3 - 1$,   b) $g(x) = x^2 - 5$,

   c) $g(x) = x + 2$,   d) $g(x) = 3x + 1$,

   e) $g(x) = 4x + 6$,   f) $g(x) = x^4 - x^2 + 1$,

   g) $g(x) = x^3 - 5x$,   h) $g(x) = x^3 - 2x^2$.

9. **Rational functions.** A **rational function** is a function which can be
expressed as a quotient of polynomials, i.e. a function whose value at a
number x is
$$\frac{f(x)}{g(x)},$$
where f, g are polynomials, and g is not the zero polynomial. Thus a
rational function is defined only for those numbers x such that $g(x) \neq 0$.
In an expression as above, we call f the numerator of the rational function,
and g its denominator. We can then work with rational functions just
as we did with rational numbers. In particular, we can put two rational
functions over a common (polynomial) denominator, and take their sum
in a manner analogous to taking the sum of rational numbers. We give
an example of this.

Example. Put the two rational functions

$$R(x) = \frac{3x^2 - 2x + 6}{x - 1} \quad \text{and} \quad S(x) = \frac{4x + 3}{x - 5}$$

over a common denominator.

This denominator will be $(x-1)(x-5)$, and the two rational functions over this common denominator are

$$R(x) = \frac{(3x^2 - 2x + 6)(x - 5)}{(x - 1)(x - 5)} \quad \text{and} \quad S(x) = \frac{(4x + 3)(x - 1)}{(x - 1)(x - 5)}.$$

Their sum is then

$$R(x) + S(x) = \frac{(3x^2 - 2x + 6)(x - 5) + (4x + 3)(x - 1)}{(x - 1)(x - 5)},$$

which we see is again a rational function, defined for all numbers $\neq 1$ or 5. We can expand the expressions in the numerator and denominator into the standard form for polynomials, and we see that

$$R(x) + S(x) = \frac{3x^3 - 13x^2 + 15x - 33}{x^2 - 6x + 5}.$$

Now you do the work, and express the sums of the following rational functions $R(x) + S(x)$ as quotients of polynomials just as we did in this example.

a) $R(x) = \dfrac{x - 4}{x + 3}$ and $S(x) = \dfrac{2x + 1}{x - 5}$

b) $R(x) = \dfrac{3x - 1}{2x + 2}$ and $S(x) = \dfrac{x - 4}{3x + 2}$

c) $R(x) = \dfrac{x^2 - 1}{x + 5}$ and $S(x) = \dfrac{3x^3 + 2}{x + 1}$

d) $R(x) = \dfrac{x^2 - x + 1}{3x - 4}$ and $S(x) = \dfrac{x + 3}{x^2 + 2}$

e) $R(x) = \dfrac{x^3 + 1}{x + 4}$ and $S(x) = \dfrac{x^4 - 2}{3x + 1}$

f) $R(x) = \dfrac{x^4 - 1}{x}$ and $S(x) = \dfrac{x - 1}{x^2 + 2}$

g) $R(x) = \dfrac{x^3 - 2x}{x^2}$ and $S(x) = \dfrac{x^2 + 1}{x - 1}$

h) $R(x) = \dfrac{2x^3 - 1}{x^2 + 2}$ and $S(x) = \dfrac{x - 1}{x + 1}$

§3. GRAPHS OF FUNCTIONS

Let f be a function, defined on a set of numbers S. By the **graph** of the function f, we shall mean the set of all points

$$(x, f(x)),$$

i.e. the set of all points whose first coordinate is x, and whose second coordinate is $f(x)$.

Example. Let $f(x) = x^2$. Then the graph of f is the graph of the equation $y = x^2$, and is a parabola, as discussed in Chapter 7.

Example. An object moves along the positive x-axis, subjected to a force inversely proportional to the square of the distance from the origin. We can then write this force as

$$y = \frac{k}{x^2},$$

where k is some constant. Suppose that $k = 3$, so that $y = 3/x^2$. It is easy to draw the graph of this equation, which is nothing but the graph of the function f such that $f(x) = 3/x^2$; see Fig. 13–1. We make a table of values, and observe that as x increases, $f(x)$ decreases.

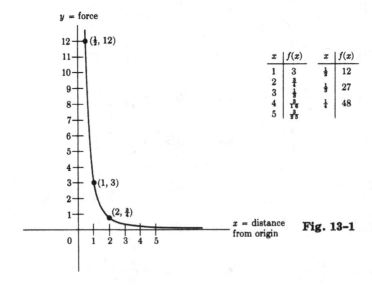

x	$f(x)$	x	$f(x)$
1	3	$\frac{1}{2}$	12
2	$\frac{3}{4}$	$\frac{1}{3}$	27
3	$\frac{1}{3}$	$\frac{1}{4}$	48
4	$\frac{3}{16}$		
5	$\frac{3}{25}$		

$x =$ distance from origin

Fig. 13–1

Example. We know that the sine is a function, associating with each number θ the value $\sin \theta$. In fact, in the chapter on trigonometry, we had already drawn its graph which looked like this (Fig. 13–2.)

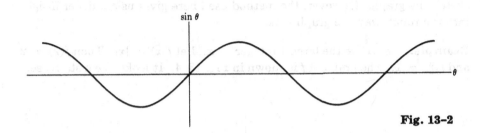

Fig. 13–2

Example. We want to sketch the graph of the function

$$y = (x - 1)(x - 2).$$

Observe that when $x = 1$ or when $x = 2$, then $y = 0$. Furthermore, there are no other values of x for which $y = 0$. Thus the graph crosses the x-axis precisely at the points $x = 1$ and $x = 2$. When $x < 1$, then $x - 1 < 0$ and $x - 2 < 0$. Hence when $x < 1$, we see that $y > 0$. Similarly, when $x > 2$, we see that $y > 0$. Finally, when

$$1 < x < 2,$$

then $x - 1 > 0$ and $x - 2 < 0$. Hence when $1 < x < 2$, we see that y is negative. Finally, when $x = 0$, we see that

$$y = (-1)(-2) = 2.$$

The graph of our function looks like this (Fig. 13–3):

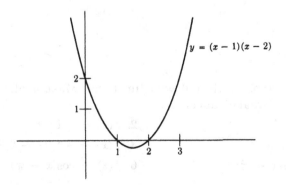

$y = (x - 1)(x - 2)$

Fig. 13–3

This graph is of course the graph of a parabola, since it is the graph of

$$y = x^2 - 3x + 2.$$

We can complete the square, and use the same method as in Chapter 12 to sketch the graph. However, the method used here gives us a quicker insight into the rough way the graph looks.

Example. Let $[x]$ be the largest integer $\leq x$. Let $f(x) = [x]$. Then $f(2) = 2$ and $f(\frac{3}{2}) = 1$. The graph of f is shown in Fig. 13–4. It looks like a staircase.

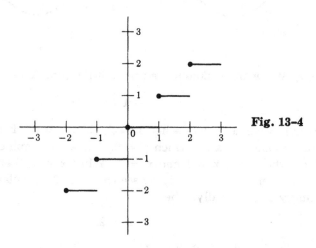

Fig. 13–4

EXERCISES

Sketch the graphs of the following functions. Make small tables of values to get an idea of what's happening.

1. $f(x) = 1/x^3$ 2. $f(x) = -1/x^3$

3. $f(x) = -1/x^2$ 4. $f(x) = -1/(x - 2)^2$

5. $f(x) = \sin(x - \pi)$ 6. $f(x) = \cos(x - \pi)$

7. $f(x) = \sin\left(x - \dfrac{\pi}{4}\right)$ 8. $f(x) = \cos\left(x - \dfrac{\pi}{4}\right)$

9. $f(x) = |x|$ 10. $f(x) = x + |x|$

11. $f(x) = x - |x|$ 12. $f(x) = 2 + |x|$

13. $f(x) = |x| - x$ 14. $f(x) = |x - 1|$

15. $f(x) = |x + 1|$ 16. $f(x) = |x - 2|$

17. $f(x) = |x + 2|$ 18. $f(x) = |x + 3|$

19. $f(x) = (x + 1)(x - 3)$ 20. $f(x) = (x - 2)(x - 5)$

21. $f(x) = (x + 2)(x - 1)$ 22. $f(x) = (x + 3)x$

23. $f(x) = (x + 2)(x + 1)$ 24. $f(x) = (x + 4)(x - 3)$

25. $f(x) = [x] + 1$ 26. $f(x) = [x + 1]$

27. $f(x) = [x + 2]$ 28. $f(x) = [x] - x$

29. $f(x) = x - [x]$ 30. $f(x) = [x] + x$

§4. EXPONENTIAL FUNCTION

We have already discussed powers like

$$a^{m/n},$$

where a is a positive number, and m, n are integers, i.e. fractional powers.

It is difficult to give an analytic development of the theory of powers

$$a^x$$

when x is not a fraction, but at least we can state the basic properties, which are intuitively clear, and then use them in applications.

Let a be a number > 0. To each number x we can associate a number denoted by a^x, such that when $x = m/n$ is a quotient of integers ($n \neq 0$) then $a^{m/n}$ is the ordinary fractional power discussed in Chapter 3, §3, and such that the function

$$x \mapsto a^x$$

has the following properties.

EXP 1. *For any numbers x, y, we have*

$$a^{x+y} = a^x a^y.$$

EXP 2. *For all numbers x, y, we have*

$$(a^x)^y = a^{xy}.$$

EXP 3. *If a, b are positive, then*

$$(ab)^x = a^x b^x.$$

EXP 4. *Assume that a > 1. If x < y, then $a^x < a^y$.*

The function

$$x \mapsto a^x$$

is called the **exponential function to the base** a. The proof given in Chapter 3, §3 that $a_0 = 1$ is valid here. From this and **EXP 1** we conclude that

$$a^{-x} = \frac{1}{a^x}.$$

This is because

$$a^{-x}a^x = a^{x-x} = a^0 = 1.$$

Remark. The values of the exponential function are always positive. In fact, if $x > 0$, and $a > 1$, then $a^x > 1$ because $a^0 = 1$ and the exponential function is increasing according to **EXP 4**. If $x < 0$, say $x = -z$ where z is positive, then

$$a^x = a^{-z} = \frac{1}{a^z},$$

and we see again that a^x is positive. However, in this case, we have $a^x < 1$.

Question. If we allow $0 < a < 1$ instead of $a > 1$, how do you have to adjust property **EXP 4** to make it valid?

It is now easy to sketch the graph of an exponential function.

Example. Let $f(x) = 2^x$. We make a small table of values, with rational numbers and integers for x.

x	$f(x)$		x	$f(x)$
0	1		-1	$\frac{1}{2}$
1	2		-2	$\frac{1}{4}$
2	4		-3	$\frac{1}{8}$
3	8		-4	$\frac{1}{16}$
4	16		-5	$\frac{1}{32}$
5	32			

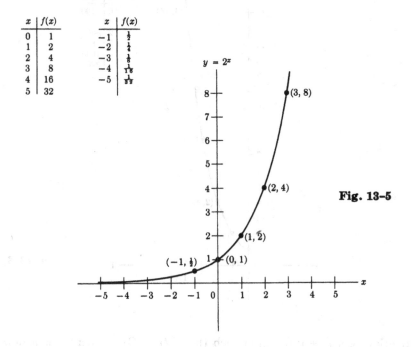

Fig. 13–5

Note that the graph climbs very steeply to the right of 0, and that it becomes very flat very fast to the left of 0.

Example. The population of a city doubles every year, and at time $t = 0$ it is equal to 100 persons. We can then express this population in the form

$$P(t) = 100 \cdot 2^t,$$

when t is an integer. Namely, when $t = 0$, we get

$$P(0) = 100.$$

Each time that t increases by 1, we see that $P(t)$ is multiplied by 2. Thus the population P is a function of time t, and again we draw the graph by making a table of values as shown in Fig. 13–6, and noticing that $P(t)$ increases when t increases.

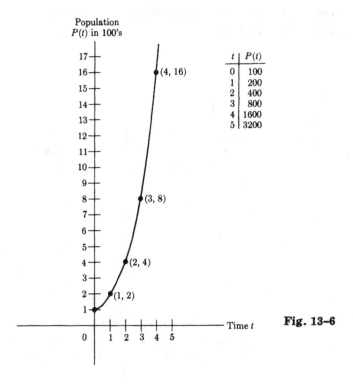

Fig. 13–6

Example. Let f be a function such that $f(t) = Ca^{5t}$, where C is a number. Such a function is said to have exponential growth. Note that when $t = 0$, we get

$$f(0) = C,$$

so that C is the value of f at $t = 0$. If f represents a population, given as a function of time, then C is the original number of persons at time $t = 0$.

Example. Certain substances disintegrate at a rate proportional to the amount of substance present. If f denotes this amount, then it is known that f as a function of time t is given by the formula

$$f(t) = Ca^{Kt},$$

where C, K are constants. Again note that C is the amount of substance present at time $t = 0$. Since the amount of substance decreases, the constant K is negative. For instance, if

$$f(t) = 3a^{-2t},$$

if $a = 2$, and if f gives the number of grams of radium at a certain place as a function of time, then there are 3 g of radium present at the beginning when $t = 0$. If the time is measured in years, then after 4 years, there are

$$f(4) = 3 \cdot a^{-8} = \frac{3}{256}$$

grams of radium left.

EXERCISES

1. Sketch the graph of the function f such that:
   a) $f(t) = 3^t$,   b) $f(t) = 4^t$ c) $f(t) = 5^t$

2. Sketch the graph of the function f such that:
   a) $f(t) = 2^{-t}$ b) $f(t) = 3^{-t}$ c) $f(t) = 5^{-t}$

3. If $1 < a < b$, which is steeper: the graph of a^x or the graph of b^x?

4. Let $a = 5$. At time $t = 0$, there are 4 g of radium in a cave. The amount of radium given as a function of time is

$$f(t) = 5a^{-3t}.$$

   How much radium is left
   a) at $t = 1$?   b) at $t = 2$?   c) at $t = 3$?

5. The population of a city triples every 50 years. At time $t = 0$, this population is 100,000. Give a formula for the population $P(t)$ as a function of t. What is the population after
   a) 100 years?   b) 150 years?   c) 200 years?

6. Bacteria in a solution double every 3 min. If there are 10^4 bacteria at the beginning, give a formula for the number of bacteria at time t. How many bacteria are there after
   a) 3 min?   b) 9 min?   c) 27 min?   d) one hour?

7. The function $f(t) = 4 \cdot 16^t$ describes the growth of bacteria.

    a) How many bacteria are present at the beginning, when $t = 0$?

    b) After $\frac{1}{2}$ hr, how many bacteria are there?

    c) Same question, after $\frac{1}{4}$ hr.

    d) Same question after 1 hr.

8. A radioactive element decays so that the amount $f(t)$ left at time t satisfies the formula

$$f(t) = 60 \cdot 2^{-.02t}.$$

    a) What is the initial quantity of this element at $t = 0$?

    b) How much is left after 500 years?

    c) How much is left after 1,000 years?

    d) How much is left after 2,000 years?

§5. LOGARITHMS

Let a be a number > 1. If $y = a^x$, then we shall say that

x is the log of y to the base a, and write $x = \log_a y$.

Example. Since $8 = 2^3$, we see that 3 is the log of 8 to the base 2. We write $3 = \log_2 8$.

Example. Since $27 = 3^3$, we see that 3 is the log of 27 to the base 3. We write $3 = \log_3 27$.

The log is a function, defined for all positive numbers. We shall not prove this, but assume that it is a basic property of numbers. In other words, we assume that:

Given a number $y > 0$, there exists a number x such that

$$a^x = y.$$

We can now prove properties of the log which are similar to those of the exponential function.

LOG 1. *For any numbers x, y, we have*

$$\log_a(xy) = \log_a x + \log_a y.$$

LOG 2. *We have*

$$\log_a 1 = 0.$$

LOG 3. *If $x < y$, then $\log_a x < \log_a y$.*

These properties can be *proved* from the corresponding properties of the exponential function. We now do this.

Proof of **LOG 1.** Let $u = \log_a x$ and $v = \log_a y$. This means that

$$x = a^u \qquad \text{and} \qquad y = a^v.$$

Hence

$$xy = a^u a^v = a^{u+v}.$$

By definition,

$$\log_a(xy) = u + v = \log_a x + \log_a y.$$

Proof of **LOG 2.** By definition, since $a^0 = 1$, this means that

$$0 = \log_a 1.$$

Proof of **LOG 3.** Let $x < y$. Let $u = \log_a x$ and $v = \log_a y$. Then

$$a^u = x \qquad \text{and} \qquad a^v = y.$$

If $u = v$, then $x = y$, which is impossible. If $v < u$, then by **EXP 4** we find $y < x$, which is also impossible. Hence we must have $u < v$, thereby proving our property **LOG 3.**

It is easy to draw a graph for the log. We leave this as an exercise. We also let you go through the properties of the log when $a < 1$, but $a > 0$. Which ones are still valid, and which type should be changed?

We can use the log to solve equations of the exponential type.

Example. Let $f(t) = 10 \cdot 2^{kt}$, where k is constant. Suppose that $f(\frac{1}{2}) = 3$. Find k.

We have

$$3 = 10 \cdot 2^{k/2}.$$

Taking the log to the base 2 on both sides, we find that

$$\log_2 3 = \log_2 10 + \frac{k}{2}.$$

Hence

$$k = 2(\log_2 3 - \log_2 10).$$

This argument can be used in a laboratory. For instance, suppose that we leave 10 g of a substance to decompose at time $t = 0$. We know that the amount of substance is given by the formula in the example above, with an unspecified constant k, and we wish to determine the value of k. If after half an hour we have 3 g of the substance left, then we get our value of k as in the example. Tables of logarithms or a computing machine then give us a decimal value for k.

Example. A radioactive substance disintegrates according to the formula

$$r(t) = C3^{-5t},$$

where C is a constant $\neq 0$. At what time will there be exactly one-third of the original amount left?

At time $t = 0$ we have $r(0) = C$ amount of substance. We must find that value of t such that

$$r(t) = \frac{1}{3}C,$$

or in other words,

$$\frac{1}{3}C = C3^{-5t}.$$

Note that we can cancel C.

Take the log to the base 3. Then the previous equation is equivalent with

$$\log_3 \left(\frac{1}{3}\right) = -5t,$$

or equivalently,

$$-1 = -5t.$$

Thus

$$t = \frac{1}{5}.$$

Observe how in this example the unspecified constant C does not appear in the final answer.

EXERCISES

1. Sketch the graph of the function g such that:

a) $g(x) = \log_2 x$　　　　　　　　　b) $g(x) = \log_3 x$.

Make a table of values to help you draw these graphs. For instance, in (a) use $x = 2, 4, 8, 16, \frac{1}{2}, \frac{1}{4}, \frac{1}{8}, \frac{1}{16}$.

2. Find the following values.

a) $\log_2 64$　　　　　　　　　　　　b) $\log_3 \left(\dfrac{1}{27}\right)$

c) $\log_5 25$　　　　　　　　　　　　d) $\log_5 \left(\dfrac{1}{25}\right)$

e) $\log_2 \left(\dfrac{1}{64}\right)$　　　　　　　　　f) $\log_3 \left(\dfrac{1}{81}\right)$

3. Let e be a fixed number > 1 and abbreviate $\log_e$ by $\log$. If a is > 1 and x is an arbitrary number, prove that

$$\log a^x = x \log a.$$

[*Hint:* Consider $e^{(\log a)x}$, and use **EXP 2**.] For instance,

$$\log 10^{2/3} = \tfrac{2}{3} \log 10.$$

For the next exercises, assume that the number e has been fixed, and that $\log = \log_e$. Thus your answers will be expressed in terms of $\log_e$. There are tables for various values of e which can be used if you want decimal answers, but we are not concerned with this here. Remember that $\log_e e = 1$. Among all numbers e which we may take as the base for logarithms, there is one which is the most useful, and you will understand why when you take calculus. For our purposes here, we want you mainly to learn to operate with the formalism of the logarithm. However, work out the next computations. Suppose that e is chosen so that $\log 2 = 0.6$. Find a decimal for:

a) $\log 2^{3/5}$　　　　　b) $\log 2^{5/2}$　　　　　c) $\log 2^{1/6}$

d) $\log 2^{2/3}$　　　　　e) $\log 8$　　　　　　f) $\log \dfrac{1}{2}$

g) $\log \dfrac{1}{16}$

4. Let $f(t) = Ce^{2t}$. Suppose that you know that $f(2) = 5$. Determine the constant C.

5. Radium disintegrates according to the formula

$$f(t) = Ce^{-5t},$$

where C is a constant. At what time will there be exactly one-half of the original amount left?

6. Bacteria increase according to the formula

$$B(t) = Ce^{kt},$$

where C and k are constants, and $B(t)$ gives the number of bacteria as a function of time t in min. At time $t = 0$, there are 10^6 bacteria. How long will it take before they increase to 10^7 if it takes 12 min to increase to 2×10^6?

7. A radioactive substance disintegrates according to the formula

$$r(t) = Ce^{-7t},$$

where C is a constant. At what time will there be exactly one-third of the original amount left?

8. A substance decomposes according to the formula

$$S(t) = Ce^{-kt},$$

where C, k are constants. At the end of 3 min, 10% of the original substance has decomposed. When will one-half of the original amount have decomposed?

9. In 1900 the population of a city was 50,000. In 1950, it was 100,000. Assume that the population as a function of time is given by the formula

$$P(t) = Ce^{kt},$$

where C, k are constants. What will be the population in 1984? In what year will it be 200,000?

10. The atmospheric pressure as a function of height is given by the formula

$$p = Ce^{-kh},$$

where C, k are constants, p is the pressure, and h is the height. If the barometer reads 30 at sea level, and 24 at 6,000 ft above sea level, find the barometric reading 10,000 ft above sea level.

11. Sugar in water decomposes according to the formula

$$S = Ce^{-kt},$$

where C, k are constants. If 30 lb of sugar reduces to 10 lb in 4 hr, when will 95% of the sugar be decomposed?

12. A particle moves with speed given by

$$s(t) = Ce^{-kt},$$

where C, k are constants. If the initial speed at $t = 0$ is 16 units/min, and if the speed is halved in 2 min, find the value of t when the speed is 10 units/min.

13. Assume that the difference d between the temperature of a body and that of surrounding air is given by the formula

$$d(t) = Ce^{-kt},$$

where C, k are constants. Let $d = 100°$, when $t = 0$, and $d = 40°$ when $t = 40$ min. Find t:

a) when $d = 70°$,

b) when $d = 16°$.

c) Find the value of d when $t = 20$ min.

14. In 1800 the population of a city was 100,000. In 1900 it was 500,000. Assume that the population as a function of time is given by the formula

$$P(t) = Ce^{kt},$$

where C, k are constants. What will be the population in the year 2,000? In what year will it be 1,000,000?

14 Mappings

§1. DEFINITION

We note that a function is an association. We have already seen other types of associations, namely mappings of the plane into itself. It is therefore convenient to define a general notion which covers both cases, and any other case like them.

We have also seen functions which are not defined for all numbers, and for which it was necessary to specify those numbers where it is defined.

Similarly, a function like $x \mapsto x^2$ does not take on all real numbers as values, only numbers ≥ 0. Thus it is convenient to specify which values a function might take. We therefore make the following definition.

Let S, S' be sets. A **mapping** from S into S' is an association

$$f: S \to S'$$

which to each element x in S associates an element $f(x)$ of S'. We call $f(x)$ the **value** of the mapping f at x, or the **image** of x under f.

Functions are special cases of mappings, and so are mappings of the plane into itself. In the latter case, $S = \mathbf{R}^2 = S'$. We have other examples. Disregard those for which you have not read the corresponding section in the book.

Example. Let $P = (1, 2)$ and $A = (-3, 5)$. The association

$$t \mapsto P + tA = (1 - 3t, 2 + 5t)$$

is a mapping from the real numbers into the plane. It is our old parametrization of a line.

Example. The association

$$\theta \mapsto (\cos \theta, \sin \theta), \qquad\qquad \theta \text{ in } \mathbf{R},$$

is a mapping from the real numbers into the plane. In fact, it is a mapping into the circle of radius 1 centered at the origin.

If $f: S \to S'$ is a mapping, and T is a subset of S, then the set of all values $f(t)$ with t in T is called the **image** of T under f, and is denoted by $f(T)$.

Example. The image of the mapping

$$\theta \mapsto (\cos \theta, \sin \theta)$$

is the circle of radius 1 centered at the origin.

Example. The image of the mapping

$$x \mapsto (x, x^2), \qquad\qquad\qquad x \text{ in } \mathbf{R},$$

is the graph of the function $x \mapsto x^2$, and is a parabola.

Example. Let a, b be positive numbers. Let

$$F_{a,b}: \mathbf{R}^2 \to \mathbf{R}^2$$

be the mapping such that

$$F_{a,b}(x, y) = (ax, by).$$

The image of the circle of radius 1 centered at the origin is an ellipse.

Example. Consider the mapping

$$f: \mathbf{R} \to \mathbf{R}^2$$

given by

$$f(t) = (t^2, t^3).$$

We view this as a curve, whose coordinates are given as a function of t. Thus we also write

$$x(t) = t^2 \qquad \text{and} \qquad y(t) = t^3$$

to denote the dependence of the coordinates on t. We view such a mapping as a parametrization of a curve. We can draw the image of this mapping in the plane. We make a small table of values as usual.

t	$x(t)$	$y(t)$		t	$x(t)$	$y(t)$
1	1	1		-1	1	-1
2	4	8		-2	4	-8
3	9	27		-3	9	-27

For any t, the value $x(t)$ is ≥ 0, and increases with t. When t is positive, $y(t)$ is positive. When t is negative, $y(t)$ is negative. Thus we see in Fig. 14–1 what the image of the mapping looks like.

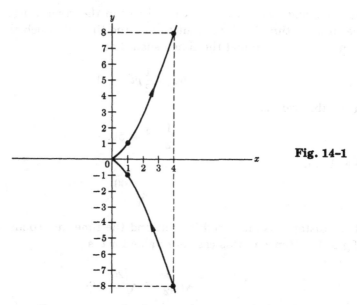

Fig. 14-1

Example. Suppose that a stone is thrown from a tall building, in the horizontal direction. Then gravity pulls the stone down. We want to give the coordinates $(x(t), y(t))$ of the stone as a function of time t. Let the building be 50 ft tall. Then

$$x(t) = ct,$$

where c is some positive constant depending on the strength of the throw. The action of gravity determines the second coordinate to be

$$y(t) = 50 - \frac{1}{2} g t^2$$

where g is constant. Thus we have a mapping

$$t \mapsto \left(ct, 50 - \frac{1}{2} g t^2 \right),$$

defined for $t \geq 0$ and for $t \leq t_0$, where t_0 is the time at which the stone strikes the ground. See the picture in Fig. 14–2.

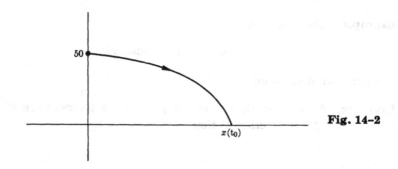

Fig. 14-2

Our mapping associates with each time t in the given range the position of the stone at time t. If we want to find the time at which the stone hits the ground, we have to find the time t such that

$$50 - \frac{1}{2} g t^2 = 0,$$

or in other words,

$$\frac{1}{2} g t^2 = 50.$$

We obtain

$$t = \sqrt{\frac{100}{g}}.$$

If the distance is measured in feet, and the time in seconds, then the value of g is 32. Hence in this case our value for t is

$$\sqrt{\frac{100}{32}} = \sqrt{\frac{25}{8}} \text{ sec.}$$

Example. Let $S = \mathbf{R}^2$. Then the association

$$X \mapsto d(X, O)$$

which to each point X associates its distance from the origin is a mapping, whose values are real numbers $\geqq 0$.

In general, we shall follow the convention that a mapping from a set S into the real numbers is called a **function.** Thus

$$X \mapsto d(X, O)$$

is called the **distance function.** Of course, given a point P, we obtain another function

$$X \mapsto d(X, P),$$

whose value at X is the distance between X and P.

Example. The mapping

$$(x, y) \mapsto x^2 + \cos(xy)$$

is a function, defined on $\mathbf{R}^2$.

Example. At a given time t, we let $f(t)$ be the temperature of a certain body. Then f is a function of time.

EXERCISES

1. Draw the image of the mapping $\mathbf{R} \to \mathbf{R}^2$ given by:
   a) $\theta \mapsto (\cos 2\theta, \sin 2\theta)$,   b) $\theta \mapsto (5 \cos \theta, 5 \sin \theta)$,
   c) $\theta \mapsto (2 \cos \theta, 3 \sin \theta)$,   d) $\theta \mapsto (4 \cos \theta, 5 \sin \theta)$,
   e) $\theta \mapsto (a \cos \theta, b \sin \theta)$ if a, b are positive numbers.

2. Draw the image of the following mappings of $\mathbf{R}$ into $\mathbf{R}^2$.
   a) $t \mapsto (1 - t^2, t)$ b) $t \mapsto (t^2, t^4)$
   c) $t \mapsto (t^4, t^8)$ d) $t \mapsto (t^2, -t^4)$
   Show in each case that the image is part of a parabola.

3. A stone is thrown from a building 50 ft tall in such a way that its co-ordinates $(x(t), y(t))$ at time t, until it hits the ground, are given by:
$$x(t) = 3t, \qquad y(t) = 50 - 16t^2.$$
   a) Find the time at which it hits the ground.
   b) Find the distance of the point where it hits the ground from the origin.
   c) Show that the path of the stone is part of a parabola and give the equation of this parabola.
   d) Find the time at which the stone is 30 ft above the ground.
   e) Find the time at which the stone is 20 ft above the ground.

4. A particle starts from a point $(0, 6)$ in the plane. It is attracted by a magnet along the x-axis, and repelled by a magnet along the y-axis in such a way that its coordinates are given by
$$x(t) = 2t \qquad \text{and} \qquad y(t) = 6 - 4t^2.$$
   Sketch the path of the particle.
   a) Find the time at which it stands at a distance of 2 units above the x-axis.
   b) Show that the path is part of a parabola and give the equation of this parabola.
   c) Find the time t_0 at which the particle hits the x-axis.
   d) Find the distance of the point at which the particle hits the x-axis from the origin.

5. A particle starts from the point $(0, 6)$ in the plane. It is attracted by a magnet below the x-axis, and repelled by a magnet along the y-axis in such a way that its coordinates are given as a function of t by
$$x(t) = 3t \qquad \text{and} \qquad y(t) = 6 - 15t^3.$$

a) Find the time at which the particle hits the x-axis.

b) Give a simple equation in terms of x and y such that the coordinates $(x(t), y(t))$ of the particle satisfy this equation. Sketch the graph of this equation.

c) Find the distance of the point at which the particle hits the x-axis from the origin.

d) Find the time at which the particle is at a distance of 2 units from the x-axis, and below the x-axis.

e) Find the time at which the particle is at a distance of 5 units below the x-axis.

f) Find the time at which the particle is at a distance of 7 units below the x-axis.

6. Draw the image of the straight line $y = 2$ under the mapping

$$(x, y) \mapsto (2^y \cos x, 2^y \sin x).$$

In general what will be the image of a line $y = $ constant under this mapping? What is the image of a line $x = $ constant under this mapping? Draw these images for

a) $y = 3$,   b) $y = -4$,

c) $x = \pi/4$,   d) $x = \pi/3$,

e) $x = -3\pi/2$,   f) $x = 2\pi/3$.

7. Let $0 \leq r_1 < r_2$ and let $0 \leq \theta_1 < \theta_2 \leq 2\pi$. Let S be the set of points in $\mathbf{R}^2$ with coordinates (r, θ) such that

$$r_1 \leq r \leq r_2 \quad \text{and} \quad \theta_1 \leq \theta \leq \theta_2.$$

Thus S is a rectangle. Let $F: \mathbf{R}^2 \to \mathbf{R}^2$ be the mapping given by

$$F(r, \theta) = (r \cos \theta, r \sin \theta).$$

a) Draw the image of S under F.

b) Draw the special case when $r_1 = 0$.

c) Draw the special case when $\theta_1 = 0$ and $\theta_2 = \pi/2$.

d) Draw the special case when $\theta_1 = 0$, $\theta_2 = 2\pi$, and $0 < r_1 < r_2$.

e) Draw the special case when $r_1 = 2$, $r_2 = 3$, $\theta_1 = \pi/4$, $\theta_2 = 3\pi/4$.

Fool around with other special cases so that you get the hang of this system. The map F in this exercise is called the **polar coordinate map.**

§2. FORMALISM OF MAPPINGS

Mappings in general satisfy a formalism similar to that developed for mappings of the plane into itself. We shall now repeat this formalism briefly.

We have the **identity mapping** I_S for any set S, such that

$$I_S(x) = x \text{ for all } x \text{ in } S.$$

Sometimes we omit the subscript S on I if the reference to S is made clear by the context.

Let

$$f: S \to T \quad \text{and} \quad g: U \to V$$

be mappings. Assume that T is a subset of U. Then we can form the **composite** mapping $g \circ f$, whose value at an element x of S is

$$(g \circ f)(x) = g(f(x)).$$

Note that $f(x)$ is an element of T, and since T is a subset of U, we can take $g(f(x))$, whose value is now in V.

Example. Let f and g be given by

$$f(x) = x^2 \quad \text{and} \quad g(u) = \sin u.$$

Then

$$g(f(x)) = g(x^2) = \sin(x^2).$$

Example. Let f and g be given by

$$f(x) = 3x - 2 \quad \text{and} \quad g(u) = \frac{1}{\cos u}.$$

Then

$$g\bigl(f(x)\bigr) = \frac{1}{\cos(3x - 2)}.$$

Observe that g is defined only for those numbers u such that $u \neq \dfrac{\pi}{2} + n\pi$, and n is an integer. Thus the composite function is defined only for those numbers x such that

$$3x - 2 \neq \frac{\pi}{2} + n\pi,$$

and n is an integer.

Example. Let $I: S \to S$ be the identity mapping, and let

$$f: S \to S$$

be any mapping. Then

$$I \circ f = f \circ I = f.$$

This is clear, because, for any x in S, we have

$$(I \circ f)(x) = I\bigl(f(x)\bigr) = f(x),$$

and similarly,

$$(f \circ I)(x) = f\bigl(I(x)\bigr) = f(x).$$

We note that the identity mapping behaves like the number 1 in a context where composition of mappings behaves like multiplication. This analogy is again evident in our next assertion.

Composition of mappings is associative. This means: If

$$f: S \to T, \qquad g: T \to U, \qquad h: U \to V$$

are mappings, then

$$(h \circ g) \circ f = h \circ (g \circ f).$$

The proof is exactly the same as the proof for the associativity of isometries of the plane into itself. Namely for any x in S, we have

$$((h \circ g) \circ f)(x) = (h \circ g)\bigl(f(x)\bigr) = h\bigl(g(f(x))\bigr)$$
$$(h \circ (g \circ f))(x) = h\bigl((g \circ f)(x)\bigr) = h\bigl(g(f(x))\bigr).$$

Let

$$f: S \to T$$

be a mapping. By an **inverse mapping** for f, we mean a mapping

$$g: T \to S$$

such that the composites of f and g are the identities of S and T, respectively; that is,

$$g \circ f = I_S \quad \text{and} \quad f \circ g = I_T.$$

We have had many examples of inverse mappings with translations, rotations, and the like. We also have examples with functions.

Example. Let $\mathbf{R}^+$ denote the positive real numbers. Let

$$f: \mathbf{R}^+ \to \mathbf{R}^+$$

be the square, i.e. $f(x) = x^2$. Then f has an inverse function, namely the square root. If $x > 0$ and g is the square root, then $g(x^2) = x$ and $f(\sqrt{x}) = x$. Thus f, g are inverse to each other.

Example. The exponential function and the logarithm are inverse to each other (when taken to the same base). In other words,

$$a^{\log_a x} = x \quad \text{and} \quad \log_a(a^x) = x.$$

This is merely the definition of the logarithm.

Thus we see that the notion of inverse mapping or inverse function includes many special cases already considered.

If $f: S \to T$ is a mapping having an inverse mapping, then we denote this inverse by

$$f^{-1}: T \to S.$$

The elementary proof concerning the inverse of a number can be applied to give the uniqueness of the inverse mapping. Indeed, suppose that

$$g: T \to S \quad \text{and} \quad h: T \to S$$

are inverse mappings for f. Then we must have $g = h$. We leave the proof as an exercise.

The section on permutations will provide further examples for mappings, as well as an extension of the theory in an interesting direction. You may well want to read our next remarks in connection with permutations, and their iterations.

Let $f: S \to S$ be a mapping of a set into itself. We may then iterate f. If x is an element of S, then we can form the iterates of the values,

$$f(x), \quad f(f(x)), \quad f(f(f(x))), \ldots.$$

We use the exponential notation, and write

$$f^2(x) = f(f(x))$$
$$f^3(x) = f(f(f(x)))$$
$$\vdots$$

In general, we write

$$f^k(x) = f(f(f(\ldots f(x) \ldots)))$$

for the iteration of f taken k times, applied to x. Thus f^k is again a function defined on S, with values in S. The value of f^k at x is $f^k(x)$. Observe that $f^{k+1} = f \circ f^k$, or, in terms of x,

$$f^{k+1}(x) = f(f^k(x)).$$

Example. Let $f: \mathbf{R} \to \mathbf{R}$ be the function such that $f(x) = x + 1$. Then

$$f^2(x) = f(f(x)) = f(x + 1) = x + 2.$$

Similarly,

$$f^3(x) = f(f^2(x)) = f(x + 2) = x + 3.$$

In general, we see that

$$f^k(x) = x + k$$

for any positive integer k.

Example. Let $F: \mathbf{R}^2 \to \mathbf{R}^2$ be the mapping given by

$$F(x, y) = (x + 3, y - 4).$$

If we denote (x, y) by X and if we let $A = (3, -4)$, then we can abbreviate the formula describing F as follows:

$$F(X) = X + A.$$

Thus we see that F is translation by A. Iterating F is analogous to the process used in the preceding example. For instance,

$$F^2(X) = F(X + A) = X + A + A = X + 2A,$$

and in general, for any positive integer k, we have

$$F^k(X) = X + kA.$$

Example. Let $f: \mathbf{R} \to \mathbf{R}$ be the map given by $f(x) = x^3$. Then

$$f^2(x) = f(f(x)) = f(x^3) = (x^3)^3 = x^9.$$

Also,

$$f^3(x) = f(f^2(x)) = f(x^9) = (x^9)^3 = x^{27}.$$

In general, for any positive integer k, we have

$$f^k(x) = x^{3^k}.$$

Just as with mappings of the plane into itself, we have a general rule of exponent for mappings; namely, if m, n are positive integers and

$$f: S \to S$$

is a mapping, then

$$\boxed{f^{m+n} = f^m \circ f^n.}$$

This simply means that if we iterate the mapping $m + n$ times, it amounts to iterating it first n times and then m times. Thus our formula is clear. We define

$$f^0 = I \quad (= I_S, \text{ identity mapping})$$

and we see that our formula also holds if m or n is 0. This is an immediate consequence of the definitions.

In fact, this formula also holds for negative integers m or n, if we make the appropriate definition. Namely, assume that f^{-1} exists. Then we define

$$f^{-k} = (f^{-1})^k$$

to be the composite of f^{-1} with itself k times. Note that

$$\boxed{f^k \circ f^{-k} = f^0 = I.}$$

This just means that if we apply f^{-1} to an element k times, and then apply f itself k times, we get that element. We can draw the picture as follows. Let x be an element of S. Then

$$f^k \circ f^{-k}(x) = \underbrace{f(f(\ldots f(}_{k \text{ times}}\underbrace{f^{-1}(f^{-1}(\ldots f^{-1}(x))))))}_{k \text{ times}}.$$

But the $f \circ f^{-1}$ in the middle is the identity I. Similarly, we combine $f \circ f^{-1}$ to get I, and we do this k times to see that

$$f^k \circ f^{-k}(x) = x.$$

Note. Later we shall discuss a more formal way of giving the above proof, namely induction. After you have read about induction, you will then understand the following argument which proves that

$$f^k \circ f^{-k} = I$$

by induction. It is true for $k = 1$ by definition. Assume it for a positive integer k. Then

$$f^{k+1} \circ f^{-(k+1)} = f \circ f^k \circ (f^{-1})^k \circ f^{-1}$$

$$= f \circ I \circ f^{-1} \qquad \text{(by induction)}$$

$$= f \circ f^{-1} = I.$$

Example. Let $f: \mathbf{R} \to \mathbf{R}$ be the map such that $f(x) = x + 3$. Then

$$f^k(x) = x + 3k \qquad \text{and} \qquad f^{-k}(x) = x - 3k.$$

Example. Let $A = (3, -4)$ and let $F: \mathbf{R}^2 \to \mathbf{R}^2$ be the map such that

$$F(X) = X + A.$$

Then

$$F^k(X) = X + kA \qquad \text{and} \qquad F^{-k}(X) = X - kA.$$

Example. Let $f: \mathbf{R} \to \mathbf{R}$ be the map such that $f(x) = x^3$. Then

$$f^k(x) = x^{3^k} \qquad \text{and} \qquad f^{-k}(x) = x^{3^{-k}}.$$

In general, for any integers m, n (positive, negative, or zero) and any map $f: S \to S$ which has an inverse, we have the formula

$$\boxed{f^{m+n} = f^m \circ f^n.}$$

This can also be proved by induction, but we omit the proof.

Example. Inserting special numbers into the above formula, we have

$$f^5 \circ f^{-3} = f^2.$$

Example. Suppose that $f^m = f^n$ for some integers m, n, and assume that f has an inverse mapping. Then, composing both sides with f^{-n}, we obtain

$$f^{m-n} = I.$$

Observe again how composition of mappings is analogous to multiplication.

Example. If f has an inverse mapping, and if $f^4 = f^{-5}$, then

$$f^9 = I.$$

Example. Let $f_1, \ldots, f_m$ be mappings of a set S such that

$$f_i^2 = I$$

for each $i = 1, \ldots, m$. Let f be a map such that

$$f_1 \circ f_2 \circ \cdots \circ f_m \circ f = I.$$

Compose on the left with f_1, which is nothing but f_1^{-1}. We obtain

$$f_2 \circ \cdots \circ f_m \circ f = f_1.$$

Now compose with f_2 on the left. We get

$$f_3 \circ \cdots \circ f_m \circ f = f_2 \circ f_1.$$

Proceeding in this way, composing with $f_3, \ldots, f_m$ successively, we find that

$$f = f_m \circ f_{m-1} \circ \cdots \circ f_2 \circ f_1.$$

This procedure will be used in the next section when we deal with permutations.

EXERCISES

1. Let $f: S \to T$ and $g: S \to T$ be mappings. Let

$$h: T \to U$$

be a mapping having an inverse mapping denoted by

$$h^{-1}: U \to T.$$

If

$$h \circ f = h \circ g,$$

prove that $f = g$. This is the **cancellation law** for mappings.

2. Let $f: S \to T$ be a mapping having an inverse mapping. Prove the following statements.

   a) If x, y are elements of S and $f(x) = f(y)$, then $x = y$.

   b) If z is an element of T, then there exists an element x of S such that $f(x) = z$.

3. Let a, b be non-zero numbers. Let $F_{a,b}: \mathbf{R}^2 \to \mathbf{R}^2$ be the mapping such that

$$F_{a,b}(x, y) = (ax, by).$$

   Show that $F_{a,b}$ has an inverse mapping.

4. Let $f: \mathbf{R}^2 \to \mathbf{R}^2$ be the mapping defined by

$$f(x, y) = (2x - y, y + x).$$

   Show that f has an inverse mapping. [*Hint:* Let $u = 2x - y$ and $v = y + x$. Solve for x, y in terms of u and v.]

5. Let $f: \mathbf{R}^2 \to \mathbf{R}^2$ be the mapping defined by

$$f(x, y) = (3x + y, 2x - 4y).$$

   Show that f has an inverse mapping.

6. Let $f: S \to S$ be a mapping which has an inverse mapping.

   a) If $f^3 = I$ and $f^5 = I$, show that $f = I$.

   b) If $f^2 = I$ and $f^7 = I$, show that $f = I$.

   c) If $f^4 = I$ and $f^{11} = I$, show that $f = I$.

7. Let f, g be mappings of a set S into itself, and assume that they have inverse mappings. Assume also that $f \circ g = g \circ f$. Express each one of the following in the form $f^m \circ g^n$ where m, n are integers.

   a) $f^3 \circ g^2 \circ f^{-2} \circ f^5 \circ g^{-5}$ b) $f^7 \circ g \circ g^4 \circ f^{-6} \circ g^3$

   c) $f^4 \circ g^5 \circ f^{-5} \circ g^{-7} \circ g^2 \circ f^2$ d) $f^4 \circ f^{-8} \circ g^2 \circ f^3 \circ g^3 \circ f^{-2}$

8. a) Let f, g be mappings of a set S into itself, and assume that they have inverse mappings. Prove that $f \circ g$ has an inverse mapping, and express it in terms of f^{-1}, g^{-1}.

   b) Let $f_1, \ldots, f_m$ be maps of S into itself, and assume that each f_i has an inverse mapping. Show that the composite

$$f_1 \circ f_2 \circ \cdots \circ f_m$$

   has an inverse mapping, and express this inverse mapping in terms of the maps f_i^{-1}.

9. Let f be a mapping of a set S into itself, and assume that f has an inverse mapping.

  a) If $f^5 = I$, express f^{-1} as a positive power of f.

  b) In general, if $f^n = I$, for some positive power of f, express f^{-1} as a positive power of f.

10. Let f, g be mappings of a set S into itself. Assume that $f^2 = g^2 = I$ and that $f \circ g = g \circ f$. Prove that $(f \circ g)^2 = I$. Prove that $(f \circ g)^3 = I$. What about $(f \circ g)^n$ for any positive integer n? What about $(f \circ g)^n$ when n is a negative integer?

§3. PERMUTATIONS

Let J_n be the set of integers $1, 2, \ldots, n$, that is the set of integers k such that $1 \leq k \leq n$. By a **permutation** of J_n, we mean a mapping

$$\sigma : J_n \to J_n$$

having the following property. If i, j are in J_n and $i \neq j$, then $\sigma(i) \neq \sigma(j)$. Thus the image of a permutation σ consists of n distinct integers

$$\sigma(1), \quad \sigma(2), \quad \ldots, \quad \sigma(n),$$

which must therefore be the integers $1, 2, \ldots, n$ in a different order. Thus we denote such a permutation σ by the symbols

$$\begin{bmatrix} 1 & 2 & 3 & \cdots & n \\ \sigma(1) & \sigma(2) & \sigma(3) & \cdots & \sigma(n) \end{bmatrix}.$$

Example. The permutation of J_4 given by

$$\sigma = \begin{bmatrix} 1 & 2 & 3 & 4 \\ 3 & 1 & 4 & 2 \end{bmatrix}$$

is such that

$$\sigma(1) = 3, \qquad \sigma(2) = 1, \qquad \sigma(3) = 4, \qquad \sigma(4) = 2.$$

If σ and σ' are permutations, we denote their composite by $\sigma\sigma'$, omitting the little circle between them for simplicity and to emphasize the analogy with "multiplication". In fact, we also call this composite the **product** of σ and σ'. Watch out! It is not always true that $\sigma\sigma' = \sigma'\sigma$.

Example. Let

$$\sigma = \begin{bmatrix} 1 & 2 & 3 & 4 \\ 2 & 3 & 4 & 1 \end{bmatrix} \quad \text{and} \quad \sigma' = \begin{bmatrix} 1 & 2 & 3 & 4 \\ 3 & 4 & 2 & 1 \end{bmatrix}.$$

To determine $\sigma'\sigma$, we just look at the effect of this composite on each one of the numbers 1, 2, 3, 4. Thus with arrows:

$$1 \xmapsto{\sigma} 2 \xmapsto{\sigma'} 4,$$
$$2 \xmapsto{\sigma} 3 \xmapsto{\sigma'} 2,$$
$$3 \xmapsto{\sigma} 4 \xmapsto{\sigma'} 1,$$
$$4 \xmapsto{\sigma} 1 \xmapsto{\sigma'} 3.$$

Thus we have

$$\sigma'\sigma = \begin{bmatrix} 1 & 2 & 3 & 4 \\ 4 & 2 & 1 & 3 \end{bmatrix}.$$

On the other hand, in a similar way, you find that

$$\sigma\sigma' = \begin{bmatrix} 1 & 2 & 3 & 4 \\ 4 & 1 & 3 & 2 \end{bmatrix}.$$

Thus $\sigma\sigma' \neq \sigma'\sigma$.

Example. Let

$$\sigma = \begin{bmatrix} 1 & 2 & 3 \\ 2 & 1 & 3 \end{bmatrix} \quad \text{and} \quad \sigma' = \begin{bmatrix} 1 & 2 & 3 \\ 3 & 1 & 2 \end{bmatrix}.$$

Then

$$\sigma\sigma'(1) = \sigma(\sigma'(1)) = \sigma(3) = 3,$$
$$\sigma\sigma'(2) = \sigma(\sigma'(2)) = \sigma(1) = 2,$$
$$\sigma\sigma'(3) = \sigma(\sigma'(3)) = \sigma(2) = 1,$$

so that we can write

$$\sigma\sigma' = \begin{bmatrix} 1 & 2 & 3 \\ 3 & 2 & 1 \end{bmatrix}.$$

If σ is a permutation of J_n, then we have already mentioned that

$$\sigma(1), \quad \sigma(2), \quad \ldots, \quad \sigma(n)$$

are simply the elements of J_n in a different order. Thus to each element j of J_n there exists a unique element i such that

$$\sigma(i) = j.$$

We can therefore define the **inverse permutation** of σ, as with inverse mappings, to be that permutation σ^{-1} such that

$$\sigma\sigma^{-1} = \sigma^{-1}\sigma = I \text{ (the identity permutation)}.$$

We have $\sigma(i) = j$ if and only if $\sigma^{-1}(j) = i$.

Example. Let

$$\sigma = \begin{bmatrix} 1 & 2 & 3 \\ 3 & 1 & 2 \end{bmatrix}.$$

Since $\sigma(1) = 3$ we have $\sigma^{-1}(3) = 1$. Since $\sigma(2) = 1$ we have $\sigma^{-1}(1) = 2$. Since $\sigma(3) = 2$ we have $\sigma^{-1}(2) = 3$. Hence

$$\sigma^{-1} = \begin{bmatrix} 1 & 2 & 3 \\ 2 & 3 & 1 \end{bmatrix}.$$

Example. The identity permutation is of course given by

$$I = \begin{bmatrix} 1 & 2 & 3 \\ 1 & 2 & 3 \end{bmatrix}.$$

If $\sigma_1, \ldots, \sigma_r$ are permutations of J_n, then the inverse of the composite permutation

$$\sigma_1 \cdots \sigma_r$$

is the permutation

$$\sigma_r^{-1} \cdots \sigma_1^{-1}.$$

Just multiply one with the other, and you will find that all the factors cancel out to give the identity.

For instance, with three permutations, we have

$$(\sigma_1\sigma_2\sigma_3)^{-1} = \sigma_3^{-1}\sigma_2^{-1}\sigma_1^{-1}$$

because

$$\sigma_1\sigma_2\sigma_3\sigma_3^{-1}\sigma_2^{-1}\sigma_1^{-1} = I,$$

and similarly on the other side.

There is an important special case of a permutation, namely the **transposition** which interchanges two distinct numbers $i \neq j$, and leaves the others fixed.

Example. The permutation

$$\sigma = \begin{bmatrix} 1 & 2 & 3 & 4 \\ 1 & 3 & 2 & 4 \end{bmatrix}$$

is a transposition, which interchanges 3 and 2, and leaves 1, 4 fixed.

If τ is a transposition, then it is clear that

$$\tau^2 = I.$$

Thus in this case, we have

$$\tau^{-1} = \tau.$$

Theorem 1. *Every permutation of J_n can be expressed as a product of transpositions.*

Proof. Let σ be a permutation of J_n. Let $\sigma(n) = k$. If $k = n$, we let τ_n be the identity. If $k \neq n$, we let τ_n be the transposition which interchanges k and n, and leaves the other integers fixed. Then $\tau_n \sigma$ leaves n fixed, and may therefore be viewed as a permutation of J_{n-1}. We now repeat our procedure. We let τ_{n-1} be either the identity, or a transposition of J_{n-1} such that

$$\tau_{n-1}\tau_n\sigma$$

leaves n and $n - 1$ fixed. We continue in this way, finding

$$\tau_n, \quad \tau_{n-1}, \quad \ldots, \quad \tau_2, \quad \tau_1$$

which are either the identity, or transpositions. Finally

$$\tau_1\tau_2 \cdots \tau_n\sigma$$

leaves every one of the numbers $1, \ldots, n$ fixed, and is therefore equal to the identity. Thus

$$\tau_1\tau_2 \cdots \tau_n\sigma = I.$$

Now multiply on the left by τ_1, then by τ_2, and so on. Since $\tau_i^2 = I$ for each i, we find that

$$\sigma = \tau_n \cdots \tau_1.$$

In this product, we may omit those τ_i which are the identity, and we find that σ has been expressed as a product of transpositions.

Note. It may happen, of course, as when σ is the identity already, that every τ_i is the identity, so that in our product, we have simply $\sigma = I$. Thus, in a sense, there is no transposition in this product. It is a matter of convention about our use of language how we deal with this case. The best convention is to agree that I is the product of zero transpositions, and allow this possibility to be included in our expression "product of transpositions". After all, we did not say how many transpositions, and it could have been zero. With this convention, the formulation we have given to Theorem 1 is correct.

The procedure used to prove Theorem 1 gives us an effective way in practice to express a permutation as a product of transpositions.

Example. Let

$$\sigma = \begin{bmatrix} 1 & 2 & 3 \\ 3 & 1 & 2 \end{bmatrix}.$$

We want to express σ as a product of transpositions. Let τ be the transposition which interchanges 3 and 1, and leaves 2 fixed. Then we find that

$$\tau\sigma = \begin{bmatrix} 1 & 2 & 3 \\ 1 & 3 & 2 \end{bmatrix}$$

so that $\tau\sigma$ is a transposition, which we denote by τ'. We can then write $\tau\sigma = \tau'$, so that composing on the left with τ yields

$$\sigma = \tau\tau'$$

because $\tau^2 = I$. This is the desired product.

Example. We want to express the permutation

$$\sigma = \begin{bmatrix} 1 & 2 & 3 & 4 \\ 2 & 3 & 4 & 1 \end{bmatrix}$$

as a product of transpositions. Let τ_1 be the transposition which interchanges 1 and 2, and leaves 3, 4 fixed. Then

$$\tau_1\sigma = \begin{bmatrix} 1 & 2 & 3 & 4 \\ 1 & 3 & 4 & 2 \end{bmatrix}.$$

Now let τ_2 be the transposition which interchanges 2 and 3, and leaves 1, 4 fixed. Then

$$\tau_2\tau_1\sigma = \begin{bmatrix} 1 & 2 & 3 & 4 \\ 1 & 2 & 4 & 3 \end{bmatrix},$$

and we see that $\tau_2\tau_1\sigma$ is a transposition, which we may denote by τ_3. Then we get $\tau_2\tau_1\sigma = \tau_3$ so that multiplying on the left with τ_2 and τ_1 successively we get

$$\sigma = \tau_1\tau_2\tau_3.$$

This is the desired expression.

In expressing a permutation as a product of transpositions, there is of course no uniqueness. This is possible in many ways.

Example. Take the permutation σ of the preceding example. Let τ_4 be the transposition which interchanges 1 and 3. Then

$$\tau_4\sigma = \begin{bmatrix} 1 & 2 & 3 & 4 \\ 2 & 1 & 4 & 3 \end{bmatrix}.$$

If again τ_1 is the transposition which interchanges 1 and 2, and τ_3 is the transposition which interchanges 3 and 4, we see that

$$\tau_4\sigma = \tau_1\tau_3.$$

Hence composing on the left with τ_4 yields

$$\sigma = \tau_4\tau_1\tau_3.$$

Even though we don't have uniqueness of the expression of σ as a product of transpositions, still an interesting phenomenon occurs. Suppose that we have written σ as a product of transpositions in two ways:

$$\sigma = \tau_1 \cdots \tau_r = \tau_1' \cdots \tau_s'.$$

The number of transpositions occurring in these expressions may not be the same, but it turns out that if r is even, then s must also be even, and if r is odd, then s must also be odd. In other words, we have:

Theorem 2. *Let σ be a permutation of J_n. In any expressions of σ as a product of transpositions, the number of transpositions occurring in such a product is either always even or always odd.*

The proof for Theorem 2 will involve a little more theory about permutations than we have right now, and we postpone it to the end. However, to use Theorem 2 is very easy, and we make some comments about that.

Suppose that a permutation σ can be expressed as a product of an even number of transpositions. Then we say that σ is an **even permutation**. If it can be expressed as a product of an odd number of transpositions, then we say that σ is an **odd permutation**. If

$$\sigma = \tau_1 \cdots \tau_m$$

is an expression of σ as a product of transposition, then we call

$$(-1)^m$$

the **sign** of σ. This sign is 1 or -1 according as σ is even or odd.

Example. The sign of the permutation

$$\sigma = \begin{bmatrix} 1 & 2 & 3 \\ 3 & 1 & 2 \end{bmatrix}$$

is 1, and σ is an even permutation (cf. a previous example where we expressed this permutation as a product of 2 transpositions).

Example. The sign of the permutation

$$\sigma = \begin{bmatrix} 1 & 2 & 3 & 4 \\ 2 & 3 & 4 & 1 \end{bmatrix}$$

is -1, and σ is an odd permutation (cf. a previous example where we expressed this permutation as a product of 3 transpositions).

If you are not interested in the proof of Theorem 2, or in the further theory of permutations, you may omit the rest of this section.

We now start on the discussion which will lead to the proof of Theorem 2. It involves looking more closely into the structure of permutations.

Let σ be a permutation of J_n and let i be one of the integers in J_n. Let us look at what happens to i when we apply successive powers of σ. We obtain

$$i, \quad \sigma(i), \quad \sigma^2(i), \quad \ldots.$$

Since J_n has only a finite number of elements, it follows that these numbers cannot be all distinct. Let us see this on an example.

Example. Let

$$\sigma = \begin{bmatrix} 1 & 2 & 3 & 4 & 5 \\ 3 & 5 & 4 & 1 & 2 \end{bmatrix}.$$

Let us start with $i = 1$. Then

$$\sigma(1) = 3,$$
$$\sigma^2(1) = \sigma(3) = 4,$$
$$\sigma^3(1) = \sigma(4) = 1.$$

Now we see that we start over again, namely

$$\sigma^4(1) = \sigma(\sigma^3(1)) = \sigma(1) = 3,$$
$$\sigma^5(1) = 4,$$
$$\sigma^6(1) = 1.$$

We see that we are going around in a circle, and we can even represent the numbers 1, 3, 4 on a circle as in Fig. 14–3.

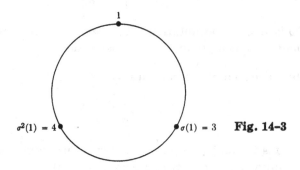

Fig. 14–3

In general, starting with any i, we have an analogous situation. Since the set J_n has only a finite number of elements, it follows that the numbers

$$i, \quad \sigma(i), \quad \sigma^2(i), \quad \sigma^3(i), \quad \ldots$$

are not all distinct. Hence there exist two positive integers r, s with $r < s$ such that $\sigma^r(i) = \sigma^s(i)$. Applying σ^{-r} to both sides, we get

$$i = \sigma^{s-r}(i).$$

Thus some power of σ leaves i fixed. Let k be the smallest positive integer such that $\sigma^k(i) = i$. *We contend that the numbers*

$$i, \quad \sigma(i), \quad \sigma^2(i), \quad \ldots, \quad \sigma^{k-1}(i)$$

are distinct.

    Proof. If $\sigma^m(i) = \sigma^n(i)$ with two integers m, n such that

$$1 \leq m < n \leq k,$$

then $\sigma^{n-m}(i) = i$, and $n - m < k$. This contradicts our assumption that k is the smallest positive integer such that $\sigma^k(i) = i$, and therefore proves our contention.

    We now see that the general situation is analogous to that of the example, and we can represent the numbers

$$i, \quad \sigma(i), \quad \sigma^2(i), \quad \ldots, \quad \sigma^{k-1}(i)$$

as going around a circle as illustrated in Fig. 14–4. We shall call the set consisting of

$$i, \quad \sigma(i), \quad \sigma^2(i), \quad \ldots, \quad \sigma^{k-1}(i)$$

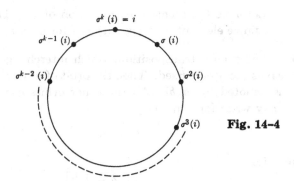

Fig. 14–4

an **orbit** of σ, or, specifically, the orbit to which i belongs. The effect of σ on this orbit will be denoted by square brackets,

$$\gamma = [i, \sigma(i), \ldots, \sigma^{k-1}(i)],$$

and will be called the **cycle** of the orbit. We call k the **period** of i under σ, or also the **period** of the cycle, or the **length** of the cycle.

Example. Consider again the permutation

$$\sigma = \begin{bmatrix} 1 & 2 & 3 & 4 & 5 \\ 3 & 5 & 4 & 1 & 2 \end{bmatrix}.$$

Then the orbit of 1 consists of 1, 3, 4, and its cycle is denoted by

$$[1, 3, 4].$$

The cycle of 2 is given by

$$[2, 5]$$

because

$$\sigma(2) = 5 \quad \text{and} \quad \sigma(5) = 2.$$

According to our terminology, we see that 1 has period 3 under σ, whereas 2 has period 2. The cycle to which 1 belongs has length 3, and the cycle to which 5 belongs has length 2.

Note that in the orbit of 1, we could have selected 3 and started considering

$$3, \quad \sigma(3), \quad \sigma^2(3), \quad \sigma^3(3) = 3,$$

so that the cycle [1, 3, 4] can also be written

$$[3, 4, 1].$$

Similarly, we have $[2, 5] = [5, 2]$.

Example. Let σ be the identity permutation of J_n. Then each orbit of σ consists of a single element, because $\sigma(i) = i$ for each i.

Example. Let τ be a transposition, which interchanges the two numbers a, b and leaves the others fixed. Then the orbit of a (or b) consists of a, b and its cycle is denoted by $[a, b]$. All the other orbits consist of one element. Thus we may write for simplicity

$$\tau = [a, b].$$

Example. Let

$$\sigma = \begin{bmatrix} 1 & 2 & 3 & 4 & 5 \\ 2 & 3 & 4 & 5 & 1 \end{bmatrix}.$$

Then σ has just one orbit, namely J_5 itself, and its cycle is

$$[1, 2, 3, 4, 5].$$

We have the usual picture representing σ in Fig. 14–5.

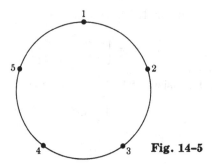

Fig. 14–5

As with transpositions, it will be useful to write

$$\sigma = [3, 4, 5, 1, 2].$$

Example. Let

$$\sigma = \begin{bmatrix} 1 & 2 & 3 & 4 & 5 \\ 3 & 5 & 4 & 1 & 2 \end{bmatrix}.$$

Then σ has two orbits, whose cycles are

$$[1, 3, 4] \qquad \text{and} \qquad [2, 5].$$

Observe that the two orbits have no elements in common. This is a special case of a general fact, which we shall state and prove below. The effect of

σ is determined by its effect on the orbits, and hence we shall use the notation

$$\sigma = [1, 3, 4][2, 5].$$

We shall call this the **orbit decomposition** of σ.

   Theorem 3. *If a, b are elements of J_n and σ is a permutation of J_n, then the orbits to which a and b belong either coincide, or have no element in common, i.e. are disjoint.*

   Proof. Suppose that the orbits have some element in common. This means that we have

$$\sigma^r(a) = \sigma^s(b)$$

for some positive integers r, s. Then

$$a = \sigma^{s-r}(b),$$

and thus we see that a occurs as some $\sigma^m(b)$. But then all the numbers

$$\sigma(a) = \sigma^{m+1}(b), \qquad \sigma^2(a) = \sigma^{m+2}(b), \ldots$$

occur in the orbit of b, which means that the orbit of a is contained in the orbit of b. Conversely, we see in the same way that the orbit of b is contained in the orbit of a, and thus the two orbits must be equal. This proves our theorem.

   Let S be a given orbit of σ and let i be any element of S. Let us denote the cycle of the orbit by

$$\gamma = [i, \sigma(i), \ldots, \sigma^{k-1}(i)].$$

If $S_1, \ldots, S_m$ are the disjoint orbits of σ, then they have corresponding cycles $\gamma_1, \ldots, \gamma_m$ and we write symbolically

$$\sigma = \gamma_1 \gamma_2 \cdots \gamma_m.$$

We call this the **orbit decomposition** of σ, as in our example above.

Example. Let

$$\sigma = \begin{bmatrix} 1 & 2 & 3 & 4 & 5 & 6 \\ 5 & 6 & 2 & 1 & 4 & 3 \end{bmatrix}.$$

Then the two orbit cycles are

$$[1, 5, 4] \qquad \text{and} \qquad [2, 6, 3].$$

Thus we get the orbit decomposition of σ; namely,

$$\sigma = [1, 5, 4][2, 6, 3].$$

In this case, both cycles have length 3, and every element of J_6 has period 3 under σ.

Example. We can write the orbit decomposition of the identity of J_n as

$$I = [1][2] \cdots [n].$$

Let σ be a permutation of J_n and let $S_1, \ldots, S_m$ be its orbits. We then have the representation of σ as a product of cycles

$$\sigma = \gamma_1 \cdots \gamma_m$$

corresponding to these orbits. We ask: What are the orbits of $\tau\sigma$, if τ is a transposition? Two cases arise. Let $\tau = [a, b]$.

Case 1. The two numbers a, b belong to the same orbit of σ, say S_1. Let

$$\gamma_1 = [i_1, \ldots, i_k]$$

be the cycle of this orbit, and suppose that

$$a = i_r \quad \text{and} \quad b = i_s$$

with $r < s$. Thus we can express symbolically the effect of $\tau\sigma$ on the orbit S_1 by

$$\tau\gamma_1 = [i_r, i_s][i_1, \ldots, i_{r-1}, i_r, \ldots, i_{s-1}, i_s, \ldots, i_k].$$

Now follow through the effect of $\tau\gamma_1$ on each one of the numbers above. We see that under $\tau\gamma_1$, we have:

$$i_1 \mapsto i_2 \mapsto \cdots \mapsto i_{r-1} \mapsto i_s \mapsto i_{s+1} \mapsto \cdots \mapsto i_k \mapsto i_1,$$

thus forming a cycle

$$\gamma_1' = [i_1, i_2, \ldots, i_{r-1}, i_s, i_{s+1}, \ldots, i_k].$$

On the other hand, we see that under $\tau\gamma_1$, we have:

$$i_r \mapsto i_{r+1} \mapsto \cdots \mapsto i_{s-1} \mapsto i_r,$$

thus forming another cycle

$$\gamma_1'' = [i_r, i_{r+1}, \ldots, i_{s-1}]$$

which is disjoint from γ_1'. Furthermore, τ has no effect on the cycles $\gamma_2, \ldots, \gamma_m$. Hence

$$\tau\sigma = \tau\gamma_1\gamma_2 \cdots \gamma_m = \gamma_1'\gamma_1''\gamma_2 \cdots \gamma_m.$$

Thus τ splits the orbits S_1 into two distinct orbits S_1' and S_1'' while leaving all other orbits fixed. **In the present case, we see that $\tau\sigma$ has one more orbit than σ.**

Case 2. The two numbers a, b belong to distinct orbits of σ, say S_1 and S_2.

If you work out the same type of argument as in Case 1, you will see that the effect of τ on the orbits of σ is to join S_1 and S_2, so that $\tau\sigma$ has one less orbit than σ. You just have to follow through what happens to each one of the integers in the cycles

$$\gamma_1 = [i_1, \ldots, i_r] \quad \text{and} \quad \gamma_2 = [j_1, \ldots, j_s]$$

corresponding to the two orbits S_1 and S_2. You will see that τ has the effect of joining these orbits into a single orbit for $\tau\sigma$. Hence in this case, $\tau\sigma$ **has one less orbit than σ.**

How does all this apply to the proof of Theorem 2? Very simply. Suppose that a given permutation σ of J_n can be written as a product of an odd number of transpositions, say

$$\sigma = \tau_1 \cdots \tau_p,$$

where p is odd. Let us start with the identity I, which has n orbits. As we multiply I successively by $\tau_p, \tau_{p-1}, \ldots, \tau_1$, we either increase the number of orbits by 1, or decrease it by 1, each time. Consequently, we obtain the relation:

Number of orbits of σ = n + an odd integer.

If it were possible to express σ as a product

$$\sigma = \tau_1' \cdots \tau_q',$$

where q is even, then arguing similarly, we would obtain the relation:

Number of orbits of σ = n + an even integer.

However the integer

(Number of orbits of σ) − n

is either odd or even, and cannot be both. Consequently we cannot express σ both as a product of an even number of transpositions and an odd number of transpositions. This concludes the proof of Theorem 2.

The general rule following from this discussion can be stated as follows.

Theorem 4. *Let σ be a permutation of J_n and suppose that σ has k orbits. If $n - k$ is even, the σ is an even permutation. If $n - k$ is odd, then σ is an odd permutation.*

Example. Theorem 4 gives us another method for determining the sign of a permutation. For instance, take $n = 6$. In a previous example we have seen that the permutation

$$\sigma = \begin{bmatrix} 1 & 2 & 3 & 4 & 5 & 6 \\ 5 & 6 & 2 & 1 & 4 & 3 \end{bmatrix}$$

has two orbits, whose cycles are

$$[1, 5, 4] \quad \text{and} \quad [2, 6, 3].$$

Since $6 - 2 = 4$ is even, it follows that σ is an even permutation.

Example. Let $n = 5$. The permutation

$$\sigma = \begin{bmatrix} 1 & 2 & 3 & 4 & 5 \\ 3 & 5 & 4 & 1 & 2 \end{bmatrix}$$

has two orbits, whose cycles are

$$[1, 3, 4] \quad \text{and} \quad [2, 5].$$

Since $5 - 2 = 3$ is odd, it follows that σ is an odd permutation.

EXERCISES

1. Express the following permutations as products of transpositions, and determine their signs.

a) $\begin{bmatrix} 1 & 2 & 3 \\ 2 & 3 & 1 \end{bmatrix}$ b) $\begin{bmatrix} 1 & 2 & 3 \\ 3 & 1 & 2 \end{bmatrix}$ c) $\begin{bmatrix} 1 & 2 & 3 \\ 3 & 2 & 1 \end{bmatrix}$

d) $\begin{bmatrix} 1 & 2 & 3 & 4 \\ 2 & 3 & 4 & 1 \end{bmatrix}$ e) $\begin{bmatrix} 1 & 2 & 3 & 4 \\ 2 & 1 & 4 & 3 \end{bmatrix}$ f) $\begin{bmatrix} 1 & 2 & 3 & 4 \\ 3 & 2 & 4 & 1 \end{bmatrix}$

g) $\begin{bmatrix} 1 & 2 & 3 & 4 \\ 4 & 2 & 1 & 3 \end{bmatrix}$ h) $\begin{bmatrix} 1 & 2 & 3 & 4 \\ 3 & 1 & 4 & 2 \end{bmatrix}$ i) $\begin{bmatrix} 1 & 2 & 3 & 4 \\ 2 & 4 & 1 & 3 \end{bmatrix}$

2. Express the following permutations as products of transpositions and determine their signs.

a) $\begin{bmatrix} 1 & 2 & 3 & 4 & 5 \\ 2 & 4 & 5 & 1 & 3 \end{bmatrix}$ b) $\begin{bmatrix} 1 & 2 & 3 & 4 & 5 \\ 4 & 5 & 1 & 3 & 2 \end{bmatrix}$

c) $\begin{bmatrix} 1 & 2 & 3 & 4 & 5 \\ 3 & 5 & 2 & 1 & 4 \end{bmatrix}$ d) $\begin{bmatrix} 1 & 2 & 3 & 4 & 5 \\ 4 & 5 & 3 & 1 & 2 \end{bmatrix}$

e) $\begin{bmatrix} 1 & 2 & 3 & 4 & 5 \\ 2 & 1 & 5 & 3 & 4 \end{bmatrix}$ f) $\begin{bmatrix} 1 & 2 & 3 & 4 & 5 \\ 3 & 4 & 1 & 2 & 5 \end{bmatrix}$

g) $\begin{bmatrix} 1 & 2 & 3 & 4 & 5 \\ 2 & 5 & 4 & 3 & 1 \end{bmatrix}$ h) $\begin{bmatrix} 1 & 2 & 3 & 4 & 5 \\ 4 & 1 & 5 & 2 & 3 \end{bmatrix}$

i) $\begin{bmatrix} 1 & 2 & 3 & 4 & 5 \\ 5 & 3 & 1 & 2 & 4 \end{bmatrix}$

3. Express the following permutations as products of transpositions and determine their signs.

a) $\begin{bmatrix} 1 & 2 & 3 & 4 & 5 & 6 \\ 3 & 2 & 1 & 6 & 4 & 5 \end{bmatrix}$ b) $\begin{bmatrix} 1 & 2 & 3 & 4 & 5 & 6 \\ 2 & 6 & 1 & 5 & 3 & 4 \end{bmatrix}$

c) $\begin{bmatrix} 1 & 2 & 3 & 4 & 5 & 6 \\ 3 & 5 & 4 & 1 & 6 & 2 \end{bmatrix}$ d) $\begin{bmatrix} 1 & 2 & 3 & 4 & 5 & 6 \\ 4 & 3 & 6 & 1 & 2 & 5 \end{bmatrix}$

e) $\begin{bmatrix} 1 & 2 & 3 & 4 & 5 & 6 \\ 5 & 3 & 1 & 2 & 4 & 6 \end{bmatrix}$ f) $\begin{bmatrix} 1 & 2 & 3 & 4 & 5 & 6 \\ 2 & 5 & 4 & 3 & 6 & 1 \end{bmatrix}$

g) $\begin{bmatrix} 1 & 2 & 3 & 4 & 5 & 6 \\ 2 & 1 & 4 & 5 & 6 & 3 \end{bmatrix}$ h) $\begin{bmatrix} 1 & 2 & 3 & 4 & 5 & 6 \\ 3 & 4 & 5 & 1 & 2 & 6 \end{bmatrix}$

i) $\begin{bmatrix} 1 & 2 & 3 & 4 & 5 & 6 \\ 2 & 4 & 6 & 5 & 3 & 1 \end{bmatrix}$

4. In each one of the cases of Exercise 1, write the inverse of the permutation.

5. In each one of the cases of Exercise 2, write the inverse of the permutation.

6. In each one of the cases of Exercise 3, write the inverse of the permutation.

7. In each one of the cases of Exercise 1, write the orbit decomposition of the permutation.

8. In each one of the cases of Exercise 2, write the orbit decomposition of the permutation.

9. In each one of the cases of Exercise 3, write the orbit decomposition of the permutation.

10. Prove that the number of odd permutations of J_n for $n \geq 2$ is equal to the number of even permutations. [*Hint:* Let $\sigma_1, \ldots, \sigma_m$ be all the distinct even permutations. Let τ be a transposition. Prove that

$$\tau\sigma_1, \quad \ldots, \quad \tau\sigma_m$$

are all distinct, and constitute all the odd permutations.]

11. After you have read the section on induction, prove that the number of permutations of J_n is equal to $n!$. [*Hint:* By induction. It is true for $n = 1$. Assume it for n. Let τ_k $(k = 1, \ldots, n)$ be the transposition which interchanges $n + 1$ and k for $k = 1, \ldots, n$. Let S_n be the set of permutations of J_n. Show that the permutations

$$I\sigma, \tau_1\sigma, \ldots, \tau_n\sigma$$

for σ in S_n give all distinct permutations of J_{n+1}. Hence the number of elements in S_{n+1} is equal to $n + 1$ times the number of elements in S_n, namely $(n + 1)n! = (n + 1)!$.]

12. In Exercises 1, 2, and 3, find the sign of the permutation by the orbit method of the last two examples of this section, and Theorem 4. Which do you think is the faster way of determining the sign of the permutation? This way, or the old way, expressing σ as explicitly as a product of transpositions?

15 Complex Numbers

§1. THE COMPLEX PLANE

The set of complex numbers is a set whose elements can be added and multiplied, so that the sum of complex numbers is a complex number, the product of complex numbers is a complex number, and so that addition and multiplication satisfy the following properties:

Addition is commutative and associative.

Multiplication is commutative, associative, and distributive with respect to addition.

Every real number is a complex number, and if a, b are real numbers, then their sum and product as complex numbers are the same as their sum and product as real numbers, respectively.

If 1 is the real number one, then $1z = z$ for every complex number z. Similarly, $0z = 0$.

Each complex number z has an additive inverse, namely $(-1)z$, so that

$$z + (-1)z = 0.$$

There exists a complex number i such that $i^2 = -1$.

Every complex number can be written in the form

$$a + bi,$$

where a, b are real numbers.

Thus in the complex numbers we can take a square root of -1. Remember the arguments we gave when we discussed square roots in the real numbers. These arguments apply here, to show that

$$i, \quad -i$$

are the only two complex numbers whose square is -1.

Just as we represented real numbers on a line, we can represent complex numbers in the plane. Namely, if $z = a + bi$ is a complex number, we view z as the point (a, b) in the plane. Thus i is represented by the point $(0, 1)$ as shown in Fig. 15–1.

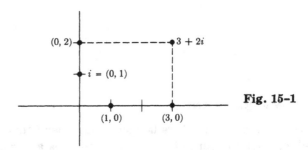

Fig. 15–1

We could, in fact, define a complex number to be a pair of real numbers (a, b), and define i to be the pair $(0, 1)$. Then we could define addition of complex numbers just as we defined addition of points in the plane, componentwise. Thus if

$$z = (x, y) \qquad \text{and} \qquad w = (u, v)$$

with x, y, u, v real, then

$$z + w = (x + u, y + v).$$

With this definition, we identify a real number x with the point $(x, 0)$ in the plane. If c is a real number, we keep our old definition of multiplication, and define

$$cz = (cx, cy).$$

Then we see that every complex number z as above can be written as

$$x + iy.$$

We would still have to define multiplication of complex numbers. If we assume that complex numbers exist satisfying our list of properties, then the product of two complex numbers

$$z = x + iy \qquad \text{and} \qquad w = u + iv$$

is given by

$$
\begin{aligned}
(x + iy)(u + iv) &= x(u + iv) + iy(u + iv) \\
&= xu + ixv + iyu + i^2yv \\
&= xu - yv + (xv + yu)i.
\end{aligned}
$$

If we want to define the multiplication of points

$$(x, y)(u, v)$$

to satisfy the properties listed above, then we must take the definition that this product is equal to the point

$$(xu - yv, xv + yu).$$

We could then verify by brute force that all the properties listed above are valid, having defined multiplication this way. This would be boring, so we omit it, and just assume all the properties.

From now on, we assume that complex numbers exist, and can be represented as points in the plane. We identify a real number a with the point $(a, 0)$. For this reason, the plane is sometimes called the **complex plane**.

In multiplying complex numbers, we use the rule $i^2 = -1$ to simplify a product and to put it in the form $a + bi$.

Example. Let

$$z = 2 + 3i \quad \text{and} \quad w = 1 - i.$$

Then

$$zw = (2 + 3i)(1 - i) = 2(1 - i) + 3i(1 - i)$$
$$= 2 - 2i + 3i - 3i^2$$
$$= 2 + 3 + i$$
$$= 5 + i.$$

Let $z = a + bi$ be a complex number. We define its **complex conjugate** (or simply its **conjugate**) to be the complex number

$$\bar{z} = a - bi.$$

Thus if $z = 2 + 3i$, then $\bar{z} = 2 - 3i$. We see at once that

$$z\bar{z} = a^2 + b^2.$$

Hence viewing z as a point in the complex plane, we see that $z\bar{z}$ is the square of the distance of this point from the origin $(0, 0)$.

Let $z = a + bi$ be a complex number $\neq 0$. Let

$$w = \frac{\bar{z}}{a^2 + b^2}.$$

Then, using the rules for multiplication, we obtain

$$zw = wz = 1,$$

because

$$z\frac{\bar{z}}{a^2 + b^2} = \frac{z\bar{z}}{a^2 + b^2} = 1.$$

This number w is called the **(multiplicative) inverse** of z, and is denoted by z^{-1}, or $1/z$. The proof given before that the multiplicative inverse is unique applies now. In other words, there exists one and only one complex number w such that $wz = 1$.

If z, w are complex numbers and $z \neq 0$, then we write w/z instead of $z^{-1}w$, just as we did with real numbers.

Example. To find the inverse of $1 + i$, we note that the conjugate of $1 + i$ is $1 - i$, and that

$$(1 + i)(1 - i) = 2.$$

Hence

$$(1 + i)^{-1} = \frac{1 - i}{2}.$$

Theorem 1. *Let z, w be complex numbers. Then*

$$\overline{zw} = \bar{z}\,\bar{w}, \qquad \overline{z + w} = \bar{z} + \bar{w}, \qquad \bar{\bar{z}} = z.$$

Proof. The proofs follow immediately from the definitions of addition, multiplication, and the complex conjugate. We leave them to you as exercises.

Let $z = x + yi$ be a complex number, with real x, y. We shall call x the **real part** of z, and y the **imaginary part** of z. These are denoted by $\text{Re}(z)$ and $\text{Im}(z)$, respectively.

Thus if $z = x + iy$, then

$$z + \bar{z} = 2x = 2\,\text{Re}(z).$$

Similarly,

$$z - \bar{z} = 2y = 2\,\text{Im}(z).$$

We define the **absolute value** of a complex number $z = x + iy$ to be

$$|z| = \sqrt{x^2 + y^2}.$$

Thus the absolute value of z is simply the distance from the origin to the point (x, y) in the plane. In terms of the absolute value, we can write

$$z^{-1} = \frac{\bar{z}}{|z|^2},$$

provided $z \neq 0$. Indeed, we observe that

$$z\bar{z} = |z|^2.$$

Theorem 2. *The absolute value of complex numbers satisfies the following properties. If z, w are complex numbers, then*

$$|zw| = |z|\,|w|$$
$$|z + w| \leq |z| + |w|.$$

Proof. We have:

$$|zw|^2 = zw\overline{zw} = z\bar{z}w\bar{w} = |z|^2|w|^2.$$

Taking the square root we conclude that $|zw| = |z|\,|w|$, thus proving the first assertion. As for the second, it is simply the triangle inequality. A simple algebraic proof can be given, but we omit it. Draw the picture of the points z, w, $z + w$ in the plane to see the parallelogram and the triangle.

EXERCISES

1. Express the following complex numbers in the form $x + iy$ where x and y are real numbers.

    a) $(-1 + 3i)^{-1}$ b) $(1 + i)(1 - i)$

    c) $(1 + i)(2 - i)$ d) $(i - 1)(2 - i)$

    e) $(7 + \pi i)(\pi + i)$ f) $(2i + 1)\pi i$

    g) $(\sqrt{2}\,i)(\pi + 3i)$ h) $(i + 1)(i - 2)(i + 3)$

2. Express the following complex numbers in the form $x + iy$, where x, y are real numbers.

    a) $(1 + i)^{-1}$ b) $\dfrac{1}{3 + i}$ c) $\dfrac{2 + i}{2 - i}$ d) $\dfrac{1}{2 - i}$

    e) $\dfrac{1 + i}{i}$ f) $\dfrac{i}{1 + i}$ g) $\dfrac{2i}{3 - i}$ h) $\dfrac{1}{-1 + i}$

3. Let z be a complex number $\neq 0$. What is the absolute value of $z/\bar{z}$?

4. Prove the statements of Theorem 1.

5. Show that for any complex number $z = x + iy$, with x, y real, we have

$$\operatorname{Im}(z) \leq |\operatorname{Im}(z)| \leq |z|.$$

§2. POLAR FORM

Let $z = x + iy$ be a complex number, which we view as a point (x, y) in the plane. Let $r = |z|$, Then

$$\frac{z}{|z|} = \frac{1}{r} z$$

has absolute value 1, and consequently can be viewed as a point on the circle of radius 1 centered at the origin. Hence there exists some real number θ such that

$$\frac{z}{r} = (\cos \theta, \sin \theta).$$

This shows that we can write

$$z = r(\cos \theta + i \sin \theta).$$

Every complex number can be written as a product of a real number ≥ 0 and a complex number of absolute value 1.

Example. Let $z = 1 + i\sqrt{3}$. Then $|z| = 2$. Hence

$$z = 2 (\cos \theta, \sin \theta).$$

Note that $\cos \theta = \frac{1}{2}$ and $\sin \theta = \sqrt{3}/2$. Hence $\theta = \pi/3$. We may say that the polar coordinates of z are $(2, \pi/3)$.

We define the symbols $e^{i\theta}$ by

$$e^{i\theta} = \cos \theta + i \sin \theta.$$

Then we have

$$z = x + iy = re^{i\theta}.$$

This is called the **polar form** of z, and is illustrated in Fig. 15–2. The notation $e^{i\theta}$ would be terrible if the next theorem were not true. But since the next theorem is true, the notation is quite good.

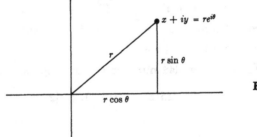

Fig. 15–2

Theorem 3. *Let θ, φ be real numbers. Then*

$$e^{i\theta}e^{i\varphi} = e^{i\theta+i\varphi}.$$

Proof. We shall see that this amounts to the addition formula for the sine and cosine. By definition, we have

$$e^{i\theta+i\varphi} = \cos(\theta + \varphi) + i\sin(\theta + \varphi).$$

Using the addition formula for the sine and cosine, we see that the preceding expression is equal to

$$\cos\theta\cos\varphi - \sin\theta\sin\varphi + i(\sin\theta\cos\varphi + \sin\varphi\cos\theta).$$

This is exactly the same complex number that we get by multiplying

$$(\cos\theta + i\sin\theta)(\cos\varphi + i\sin\varphi).$$

This proves Theorem 3.

Thus exponentiation with a complex number of the form $i\theta$ obeys the same basic rule as ordinary exponentiation.

Let z and w be complex numbers. Let us write them in polar form; namely

$$z = re^{i\theta} \qquad \text{and} \qquad w = se^{i\varphi},$$

where r, s, θ, φ are real numbers. Then their product is equal to

$$zw = re^{i\theta}se^{i\varphi} = rse^{i(\theta+\varphi)}.$$

Thus, to multiply complex numbers, we may say roughly that we multiply their absolute values and add their angles.

Example. Find a complex number whose square is $4e^{i\pi/2}$.

Let $z = 2e^{i\pi/4}$. Then, using Theorem 3, we find that

$$z^2 = 4e^{i\pi/2}.$$

Example. Let n be a positive integer. Find a complex number w such that $w^n = e^{i\pi/2}$.

It is clear that the complex number $w = e^{i\pi/2n}$ satisfies our requirements.

Example. We have

$$\boxed{e^{i\pi} = -1 \qquad \text{and} \qquad e^{2\pi i} = 1.}$$

This follows at once from the definition and the known values of the sine and cosine.

EXERCISES

1. Put the following complex numbers in polar form.

   a) $1 + i$ b) $\sqrt{3} + i$ c) -3 d) $4i$

   e) $1 - i\sqrt{3}$ f) $-5i$ g) -7 h) $-1 - i$

2. Put the following complex numbers in the ordinary form $x + iy$. Also plot them as points in the plane.

   a) $e^{3\pi i}$ b) $e^{2\pi i/3}$ c) $3e^{i\pi/4}$ d) $e^{-i\pi/3}$

   e) $e^{2\pi i/6}$ f) $e^{-i\pi/2}$ g) $e^{-i\pi}$ h) $e^{-5i\pi/4}$

3. Let z be a complex number $\neq 0$. Show that there are precisely two distinct complex numbers whose square is z.

4. Let z be a complex number $\neq 0$. Let n be a positive integer. Show that there are n distinct complex numbers w such that $w^n = z$. Write these complex numbers in polar form. The proof given that a polynomial of degree $\leq n$ has at most n roots applies to the complex case, and thus we see that there are no other complex numbers w such that $w^n = z$ other than those you have presumably written down.

5. Write in polar form the n complex numbers w such that $w^n = 1$. Plot all of these as points in the plane for $n = 2, 3, 4, 5$.

6. If θ is real, show that

$$\cos \theta = \frac{e^{i\theta} + e^{-i\theta}}{2} \qquad \text{and} \qquad \sin \theta = \frac{e^{i\theta} - e^{-i\theta}}{2i}.$$

16 Induction and Summations

§1. INDUCTION

Induction is an axiom which allows us to give proofs that certain properties are true for all integers.

Suppose that we wish to prove a certain assertion concerning positive integers n. Let $A(n)$ denote the assertion concerning the integer n. To prove it for all n, it suffices to prove the following.

IND 1. *The assertion $A(1)$ is true (i.e. the assertion concerning the integer 1 is true).*

IND 2. *Assuming the assertion proved for all positive integers $\leq n$, prove it for $n + 1$, i.e. prove that $A(n + 1)$ is true.*

The combination of these two steps is known as induction. For instance, **IND 1** gives us a starting point, and **IND 2** allows us to prove $A(2)$ from $A(1)$, then $A(3)$ from $A(2)$, and so forth, proceeding stepwise.

We shall now give examples.

Example. For all integers $n \geq 1$, we have

$$1 + 2 + 3 + \cdots + n = \frac{n(n + 1)}{2}.$$

Proof. By induction. The assertion $A(n)$ is the assertion of the theorem. When $n = 1$, it simply states that

$$1 = \frac{1(1 + 1)}{2},$$

which is clearly true.

Assume now that the assertion is true for the integer n. Then

$$1 + 2 + \cdots + n + (n + 1) = \frac{n(n + 1)}{2} + (n + 1).$$

Putting the expression on the right of the equality sign over a common denominator 2, we see that it is equal to

$$\frac{n^2 + n + 2n + 2}{2} = \frac{(n + 1)(n + 2)}{2}.$$

Hence assuming $A(n)$, we have proved that

$$1 + 2 + \cdots + (n + 1) = \frac{(n + 1)(n + 2)}{2},$$

which is none other than assertion $A(n + 1)$. This proves our result.

Notational remark. We have just written the sum of the first n positive integers above using three dots to denote intermediate integers. There is a notation which avoids the use of such dots, using a capital Greek sigma sign ($\sum$ for sum). Thus we write

$$\sum_{k=1}^{n} k = 1 + 2 + \cdots + n.$$

Similarly, we write

$$\sum_{k=1}^{n} k^2 = 1^2 + 2^2 + \cdots + n^2,$$

or also

$$\sum_{k=1}^{n} \sin k = \sin 1 + \sin 2 + \cdots + \sin n.$$

If f is any function, we would write

$$\sum_{k=1}^{n} f(k) = f(1) + f(2) + \cdots + f(n).$$

We can write the distributivity with respect to a sum of n terms with the sigma notation as follows. For any number c, we have

$$c \sum_{k=1}^{n} f(k) = \sum_{k=1}^{n} cf(k).$$

For instance,

$$5 \sum_{k=1}^{n} k^3 = \sum_{k=1}^{n} 5k^3.$$

Expanding with use of dots, we can also write this as

$$5(1^3 + 2^3 + \cdots + n^3) = 5 \cdot 1^3 + 5 \cdot 2^3 + \cdots + 5 \cdot n^3.$$

Example. Let f be a function defined for all real numbers such that

$$f(x + y) = f(x)f(y)$$

for all numbers x, y. Let $f(1) = a$. We want to prove by induction that $f(n) = a^n$ for all positive integers n. This is true for $n = 1$. Assume the result for an integer n. Then

$$f(n + 1) = f(n)f(1) = a^n a = a^{n+1},$$

thereby proving our assertion.

Example. We want a simple formula for the number of ways of selecting k objects out of a set of n objects. This number is denoted by C_k^n and is called a **binomial coefficient.** The reason for this name is simple. Suppose that we wish to expand the product

$$(x + y)^n = (x + y)(x + y) \cdots (x + y)$$

as a sum of terms involving powers of x and y. In this expansion, we select x from k factors, and we select y from $n - k$ factors. We then take the sum over all possible such selections. Thus we find that

$$(x + y)^n = \sum_{k=0}^{n} C_k^n x^k y^{n-k}.$$

It turns out that C_k^n has a simple expression. Let $n!$ denote the product of the first n integers, so that

$$1! = 1,$$
$$2! = 1 \cdot 2 = 2,$$
$$3! = 1 \cdot 2 \cdot 3 = 6,$$
$$4! = 1 \cdot 2 \cdot 3 \cdot 4 = 24,$$
$$5! = 1 \cdot 2 \cdot 3 \cdot 4 \cdot 5 = 120,$$

and so on. By convention, we *define* $0! = 1$. Then the value for C_k^n is given by

$$C_k^n = \frac{n!}{k!(n - k)!}.$$

For instance,

$$C_2^4 = \frac{4!}{2!(4 - 2)!} = \frac{24}{2 \cdot 2} = 6.$$

It is very simple to prove by induction that C_k^n has the value described above, and we shall let you have fun with that, in Exercise 9. Instead of the notation

C_k^n, one also uses the notation

$$\binom{n}{k}$$

for the binomial coefficient.

As a matter of convenience, if n is an integer ≥ 0 and k is an integer such that $k < 0$ or $k > n$, then we **define**

$$\binom{n}{k} = 0.$$

EXERCISES

1. Prove that, for all integers $n \geq 1$, we have

$$1 + 3 + 5 + \cdots + (2n - 1) = n^2.$$

2. Prove that, for all integers $n \geq 1$, we have

   a) $1^2 + 2^2 + \cdots + n^2 = \dfrac{1}{6}n(n + 1)(2n + 1)$,

   b) $1^3 + 2^3 + 3^3 + \cdots + n^3 = \left[\dfrac{n(n + 1)}{2}\right]^2.$

3. Prove that

$$1^2 + 3^2 + 5^2 + \cdots + (2n - 1)^2 = \frac{1}{3}(4n^3 - n).$$

4. Prove that $n(n^2 + 5)$ is divisible by 6 for all integers $n \geq 1$.

5. Prove that, for $x \neq 1$, we have

$$(1 + x)(1 + x^2)(1 + x^4) \cdots (1 + x^{2^n}) = \frac{1 - x^{2^{n+1}}}{1 - x}.$$

6. Let f be a function defined for all real numbers such that

$$f(xy) = f(x) + f(y)$$

   for all real numbers x, y. Show that

$$f(x^n) = nf(x)$$

   for all x.

7. Let f be a function defined for all numbers such that

$$f(xy) = f(x)f(y)$$

for all real numbers x, y. Show that $f(x^n) = f(x)^n$ for all positive integers n and all real numbers x.

8. Using Exercises 1, 2, and 3, write out simple expressions giving the values for the following sums.

a) $\displaystyle\sum_{k=1}^{2n} k^2$

b) $\displaystyle\sum_{k=1}^{2n} k^3$

c) $\displaystyle\sum_{k=1}^{2n} (2k - 1)$

d) $\displaystyle\sum_{k=1}^{m-1} k^2$

e) $\displaystyle\sum_{k=1}^{m-1} k^3$

f) $\displaystyle\sum_{k=1}^{m} (2k - 1)$

9. **Binomial coefficients.** Let

$$\binom{n}{k} = \frac{n!}{k!\,(n - k)!},$$

where n, k are integers ≥ 0, $0 \leq k \leq n$, and $0!$ is defined to be 1. Prove the following assertions.

a) $\displaystyle\binom{n}{k} = \binom{n}{n - k}$

b) $\displaystyle\binom{n}{k - 1} + \binom{n}{k} = \binom{n + 1}{k}$ (for $k > 0$)

c) Prove by induction that for all numbers x, y, we have

$$(x + y)^n = \sum_{k=0}^{n} \binom{n}{k} x^k y^{n-k}.$$

10. Write down the numerical values for the following binomial coefficients:

$$\binom{3}{1}, \binom{3}{2}, \binom{3}{3},$$

$$\binom{4}{0}, \binom{4}{1}, \binom{4}{2}, \binom{4}{3}, \binom{4}{4},$$

$$\binom{5}{0}, \binom{5}{1}, \binom{5}{2}, \binom{5}{3}, \binom{5}{4}, \binom{5}{5}.$$

11. Write out in full the expansions of:

a) $(x + y)^3$ b) $(x + y)^4$ c) $(x + y)^5$

Observe that you might already have done these in Chapter 1, §3.

12. Theorem. *All billiard balls have the same color.*

Proof: By induction, on the number n of billiard balls. Our theorem is certainly true for $n = 1$, i.e. for one billiard ball. Assume it for n billiard balls. We prove it for $n + 1$. Look at the first n billiard balls among those $n + 1$. By induction, they have the same color. Now look at the last n among those $n + 1$. They have the same color. Hence all $n + 1$ have the same color (Fig. 16–1).

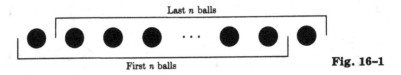

Last n balls

First n balls

Fig. 16–1

What is wrong with this "proof"?

13. Let E be a set with n elements, and let F be a set with m elements. Show that the total number of mappings from E to F is m^n. [*Hint:* Use induction on n. First take $n = 1$. How many ways are there of mapping a single element into a set with m elements? Then assume the result for n, and prove it for $n + 1$, taking into account the number of possible ways of mapping the $(n + 1)$-th element of E into F.]

§2. SUMMATIONS

We have already met simple cases of summations in the preceding section, when we computed the values of certain sums. We shall give here other applications of such summations.

We go back to the old problem of computing volumes, similar to the problem of computing areas and lengths discussed in Chapter 7. We shall treat a typical example, and let you work out others along similar lines. Our purpose is to compute the volume of a cone, of a ball, and more generally of a solid obtained by revolving a curve around an axis, which we take to be the x-axis.

We work out the volume of a cone in detail. In fact, we consider a special cone as shown on Fig. 16–2.

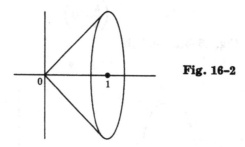

Fig. 16–2

We suppose that the height of the cone is equal to 1, and that the radius of the base is equal to 1 also. Then the cone is obtained by revolving the line $y = x$ around the x-axis, between the values $x = 0$ and $x = 1$.

We wish to prove that the volume of this cone is equal to $\pi/3$. We use the method discovered by Archimedes. We first observe that the volume of a cylinder whose base has radius r and of height h as shown in Fig. 16–3 is $\pi r^2 h$.

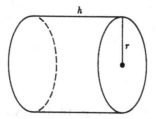

Fig. 16–3

Using this, we approximate the cone by cylinders as follows. Cut up the interval on the x-axis between 0 and 1 into n segments of equal length $1/n$. The end points of these segments are the rational numbers

$$0, \ \frac{1}{n}, \ \frac{2}{n}, \ \ldots, \ \frac{n-1}{n}, \ \frac{n}{n} = 1.$$

Using general notation, we can say that the small segments are the segments

$$\left[\frac{k}{n}, \frac{k+1}{n}\right],$$

for $k = 0, \ldots, n$. We use the notation $[a, b]$ to denote the set of all numbers x such that $a \leq x \leq b$.

We draw the cylinder whose height is $1/n$ and whose base is the disc of radius k/n centered at the point

$$\left(\frac{k}{n}, 0\right)$$

as shown in Fig. 16–4(a) and (b).

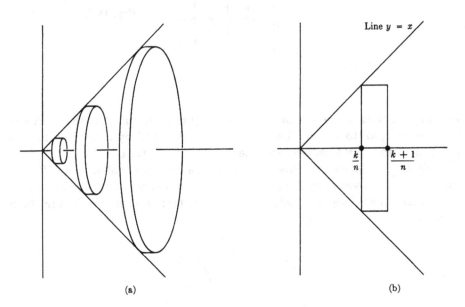

(a)   (b)

Fig. 16–4

Fig. 16–4(b) is a cross section of Fig. 16–4(a), viewed from the side. The volume of each small cylinder is equal to

(1)
$$\pi \frac{1}{n}\left(\frac{k}{n}\right)^2.$$

Hence the sum of the volumes of the small cylinders is equal to

(2)
$$\sum_{k=0}^{n} \pi \frac{1}{n}\left(\frac{k}{n}\right)^2 = \frac{\pi}{n^3}\sum_{k=0}^{n} k^2.$$

Of course, in the present case, the term in this sum corresponding to $k = 0$ is equal to 0, so our sum is really equal to the sum

(3)
$$\frac{\pi}{n^3}\sum_{k=1}^{n} k^2.$$

If we use Exercise 2 of §1, then we get a simple expression for the value of the sum, namely

(4)
$$\frac{\pi}{n^3} \frac{1}{6} n(n + 1)(2n + 1),$$

which is equal to

(5)
$$\frac{\pi}{6} 1 \left(1 + \frac{1}{n}\right)\left(2 + \frac{1}{n}\right).$$

You see this by dividing each expression n, $n + 1$, $2n + 1$ by n, thus using up the n^3 in the denominator.

Now as we let n get arbitrarily large, we see that the expression of (5) approaches the volume of the cone, and also approaches

$$\frac{\pi}{6} \cdot 2 = \frac{\pi}{3}.$$

This achieves what we wanted.

To get the volume of an arbitrary cone, we can proceed in two ways. Suppose that the cone has a base whose radius is r, and let the height be h. Thus we draw the cone as in Fig. 16–5.

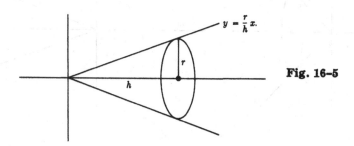

Fig. 16–5

We see that the cone is obtained by rotating the line

$$y = \frac{r}{h} x$$

around the x-axis, between the values $x = 0$ and $x = h$. We could either follow the same method we used above, decomposing the segment $[0, h]$ into n segments of length h/n, forming the cylinders as before, taking the sum, and then taking the limit. We leave this as Exercise 1. We would then get:

The volume of a cone whose base has radius r and whose height is h is equal to $\frac{1}{3}\pi r^2 h$.

But we can also proceed using mixed dilations, following the ideas of the exercises of Chapter 7, §1. Let us denote by $C(r, h)$ the cone whose base has radius r and whose height is h. It should be intuitively clear (and we shall give a coordinate argument below) that $C(r, h)$ is obtained by a mixed dilation of the cone $C(1, 1)$ whose volume was obtained above. Namely, a point in 3-space is described by three coordinates (x, y, z). If a, b, c are three positive numbers, we let

$$F_{a,b,c} : \mathbf{R}^3 \to \mathbf{R}^3$$

be the map such that

$$F_{a,b,c}(x, y, z) = (ax, by, cz).$$

Then the cone $C(r, h)$ is the image of the cone $C(1, 1)$ under the mixed dilation

$$F_{h,r,r}.$$

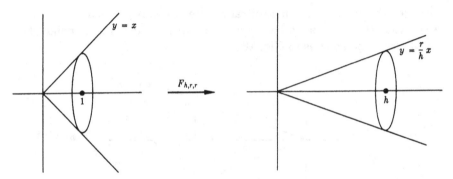

Fig. 16–6

Referring back to Chapter 7, suppose that we have a mixed dilation $F_{a,b,c}$ as above. If R is a rectangular solid whose sides have lengths t, u, v, then the dilation of R by $F_{a,b,c}$ is a rectangular solid whose sides have lengths at, bu, cv. The volume of R is tuv. and the volume of the dilated solid $F_{a,b,c}(R)$ is $atbucv = abctuv$. In other words, the volume gets multiplied by abc under the mixed dilation. If S is an arbitrary solid in 3-space, and if we approximate S by rectangular solids, then we conclude that a similar result holds for S. In other words, if V is the volume of S, then

$$\text{volume of } F_{a,b,c}(S) = abcV.$$

In the application to the cone, we have $a = h$, $b = r$, $c = r$, so that

$$abc = r^2 h.$$

We can then conclude that the volume of $C(r, h)$ is equal to hr^2 times the volume of $C(1, 1)$. Thus we obtain the value

$$\text{Volume of } C(r, h) = \frac{1}{3} \pi r^2 h.$$

Let us now look more precisely at the reason why $C(r, h)$ is the dilation of the cone $C(1, 1)$ by the mixed dilation $F_{h,r,r}$. First we must express the cone in terms of coordinates. Let (x, y, z) be the coordinates of a point in 3-space. If this point lies in the cone $C(r, h)$, then

$$0 \leq x \leq h.$$

For each such value of x, a point (x, y, z) lies in the cone if and only if

$$y^2 + z^2 \leq \left(\frac{r}{h} x\right)^2.$$

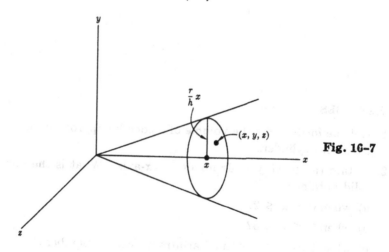

Fig. 16-7

Thus the points on the cone $C(r, h)$ can be described as those points (x, y, z) such that

$$0 \leq x \leq h,$$

(∗)
$$\left(\frac{y}{r}\right)^2 + \left(\frac{z}{r}\right)^2 \leq \left(\frac{x}{h}\right)^2.$$

Let

$$x' = \frac{x}{h}, \qquad y' = \frac{y}{r}, \qquad z' = \frac{z}{r}.$$

If (x, y, z) satisfy (∗), i.e. lie in $C(r, h)$, then (x', y', z') satisfy

(∗∗) $0 \le x'^2 \le 1,$

$$y'^2 + z'^2 \le x'^2.$$

Hence (x', y', z') is a point in the cone $C(1, 1)$. Conversely, it is also clear that if (x', y', z') is a point in $C(1, 1)$, then (x, y, z) is a point of $C(r, h)$. Hence $C(r, h)$ is the image of $C(1, 1)$ under the mixed dilation

$$F_{h,r,r}.$$

This concludes our argument.

EXERCISES

1. Get the formula for the volume of a cone by approximating an arbitrary cone with cylinders.

2. Rotate the curve $y = 3x$ about the x-axis. What is the volume of the solid obtained

   a) when $0 \le x \le 2$?

   b) when $0 \le x \le 5$?

   c) when $0 \le x \le c$ with an arbitrary positive number c?

3. Rotate the curve $y = \sqrt{x}$ about the x-axis. What is the volume of the solid obtained when $0 \le x \le h$ and h has the value:

   a) $h = 1$?   b) $h = 2$?   c) $h = 3$?   d) arbitrary h?

4. Rotate the curve $y = \sqrt{r^2 - x^2}$ about the x-axis. What is the volume of the solid obtained when $0 \le x \le r$ and r has the value:

   a) $r = 1$,   b) $r = 2$,

   c) $r = 3$,   d) r is arbitrary?

5. Look again at Exercise 4. What is the solid obtained? You should now be able to see that the volume of a spherical ball of radius r is equal to

$$\frac{4}{3} \pi r^3.$$

6. Find the area between the curve $y = x^2$ and the x-axis, from the origin to the following values of x:

a) $x = 1$,   b) $x = 2$,   c) $x = 3$,   d) $x = 4$.

Use a sum of areas of small rectangles as indicated in Fig. 16–8.

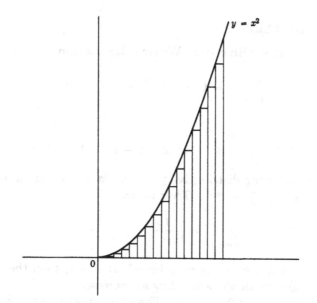

Fig. 16–8

7. Find the area between the curve $y = x^3$ and the x-axis between the origin and the following values of x:

a) $x = 1$,   b) $x = 2$,   c) $x = 3$,   d) $x = 4$.

8. Let S be the region in the plane consisting of all points (x, y) such that

$$0 \le x \le 1 \quad \text{and} \quad 0 \le y \le x^2.$$

Let T be the region in the plane consisting of all points (x, y) such that

$$0 \le x \le 1 \quad \text{and} \quad 0 \le y \le 3x^2.$$

Express T as the image of S under a mixed dilation. What is the area of T?

9. Let c be a number > 0. What is the area of the region lying between the curve $y = cx^2$ and the x-axis, from the origin to the following values of x:

a) $x = 1$,   b) $x = 2$,   c) $x = 3$.

§3. GEOMETRIC SERIES

Let c be a number with $c \neq 1$. We consider the sum

$$\sum_{k=0}^{n} c^k = 1 + c + c^2 + \cdots + c^n.$$

Observe that

$$(1 + c + c^2 + \cdots + c^n)(1 - c) = 1 - c^{n+1}.$$

This is immediate, using distributivity. There is a cancellation of all terms in the product, except $1 - c^{n+1}$. Thus we have

$$\sum_{k=0}^{n} c^k = \frac{1}{1 - c} - \frac{c^{n+1}}{1 - c}.$$

What happens when n becomes very large? If $c > 1$, then the power c^{n+1} becomes large, and we don't regard this as interesting.

Suppose however that $0 \leq c < 1$. Then c^{n+1} approaches 0. For instance, suppose that $c = \frac{1}{2}$. Then

$$c^2 = \frac{1}{4}, \qquad c^3 = \frac{1}{8}, \qquad c^4 = \frac{1}{16}, \qquad \cdots \cdots$$

We see that the denominator is an increasing power of 2, and that the fraction for c^{n+1} thus approaches 0. Hence

$$\frac{c^{n+1}}{1 - c}$$

approaches 0 as n becomes arbitrarily large. Thus we may say that the sum

$$1 + c + c^2 + \cdots + c^n$$

approaches

$$\frac{1}{1 - c}$$

as n becomes arbitrarily large. It is convenient to abbreviate this by the symbols

$$\sum_{k=1}^{\infty} c^k = \frac{1}{1 - c}.$$

The symbol ∞ is called **"infinity"**, and is explained only in terms of the context we have just introduced. Also, instead of saying "as n becomes arbitrarily large", we also say "as n approaches infinity".

Warning. There is no number called "infinity". The use of the word infinity is meaningful only as an abbreviation in the context just described.

Example. Let $c = \frac{1}{5}$. Then

$$1 + \frac{1}{5} + \left(\frac{1}{5}\right)^2 + \left(\frac{1}{5}\right)^3 + \cdots = \frac{1}{1 - \frac{1}{5}} = \frac{5}{4}.$$

Here, we write three dots . . . instead of using the summation sign.

The symbols

$$\sum_{k=1}^{\infty} c^k$$

are also called the **geometric series**. This symbol has a numerical value when $0 < c < 1$, and this value is then $1/(1 - c)$ as above.

EXERCISES

1. Find the value of the geometric series for the following values of c.

a) $\frac{1}{3}$ b) $\frac{1}{4}$ c) $\frac{1}{5}$ d) $\frac{1}{6}$

e) $\frac{3}{4}$ f) $\frac{2}{3}$ g) $\frac{2}{5}$ h) $\frac{3}{5}$

2. Argue in a manner similar to that of the text to give a value to the geometric series when $-1 < c \leq 0$. Give the general formula, and also

give the specific numerical values of the geometric series for the following numbers c.

a) $-\dfrac{1}{3}$ b) $-\dfrac{1}{4}$ c) $-\dfrac{1}{5}$ d) $-\dfrac{1}{6}$

e) $-\dfrac{3}{4}$ f) $-\dfrac{2}{3}$ g) $-\dfrac{2}{5}$ h) $-\dfrac{3}{5}$

3. Let c be a complex number such that $0 \le |c| < 1$. Again argue in a similar way to give a value to the geometric series.

4. What is the value of the sum

$$\sum_{k=0}^{n} c^k,$$

when c is equal to $re^{i\theta}$ and $0 \le r < 1$? Express your answer in the same form as that used to discuss the geometric series. Similarly, give the value of the series

$$\sum_{k=0}^{\infty} c^k,$$

when $c = re^{i\theta}$ and $0 \le r < 1$.

5. Let $c = e^{2\pi i/n}$ for some positive integer n. What is the value of

$$1 + c + c^2 + \cdots + c^{n-1}?$$

6. Let c be a complex number $\ne 1$, such that $c^n = 1$ for some positive integer n. What is the value of

$$1 + c + c^2 + \cdots + c^{n-1}?$$

7. Consider the sums

$$\sum_{k=1}^{n} \frac{1}{k} = 1 + \frac{1}{2} + \frac{1}{3} + \cdots + \frac{1}{n}.$$

Starting with $\frac{1}{3}$, group the terms of this sum by taking the first two, then the next four, then the next eight, and so on, by groups of 2^d where $d = 1, 2, \ldots$. In each such grouping, replace each term by the fraction furthest to the right, i.e. replace 1 by $\frac{1}{2}$, then $\frac{1}{3}$ by $\frac{1}{4}$, then $\frac{1}{5}, \frac{1}{6}, \frac{1}{7}$ by $\frac{1}{8}$, then the next group of fractions by $\frac{1}{16}$, and so on. In this way, show that these sums can be made to have arbitrarily large values, for sufficiently large n.

8. Consider the sums

$$\sum_{k=1}^{n} \frac{1}{k^2} = 1 + \frac{1}{2^2} + \frac{1}{3^2} + \cdots + \frac{1}{n^2}.$$

Group the terms differently, putting together $\dfrac{1}{2^2}$ and $\dfrac{1}{3^3}$; then $\dfrac{1}{4^2}$ up to $\dfrac{1}{7^2}$, then $\dfrac{1}{8^2}$ up to $\dfrac{1}{15^2}$, and so on. Prove that these sums are ≤ 2, no matter what value n has.

9. A ball is thrown to a height of 6 ft and falls back down. Each time it rebounds to a height equal to four-fifths of the preceding height. What distance will the ball have traveled after it touches the ground for the

   a) 3rd time?   b) 5th time?   c) 20th time?

   d) Assuming that the ball goes on bouncing forever according to the above prescription, what distance does it travel?

10. A ball falls from a height of 10 ft and rebounds each time to a height equal to one-fifth of its previous height. What distance will the ball have traveled when it touches the ground for the

   a) 5th time?   b) 20th time?

   c) Assuming that the ball goes on bouncing forever according to the above prescription, what distance will it travel?

17 Determinants

§1. MATRICES

An array of four numbers

$$A = \begin{pmatrix} a & b \\ c & d \end{pmatrix}$$

is called a 2 × 2 **matrix** (read "**two by two matrix**"). For instance

$$\begin{pmatrix} 3 & -5 \\ 2 & 7 \end{pmatrix}$$

is a 2 × 2 matrix. In a sense, a 2 × 2 matrix can be viewed as a generalization of our "pairs", and this can be done in two ways: by looking at the rows or at the columns of the matrix. In our example, we have two rows,

$$(3, -5) \quad \text{and} \quad (2, 7),$$

as well as two columns,

$$\begin{pmatrix} 3 \\ 2 \end{pmatrix} \quad \text{and} \quad \begin{pmatrix} -5 \\ 7 \end{pmatrix}.$$

These are called the first and second columns, respectively.

For this chapter, we call a pair of numbers, written either vertically or horizontally, a **vector**. If it is written vertically, we call it a **column vector**. If it is written horizontally, we call it a **row vector**.

Example. The row vectors of the matrix

$$\begin{pmatrix} -6 & 8 \\ 5 & -3 \end{pmatrix}$$

are $(-6, 8)$ and $(5, -3)$. The column vectors of this same matrix are

$$\begin{pmatrix} -6 \\ 5 \end{pmatrix} \quad \text{and} \quad \begin{pmatrix} 8 \\ -3 \end{pmatrix}.$$

We add column vectors componentwise, just as we added row vectors. For instance,

$$\begin{pmatrix} -1 \\ 6 \end{pmatrix} + \begin{pmatrix} 3 \\ -2 \end{pmatrix} = \begin{pmatrix} 2 \\ 4 \end{pmatrix}.$$

Similarly, we multiply a column vector by a number, componentwise. For instance,

$$-7 \begin{pmatrix} 2 \\ 3 \end{pmatrix} = \begin{pmatrix} -14 \\ -21 \end{pmatrix}.$$

Thus writing pairs of numbers vertically is just a notational convenience to distinguish rows and columns; it does not affect the basic nature of the algebraic operations among them.

For a general 2×2 matrix

$$\begin{pmatrix} a & b \\ c & d \end{pmatrix},$$

we call

$$\begin{pmatrix} a \\ c \end{pmatrix} \quad \text{and} \quad \begin{pmatrix} b \\ d \end{pmatrix}$$

its **first** and **second column**, respectively. The numbers a, b, c, d are called the **components** of the matrix. We write the components a, b, c, d of a 2×2 matrix also with subscripts:

$$\begin{pmatrix} a_{11} & a_{12} \\ a_{21} & a_{22} \end{pmatrix}.$$

The first subscript indicates the row, and the second subscript indicates the column. For instance, a_{12} means the component of the matrix in the first row and second column. Similarly, a_{22} means the component of the second row and second column.

Example. In the matrix

$$A = \begin{pmatrix} 3 & -5 \\ 7 & 1 \end{pmatrix},$$

we have $a_{11} = 3$, $a_{21} = 7$, $a_{12} = -5$, and $a_{22} = 1$.

With this subscript notation, we denote the rows by

$$A_1 = (a_{11}, a_{12}) \quad \text{and} \quad A_2 = (a_{21}, a_{22}).$$

We denote the columns by

$$A^1 = \begin{pmatrix} a_{11} \\ a_{21} \end{pmatrix} \quad \text{and} \quad A^2 = \begin{pmatrix} a_{12} \\ a_{22} \end{pmatrix}.$$

Thus we use a superscript like A^1, A^2 to denote columns instead of rows in a matrix A.

We can do similar things with a 3×3 matrix, which is an array of numbers as follows.

$$\begin{pmatrix} a_{11} & a_{12} & a_{13} \\ a_{21} & a_{22} & a_{23} \\ a_{31} & a_{32} & a_{33} \end{pmatrix}.$$

This 3×3 matrix has three rows denoted by A_1, A_2, A_3 and three columns, denoted by A^1, A^2, A^3. Its **components** are the numbers a_{ij}, where the indices i, j range over the integers 1, 2, 3.

Example. The three columns of the matrix

$$\begin{pmatrix} 3 & -1 & 4 \\ 2 & 5 & 7 \\ -8 & 2 & 3 \end{pmatrix}$$

are

$$A^1 = \begin{pmatrix} 3 \\ 2 \\ -8 \end{pmatrix}, \quad A^2 = \begin{pmatrix} -1 \\ 5 \\ 2 \end{pmatrix}, \quad A^3 = \begin{pmatrix} 4 \\ 7 \\ 3 \end{pmatrix}.$$

The three rows are

$$A_1 = (3, -1, 4), \quad A_2 = (2, 5, 7), \quad A_3 = (-8, 2, 3).$$

For this matrix, we have $a_{21} = 2$ and $a_{12} = -1$.

Of course, we could now generalize further, and deal with 4×4 matrices, etc., but in the rest of the chapter, we study only 2×2 and 3×3 matrices. For the higher theory, look up my book *Introduction to Linear Algebra*.

It is natural to flip a 2×2 matrix or a 3×3 matrix across the diagonal, and to change rows into columns, or columns into rows. This process is called **transposition.** Thus the **transpose** of the matrix

$$\begin{pmatrix} a & b \\ c & d \end{pmatrix}$$

is the matrix

$$\begin{pmatrix} a & c \\ b & d \end{pmatrix}.$$

To give a numerical example, the transpose of the matrix

$$\begin{pmatrix} 3 & -8 \\ 7 & 15 \end{pmatrix}$$

is the matrix

$$\begin{pmatrix} 3 & 7 \\ -8 & 15 \end{pmatrix}.$$

Similarly, the transpose of a matrix

(1)
$$\begin{pmatrix} 4 & -3 & 2 \\ -1 & 5 & 7 \\ 9 & -8 & 14 \end{pmatrix}$$

is the matrix

(2)
$$\begin{pmatrix} 4 & -1 & 9 \\ -3 & 5 & -8 \\ 2 & 7 & 14 \end{pmatrix}.$$

The **transpose** of a matrix A is denoted by ${}^tA$. The matrix

$$A = \begin{pmatrix} a_{11} & a_{12} & a_{13} \\ a_{21} & a_{22} & a_{23} \\ a_{31} & a_{32} & a_{33} \end{pmatrix}$$

is sometimes denoted by (a_{ij}), where the first index i denotes the row, and the second index j denotes the column. If $A = (a_{ij})$, then its transpose is denoted by

$${}^tA = (a_{ji}),$$

with a reversal of the indices i and j.

With this notation, we call the components a_{ii} of the matrix its **diagonal components.** Observe that the diagonal components remain unchanged when we transpose a matrix.

Example. The diagonal components of the matrix

$$\begin{pmatrix} 2 & 1 & 3 \\ -1 & 5 & 7 \\ -4 & 6 & -8 \end{pmatrix}$$

are

$$a_{11} = 2, \qquad a_{22} = 5, \qquad a_{33} = -8.$$

EXERCISES

In each of the following cases, write down the second row and first column of the indicated matrix. Also write down its transpose.

1. $\begin{pmatrix} 2 & -5 \\ -3 & -7 \end{pmatrix}$ 　　　　2. $\begin{pmatrix} 3 & -7 \\ 8 & 1 \end{pmatrix}$ 　　　　3. $\begin{pmatrix} -4 & 6 \\ 5 & -9 \end{pmatrix}$

4. $\begin{pmatrix} 3 & 5 & 6 \\ -1 & 2 & 3 \\ 7 & 3 & -2 \end{pmatrix}$ 　　5. $\begin{pmatrix} -1 & 3 & -4 \\ 2 & 1 & 6 \\ 5 & -8 & -2 \end{pmatrix}$ 　　6. $\begin{pmatrix} -3 & 4 & 6 \\ -2 & -1 & -7 \\ \frac{1}{2} & 3 & \frac{1}{3} \end{pmatrix}$

7. Find the sum of the first two columns in the matrix of Exercise 4.

8. Find the sum of the second and third rows in the matrix of
   a) Exercise 4, 　　　　b) Exercise 5, 　　　　c) Exercise 6.

9. Find the sum of the second and third columns in the matrix of
   a) Exercise 4, 　　　　b) Exercise 5, 　　　　c) Exercise 6.

10. Find the sum of the first and third columns in the matrix of
   a) Exercise 4, 　　　　b) Exercise 5, 　　　　c) Exercise 6.

§2. DETERMINANTS OF ORDER 2

Let

$$A = \begin{pmatrix} a & b \\ c & d \end{pmatrix}$$

be a 2 × 2 matrix. We define its **determinant** to be the number

$$ad - bc.$$

Example. The determinant of the matrix

$$\begin{pmatrix} 3 & 6 \\ 1 & 7 \end{pmatrix}$$

is equal to $3 \cdot 7 - 1 \cdot 6 = 15$.

Example. The determinant of the matrix

$$\begin{pmatrix} -2 & -5 \\ 4 & 8 \end{pmatrix}$$

is equal to $(-2) \cdot 8 - (-5) \cdot 4 = -16 + 20 = 4$.

The determinant of the matrix A is denoted by the symbols

$$|A| = \begin{vmatrix} a & b \\ c & d \end{vmatrix}.$$

Thus we have

$$\begin{vmatrix} -2 & -5 \\ 4 & 8 \end{vmatrix} = 4.$$

This notation is useful when we want to exhibit the components of the matrix. In other circumstances, it is useful to denote the determinant of A by the symbols

$$D(A),$$

which are short.

The determinant is an important tool for solving linear equations. We see this in the next theorem.

Theorem 1. *Let a, b, c, d, u, v be numbers. Assume that the determinant $ad - bc$ is not equal to 0. Then the system of linear equations*

$$ax + by = u,$$
$$cx + dy = v$$

has a unique solution.

Proof. The proof follows the method which we have already met in Chapter 3, consisting of eliminating a variable. For instance, multiply the first equation by d, multiply the second by b, and subtract the second from the first. The term involving y then becomes equal to 0, and we find that

$$adx - bcx = du - bv.$$

But $adx - bcx = x(ad - bc)$. Hence we find that if (x, y) is a solution, then

(1)
$$x = \frac{du - bv}{ad - bc}.$$

Eliminating x instead of y from the equations, we would find similarly that

(2)
$$y = \frac{av - cu}{ad - bc}.$$

What these arguments show is that if the equations have a solution, then this solution is given by formulas (1) and (2). But, conversely, computing $ax + by$ with these values for x and y, we find u. Similarly, computing $cx + dy$, we find v. For instance:

$$ax + by = a\frac{du - bv}{ad - bc} + b\frac{av - cu}{ad - bc} = \frac{adu - abv + bav - bcu}{ad - bc} = u,$$

because first abv cancels, and then $ad - bc$ cancels from the numerator and denominator, to yield u. This proves our theorem.

Observe that the solutions for x and y are quotients of expressions which are themselves determinants. For instance, we can rewrite the expression for x and y in the form

$$x = \frac{\begin{vmatrix} u & b \\ v & d \end{vmatrix}}{\begin{vmatrix} a & b \\ c & d \end{vmatrix}} \qquad y = \frac{\begin{vmatrix} a & u \\ c & v \end{vmatrix}}{\begin{vmatrix} a & b \\ c & d \end{vmatrix}}.$$

Note how the column

$$\begin{pmatrix} u \\ v \end{pmatrix}$$

occurs as the first column in the numerator for x, and how it occurs as the second column in the numerator for y. In both cases, the denominator is the determinant $D(A) = ad - bc$. These facts are typical of the more general case to be studied later for 3×3 determinants.

Example. The system of equations

$$3x - 2y = 5,$$
$$4x + 7y = -4$$

has a solution, given by

$$x = \frac{\begin{vmatrix} 5 & -2 \\ -4 & 7 \end{vmatrix}}{\begin{vmatrix} 3 & -2 \\ 4 & 7 \end{vmatrix}} \quad \text{and} \quad y = \frac{\begin{vmatrix} 3 & 5 \\ 4 & -4 \end{vmatrix}}{\begin{vmatrix} 3 & -2 \\ 4 & 7 \end{vmatrix}}.$$

In practice, for the 2 × 2 case, it is easier to work out the elimination method without determinants each time. However, we have described the method of determinants because in cases involving more unknowns, this method does become useful sometimes.

Theorem 2. *If A is a 2 × 2 matrix, then the determinant of A is equal to the determinant of the transpose of A. In other words,*

$$D(A) = D({}^t A).$$

Proof. This is immediate from the definition of the determinant. We have

$$|A| = \begin{vmatrix} a & b \\ c & d \end{vmatrix} \quad \text{and} \quad |{}^t A| = \begin{vmatrix} a & c \\ b & d \end{vmatrix},$$

and

$$ad - bc = ad - cb.$$

Of course, the property expressed in Theorem 2 is very simple. We give it here because it is satisfied by 3 × 3 determinants which will be studied later.

EXERCISES

1. Compute the following determinants.

a) $\begin{vmatrix} 3 & -5 \\ 4 & 2 \end{vmatrix}$ b) $\begin{vmatrix} 2 & -1 \\ -3 & 4 \end{vmatrix}$ c) $\begin{vmatrix} -3 & 4 \\ 2 & -1 \end{vmatrix}$

d) $\begin{vmatrix} -5 & 3 \\ 4 & 6 \end{vmatrix}$ e) $\begin{vmatrix} 3 & 3 \\ -7 & -8 \end{vmatrix}$ f) $\begin{vmatrix} -5 & -4 \\ 6 & 3 \end{vmatrix}$

2. Compute the determinant

$$\begin{vmatrix} \cos\theta & -\sin\theta \\ \sin\theta & \cos\theta \end{vmatrix}$$

for any real number θ.

3. Compute the determinant

$$\begin{vmatrix} \cos\theta & \sin\theta \\ \sin\theta & \cos\theta \end{vmatrix}$$

when

a) $\theta = \pi$,   b) $\theta = \pi/2$,   c) $\theta = \pi/3$,   d) $\theta = \pi/4$.

§3. PROPERTIES OF 2 × 2 DETERMINANTS

Consider a 2 × 2 matrix A with columns A^1, A^2. The determinant $D(A)$ has interesting properties with respect to these columns, which we shall describe. Thus it is useful to use the notation

$$D(A) = D(A^1, A^2)$$

to emphasize the dependence of the determinant on its columns. If the two columns are denoted by

$$B = \begin{pmatrix} b_1 \\ b_2 \end{pmatrix} \quad \text{and} \quad C = \begin{pmatrix} c_1 \\ c_2 \end{pmatrix},$$

then we would write

$$D(B, C) = \begin{vmatrix} b_1 & c_1 \\ b_2 & c_2 \end{vmatrix} = b_1 c_2 - c_1 b_2.$$

We may view the determinant as a certain type of "product" between the columns B and C. To what extent does this product satisfy the same rules as the product of numbers. Answer: To some extent, which we now determine precisely.

To begin with, this "product" satisfies distributivity. In the determinant notation, this means:

D1. *If $B = B' + B''$, i.e.*

$$\begin{pmatrix} b_1 \\ b_2 \end{pmatrix} = \begin{pmatrix} b_1' \\ b_2' \end{pmatrix} + \begin{pmatrix} b_1'' \\ b_2'' \end{pmatrix},$$

then

$$D(B' + B'', C) = D(B', C) + D(B'', C).$$

Similarly, if $C = C' + C''$, then

$$D(B, C' + C'') = D(B, C') + D(B, C'').$$

Proof. Of course, the proof is quite simple using the definition of the determinant. We have:

$$D(B' + B'', C) = \begin{vmatrix} b_1' + b_1'' & c_1 \\ b_2' + b_2'' & c_2 \end{vmatrix}$$

$$= (b_1' + b_1'')c_2 - (b_2' + b_2'')c_1$$

$$= b_1'c_2 + b_1''c_2 - b_2'c_1 - b_2''c_1$$

$$= D(B', C) + D(B'', C).$$

Distributivity on the other side is proved similarly.

D2. *If x is a number, then*

$$D(xB, C) = x \cdot D(B, C) = D(B, xC).$$

Proof. We have:

$$D(xB, C) = \begin{vmatrix} xb_1 & c_1 \\ xb_2 & c_2 \end{vmatrix} = xb_1c_2 - xb_2c_1 = x(b_1c_2 - b_2c_1)$$

$$= xD(B, C).$$

Again, the other equality is proved similarly.

The two vectors

$$E^1 = \begin{pmatrix} 1 \\ 0 \end{pmatrix} \quad \text{and} \quad E^2 = \begin{pmatrix} 0 \\ 1 \end{pmatrix}$$

will be called the **unit vectors.** The matrix formed by them, namely

$$E = \begin{pmatrix} 1 & 0 \\ 0 & 1 \end{pmatrix},$$

will be called the **unit matrix.** We have:

D3. *If E is the unit matrix, then $D(E) = D(E^1, E^2) = 1$.*

This is obvious.

D4. *If the two columns of the matrix are equal, then the determinant is equal to 0. In other words,*

$$D(B, B) = 0.$$

Proof. This is obvious, because

$$\begin{vmatrix} b_1 & b_1 \\ b_2 & b_2 \end{vmatrix} = b_1 b_2 - b_2 b_1 = 0.$$

These four basic properties are fundamental, and other properties can be deduced from them, without going back to the definition of the determinant in terms of the components of the matrix.

D5. *If we add a multiple of one column to the other, then the value of the determinant does not change. In other words, let x be a number. Then*

$$D(B + xC, C) = D(B, C) \quad \text{and} \quad D(B, C + xB) = D(B, C).$$

Written out in terms of components, the first relation reads

$$\begin{vmatrix} b_1 + xc_1 & c_1 \\ b_2 + xc_2 & c_2 \end{vmatrix} = \begin{vmatrix} b_1 & c_1 \\ b_2 & c_2 \end{vmatrix}.$$

Proof. Using **D1, D2, D4** in succession, we find that

$$D(B + xC, C) = D(B, C) + D(xC, C)$$
$$= D(B, C) + xD(C, C) = D(B, C).$$

A similar proof applies to $D(B, C + xB)$.

D6. *If the two columns are interchanged, then the value of the determinant changes by a sign. In other words, we have*

$$D(B, C) = -D(C, B).$$

Proof. Again, we use **D1, D2, D4** successively, and get:

$$
\begin{aligned}
0 = D(B + C, B + C) &= D(B, B + C) + D(C, B + C) \\
&= D(B, B) + D(B, C) + D(C, B) + D(C, C) \\
&= D(B, C) + D(C, B).
\end{aligned}
$$

This proves that $D(B, C) = -D(C, B)$, as desired.

Of course, you can also give a proof using the components of the matrix. Do this as an exercise. However, there is some point in doing it as above, because in the study of determinants in the higher-dimensional case later, a proof with components becomes much messier, while the proof following the same pattern as the one we have given remains neat.

Observe that **D6** shows that our determinant, viewed as a "product," is *not* commutative. Commutativity would mean that

$$D(B, C) = D(C, B),$$

and this is *not true*. Note that **D6** was deduced from distributivity and the special property **D4**. The property expressed by **D6** is often called **anti-commutativity**—"anti" because of the minus sign which appears.

We can find the values of x and y in Theorem 1, by using the properties we have just proved. This new argument is the one which generalizes later. It runs as follows. We wish to solve the system of linear equations

(*)
$$
\begin{aligned}
a_1 x + b_1 y &= c_1 \\
a_2 x + b_2 y &= c_2.
\end{aligned}
$$

Observe how we have numbered the coefficients in such a way that we can write the columns easily, in an abbreviated fashion. Namely, (*) can be written in the form

(**)
$$x A + y B = C,$$

where

$$A = \begin{pmatrix} a_1 \\ a_2 \end{pmatrix}, \qquad B = \begin{pmatrix} b_1 \\ b_2 \end{pmatrix}, \qquad C = \begin{pmatrix} c_1 \\ c_2 \end{pmatrix}.$$

Suppose that x, y are solutions of the system (**). Using our properties of the determinant, we have:

$$D(C, B) = D(xA + yB, B) = D(xA, B) + D(yB, B)$$
$$= xD(A, B) + yD(B, B)$$
$$= xD(A, B).$$

Hence we find that

$$x = \frac{D(C, B)}{D(A, B)} = \frac{\begin{vmatrix} c_1 & b_1 \\ c_2 & b_2 \end{vmatrix}}{\begin{vmatrix} a_1 & b_1 \\ a_2 & b_2 \end{vmatrix}}.$$

If you compare this with the solution found previously you will find, of course, that we have obtained the same value, namely a quotient of determinants whose denominator is the determinant of the coefficients of the equations. Similarly, you can find the value for y using the same method. Do it as an exercise.

EXERCISES

1. Prove the other half of **D1**, i.e. distributivity on the side other than that given in the text.

2. Prove the other half of **D2**.

3. Prove the other half of **D5**.

4. Using the same method as at the end of the section, find the value for y as a quotient of two determinants.

5. Solve the linear equations of Chapter 2, §1 by determinants.

6. Let c be a number, and let A be a 2 × 2 matrix. Define cA to be the matrix obtained by multiplying all components of A by c. How does $D(cA)$ differ from $D(A)$?

7. Let $A = (a_1, a_2)$ and $B = (b_1, b_2)$. Define their **dot product** $A \cdot B$ by the formula

$$A \cdot B = a_1 b_1 + a_2 b_2.$$

For instance, $(3, 1) \cdot (-4, 5) = -3 \cdot 4 + 1 \cdot 5 = -7$. Thus the dot product is a number. Prove that this product is commutative and distributive with respect to addition. If $A \neq (0, 0)$, prove that $A \cdot A > 0$. Give an example of $A \neq O$ and $B \neq O$ such that $A \cdot B = 0$. Compute $A \cdot B$ for the following values of A and B.

a) $A = (-4, 3)$ and $B = (5, -2)$

b) $A = (-2, -1)$ and $B = (-3, 4)$

§4. DETERMINANTS OF ORDER 3

We shall define the determinant for 3×3 matrices, and we shall see that it satisfies properties analogous to those of the 2×2 case.

Let

$$A = (a_{ij}) = \begin{pmatrix} a_{11} & a_{12} & a_{13} \\ a_{21} & a_{22} & a_{23} \\ a_{31} & a_{32} & a_{33} \end{pmatrix}$$

be a 3×3 matrix. We define its **determinant** according to the formula known as the **expansion by a row,** say the first row. That is, we define

(1)
$$D(A) = a_{11} \begin{vmatrix} a_{22} & a_{23} \\ a_{32} & a_{33} \end{vmatrix} - a_{12} \begin{vmatrix} a_{21} & a_{23} \\ a_{31} & a_{33} \end{vmatrix} + a_{13} \begin{vmatrix} a_{21} & a_{22} \\ a_{31} & a_{32} \end{vmatrix}.$$

and we denote $D(A)$ also with the two vertical bars

$$D(A) = \begin{vmatrix} a_{11} & a_{12} & a_{13} \\ a_{21} & a_{22} & a_{23} \\ a_{31} & a_{32} & a_{33} \end{vmatrix}.$$

We may describe the sum in (1) as follows. Let A_{ij} be the matrix obtained from A by deleting the i-th row and j-th column. Then the sum for $D(A)$ can be written as

$$a_{11}D(A_{11}) - a_{12}D(A_{12}) + a_{13}D(A_{13}).$$

In other words, each term consists of the product of an element of the first row and the determinant of the 2×2 matrix obtained by deleting the first row and the j-th column, and putting the appropriate sign to this term as shown.

Example. Let

$$A = \begin{pmatrix} 2 & 1 & 0 \\ 1 & 1 & 4 \\ -3 & 2 & 5 \end{pmatrix}.$$

Then

$$A_{11} = \begin{pmatrix} 1 & 4 \\ 2 & 5 \end{pmatrix}, \qquad A_{12} = \begin{pmatrix} 1 & 4 \\ -3 & 5 \end{pmatrix}, \qquad A_{13} = \begin{pmatrix} 1 & 1 \\ -3 & 2 \end{pmatrix}$$

and our formula for the determinant of A yields

$$D(A) = 2\begin{vmatrix} 1 & 4 \\ 2 & 5 \end{vmatrix} - 1\begin{vmatrix} 1 & 4 \\ -3 & 5 \end{vmatrix} + 0\begin{vmatrix} 1 & 1 \\ -3 & 2 \end{vmatrix}$$

$$= 2(5 - 8) - 1(5 + 12) + 0$$

$$= -23.$$

Thus the determinant is a number. To compute this number in the above example, we computed the determinants of the 2×2 matrices explicitly. We can also expand these in the general definition, and thus we find a six-term expression for the determinant of a general 3×3 matrix $A = (a_{ij})$, namely:

(2)
$$\begin{aligned} D(A) = {}& a_{11}a_{22}a_{33} - a_{11}a_{32}a_{23} - a_{12}a_{21}a_{33} \\ & + a_{12}a_{23}a_{31} + a_{13}a_{21}a_{32} - a_{13}a_{22}a_{31}. \end{aligned}$$

Do not memorize (2). Remember only (1), and write down (2) only when needed for specific purposes.

We could have used the other rows to expand the determinant, instead of the first row. For instance, the expansion according to the second row is given by

$$-a_{21}\begin{vmatrix} a_{12} & a_{13} \\ a_{32} & a_{33} \end{vmatrix} + a_{22}\begin{vmatrix} a_{11} & a_{13} \\ a_{31} & a_{33} \end{vmatrix} - a_{23}\begin{vmatrix} a_{11} & a_{12} \\ a_{31} & a_{32} \end{vmatrix}$$

$$= -a_{21}D(A_{21}) + a_{22}D(A_{22}) - a_{23}D(A_{23}).$$

Again, each term is the product of a_{2j} with the determinant of the 2×2 matrix obtained by deleting the second row and j-th column, together with the appropriate sign in front of each term. This sign is determined according to the pattern:

$$\begin{pmatrix} + & - & + \\ - & + & - \\ + & - & + \end{pmatrix}.$$

If you write down the two terms for each one of the 2×2 determinants in the expansion according to the second row, you will obtain six terms, and you will find immediately that they give you the same value which we wrote down in formula (2). Thus expanding according to the second row gives the same value for the determinant as expanding according to the first row.

Furthermore, we can also expand according to any one of the columns. For instance, expanding according to the first column, we find that

$$a_{11}\begin{vmatrix} a_{22} & a_{23} \\ a_{32} & a_{33} \end{vmatrix} - a_{21}\begin{vmatrix} a_{12} & a_{13} \\ a_{32} & a_{33} \end{vmatrix} + a_{31}\begin{vmatrix} a_{12} & a_{13} \\ a_{22} & a_{23} \end{vmatrix}$$

yields precisely the same six terms as in (2), if you write down each one of the two terms corresponding to each one of the 2×2 determinants in the above expression.

Example. We compute the determinant

$$\begin{vmatrix} 3 & 0 & 1 \\ 1 & 2 & 5 \\ -1 & 4 & 2 \end{vmatrix}$$

by expanding according to the second column. The determinant is equal to

$$2\begin{vmatrix} 3 & 1 \\ -1 & 2 \end{vmatrix} - 4\begin{vmatrix} 3 & 1 \\ 1 & 5 \end{vmatrix} = 2(6 - (-1)) - 4(15 - 1) = -42.$$

Note that the presence of 0 in the first row and second column eliminates one term in the expansion, since this term is equal to 0.

If we expand the above determinant according to the third column, we find the same value, namely

$$+1\begin{vmatrix} 1 & 2 \\ -1 & 4 \end{vmatrix} - 5\begin{vmatrix} 3 & 0 \\ -1 & 4 \end{vmatrix} + 2\begin{vmatrix} 3 & 0 \\ 1 & 2 \end{vmatrix} = -42.$$

Theorem 3. *If A is a 3×3 matrix, then $D(A) = D({}^tA)$. In other words, the determinant of A is equal to the determinant of the transpose of A.*

Proof. This is true because expanding $D(A)$ according to rows or columns gives the same value, namely the expression in (2).

EXERCISES

1. Write down the expansion of a 3×3 determinant according to the third row, the second column, and the third column, and verify in each case that you get the same six terms as in (2).

2. Compute the following determinants by expanding according to the second row, and also according to the third column, as a check for your computation. Of course, you should find the same value.

a) $\begin{vmatrix} 2 & 1 & 2 \\ 0 & 3 & -1 \\ 4 & 1 & 1 \end{vmatrix}$
b) $\begin{vmatrix} 3 & -1 & 5 \\ -1 & 2 & 1 \\ -2 & 4 & 3 \end{vmatrix}$
c) $\begin{vmatrix} 2 & 4 & 3 \\ -1 & 3 & 0 \\ 0 & 2 & 1 \end{vmatrix}$

d) $\begin{vmatrix} 1 & 2 & -1 \\ 0 & 1 & 1 \\ 0 & 2 & 7 \end{vmatrix}$
e) $\begin{vmatrix} -1 & 5 & 3 \\ 4 & 0 & 0 \\ 2 & 7 & 8 \end{vmatrix}$
f) $\begin{vmatrix} 3 & 1 & 2 \\ 4 & 5 & 1 \\ -1 & 2 & -3 \end{vmatrix}$

3. Compute the following determinants.

a) $\begin{vmatrix} 4 & 0 & 0 \\ 0 & 5 & 0 \\ 0 & 0 & 7 \end{vmatrix}$
b) $\begin{vmatrix} -3 & 0 & 0 \\ 0 & 5 & 0 \\ 0 & 0 & -8 \end{vmatrix}$
c) $\begin{vmatrix} 6 & 0 & 0 \\ 0 & 5 & 0 \\ 0 & 0 & -2 \end{vmatrix}$

4. Let a, b, c be numbers. In terms of a, b, c, what is the value of the determinant

$$\begin{vmatrix} a & 0 & 0 \\ 0 & b & 0 \\ 0 & 0 & c \end{vmatrix} ?$$

5. Compute the following determinants.

a) $\begin{vmatrix} 3 & 1 & -5 \\ 0 & 4 & 1 \\ 0 & 0 & -2 \end{vmatrix}$
b) $\begin{vmatrix} 4 & -6 & 7 \\ 0 & 2 & -8 \\ 0 & 0 & -9 \end{vmatrix}$

c) $\begin{vmatrix} 6 & 0 & 0 \\ -4 & -5 & 0 \\ 7 & 20 & -3 \end{vmatrix}$
d) $\begin{vmatrix} 5 & 0 & 0 \\ 4 & 2 & 0 \\ -17 & 19 & -3 \end{vmatrix}$

6. In terms of the components of the matrix, what is the value of the determinant:

a) $\begin{vmatrix} a_{11} & a_{12} & a_{13} \\ 0 & a_{22} & a_{23} \\ 0 & 0 & a_{33} \end{vmatrix} ?$
b) $\begin{vmatrix} a_{11} & 0 & 0 \\ a_{21} & a_{22} & 0 \\ a_{31} & a_{32} & a_{33} \end{vmatrix} ?$

§5. PROPERTIES OF 3 × 3 DETERMINANTS

We shall now see that 3×3 determinants satisfy the properties **D1** through **D6**, listed previously for 2×2 determinants. These properties are concerned with the columns of the matrix, and hence it is useful to use the same notation which we used before. If A^1, A^2, A^3 are the columns of the 3×3 matrix A, then we write

$$D(A) = D(A^1, A^2, A^3).$$

For the rest of this section, we assume that our column and row vectors have dimension 3; that is, that they have three components. Thus any column vector B in this section can be written in the form

$$B = \begin{pmatrix} b_1 \\ b_2 \\ b_3 \end{pmatrix}.$$

D1. *Suppose that the first column can be written as a sum,*

$$A^1 = B + C,$$

that is,

$$\begin{pmatrix} a_{11} \\ a_{21} \\ a_{31} \end{pmatrix} = \begin{pmatrix} b_1 \\ b_2 \\ b_3 \end{pmatrix} + \begin{pmatrix} c_1 \\ c_2 \\ c_3 \end{pmatrix}.$$

Then

$$D(B + C, A^2, A^3) = D(B, A^2, A^3) + D(C, A^2, A^3).$$

and the analogous rule holds with respect to the second and third columns.

Proof. We use the definition of the determinant, namely the expansion according to the first row. We see that each term splits into a sum of two terms corresponding to B and C. For instance,

$$a_{11} \begin{vmatrix} a_{22} & a_{23} \\ a_{31} & a_{33} \end{vmatrix} = b_1 \begin{vmatrix} a_{22} & a_{23} \\ a_{31} & a_{33} \end{vmatrix} + c_1 \begin{vmatrix} a_{22} & a_{23} \\ a_{31} & a_{33} \end{vmatrix},$$

$$a_{12} \begin{vmatrix} b_2 + c_2 & a_{23} \\ b_3 + c_3 & a_{33} \end{vmatrix} = a_{12} \begin{vmatrix} b_2 & a_{23} \\ b_3 & a_{33} \end{vmatrix} + a_{12} \begin{vmatrix} c_2 & a_{23} \\ c_3 & a_{33} \end{vmatrix},$$

$$a_{13} \begin{vmatrix} b_2 + c_2 & a_{22} \\ b_3 + c_3 & a_{32} \end{vmatrix} = a_{13} \begin{vmatrix} b_2 & a_{22} \\ b_3 & a_{32} \end{vmatrix} + a_{13} \begin{vmatrix} c_2 & a_{22} \\ c_3 & a_{32} \end{vmatrix}.$$

Summing with the appropriate sign yields the desired relation.

D2. *If x is a number, then*

$$D(xA^1, A^2, A^3) = x \cdot D(A^1, A^2, A^3),$$

and similarly for the other columns.

Proof. We have:

$$D(xA^1, A^2, A^3) = xa_{11} \begin{vmatrix} a_{22} & a_{23} \\ a_{32} & a_{33} \end{vmatrix} - a_{12} \begin{vmatrix} xa_{21} & a_{23} \\ xa_{31} & a_{33} \end{vmatrix} + a_{13} \begin{vmatrix} xa_{21} & a_{22} \\ xa_{31} & a_{32} \end{vmatrix}$$

$$= x \cdot D(A^1, A^2, A^3).$$

The proof is similar for the other columns.

In the 3 × 3 case, we also have the **unit vectors,** namely

$$E^1 = \begin{pmatrix} 1 \\ 0 \\ 0 \end{pmatrix}, \qquad E^2 = \begin{pmatrix} 0 \\ 1 \\ 0 \end{pmatrix}, \qquad E^3 = \begin{pmatrix} 0 \\ 0 \\ 1 \end{pmatrix},$$

and the unit **3 × 3 matrix,** namely

$$E = \begin{pmatrix} 1 & 0 & 0 \\ 0 & 1 & 0 \\ 0 & 0 & 1 \end{pmatrix}.$$

D3. *If E is the unit matrix, then* $D(E) = D(E^1, E^2, E^3) = 1$.

Proof. This is obvious from the expansion according to the first row.

D4. *If two columns of the matrix are equal, then the determinant is equal to 0.*

Proof. Suppose that $A^1 = A^2$, and look at the expansion of the determinant according to the first row. Then $a_{11} = a_{12}$, and the first two terms cancel. The third term is equal to 0 because it involves a 2 × 2 determinant whose two columns are equal.   The proof for the other cases is similar. (Other cases: $A^2 = A^3$ and $A^1 = A^3$.)

Observe that to prove our basic four properties, we needed to use the definition of the determinant, i.e. its expansion according to the first row. For the remaining properties, we can give a proof which is not based directly on this expansion, but only on the formalism of **D1** through **D4**. This has the advantage of making the arguments easier, and in fact of making them completely analogous to those used in the 2 × 2 case. We carry them out.

D5. *If we add a multiple of one column to another, then the value of the determinant does not change. In other words, let x be a number. Then for instance*

$$D(A^1, A^2 + xA^1, A^3) = D(A^1, A^2, A^3),$$

and similarly in all other cases.

Proof. We have

$$
\begin{aligned}
D(A^1, A^2, + xA^1, A^3) &= D(A^1, A^2, A^3) + D(A^1, xA^1, A^3) &\text{(by } \mathbf{D1}) \\
&= D(A^1, A^2, A^3) + x \cdot D(A^1, A^1, A^3) &\text{(by } \mathbf{D2}) \\
&= D(A^1, A^2, A^3) &\text{(by } \mathbf{D4}).
\end{aligned}
$$

This proves what we wanted. The proofs of the other cases are similar.

D6. *If two adjacent columns are interchanged, then the determinant changes by a sign. In other words, we have*

$$D(A^1, A^3, A^2) = -D(A^1, A^2, A^3),$$

and similarly in the other case.

Proof. We use the same method as before. We find:

$$
\begin{aligned}
0 &= D(A^1, A^2 + A^3, A^2 + A^3) \\
&= D(A^1, A^2, A^2 + A^3) + D(A^1, A^3, A^2 + A^3) \\
&= D(A^1, A^2, A^2) + D(A^1, A^2, A^3) + D(A^1, A^3, A^2) + D(A^1, A^3, A^3) \\
&= D(A^1, A^2, A^3) + D(A^1, A^3, A^2),
\end{aligned}
$$

using **D1** and **D4**. This proves **D6** in this case, and the other cases are proved similarly.

Using these rules, especially **D5**, we can compute determinants a little more efficiently. For instance, we have already noticed that when a 0 occurs in the given matrix, we can expand according to the row (or column) in which this 0 occurs, and it eliminates one term. Using **D5** repeatedly, we can change the matrix so as to get as many zeros as possible, and then reduce the computation to one term.

Example. Compute the determinant

$$
\begin{vmatrix}
3 & 0 & 1 \\
1 & 2 & 5 \\
-1 & 4 & 2
\end{vmatrix}.
$$

We already have 0 in the first row. We subtract two times the second row from the third row. Our determinant is then equal to

$$\begin{vmatrix} 3 & 0 & 1 \\ 1 & 2 & 5 \\ -3 & 0 & -8 \end{vmatrix}.$$

We expand according to the second column. The expansion has only one term ≠ 0, with a + sign, and that is:

$$2 \begin{vmatrix} 3 & 1 \\ -3 & -8 \end{vmatrix}.$$

The 2 × 2 determinant can be evaluated by our definition of $ad - bc$, and we find the value

$$2(-24 - (-3)) = -42.$$

Example. We compute the determinant

$$\begin{vmatrix} 4 & 7 & 10 \\ 3 & 7 & 5 \\ 5 & -1 & 10 \end{vmatrix}.$$

We subtract two times the second row from the first row, and then from the third row, yielding

$$\begin{vmatrix} -2 & -7 & 0 \\ 3 & 7 & 5 \\ -1 & -15 & 0 \end{vmatrix},$$

which we expand according to the third column, and get

$$-5(30 - 7) = -5(23) = -115.$$

Note that the term has a minus sign, determined by our usual pattern of signs.

EXERCISES

1. Write out in full and prove property **D1** with respect to the second column and the third column.

2. Same thing for property **D2**.

3. Prove the two cases not treated in the text for property **D4**.

4. Prove **D5** in the case

  a) you add a multiple of the third column to the first;

  b) you add a multiple of the second column to the first;

  c) you add a multiple of the third column to the second.

5. Prove **D6** in the second case.

6. If you interchange the first and third columns of the given matrix, how does its determinant change? What about interchanging the first and third row?

7. State **D5** and **D6** for rows.

8. Compute the following determinants, making the computation as easy as you can.

a) $\begin{vmatrix} 4 & -9 & 2 \\ 4 & -9 & 2 \\ 3 & 1 & 5 \end{vmatrix}$ b) $\begin{vmatrix} 4 & -1 & 1 \\ 2 & 0 & 0 \\ 1 & 5 & 7 \end{vmatrix}$ c) $\begin{vmatrix} 2 & -1 & 4 \\ 1 & 1 & 5 \\ 1 & 2 & 3 \end{vmatrix}$

d) $\begin{vmatrix} 3 & 1 & 1 \\ 2 & 5 & 5 \\ 8 & 7 & 7 \end{vmatrix}$ e) $\begin{vmatrix} 2 & 1 & 1 \\ 3 & 1 & 5 \\ 4 & -2 & 3 \end{vmatrix}$ f) $\begin{vmatrix} -4 & 4 & 2 \\ 5 & 1 & 3 \\ 2 & 1 & 4 \end{vmatrix}$

g) $\begin{vmatrix} 7 & 3 & 2 \\ 1 & -1 & 1 \\ 2 & 1 & 3 \end{vmatrix}$ h) $\begin{vmatrix} 3 & 2 & 1 \\ 1 & 1 & 1 \\ -1 & 3 & 4 \end{vmatrix}$ i) $\begin{vmatrix} -2 & -1 & 1 \\ 3 & 1 & -1 \\ -1 & 2 & 3 \end{vmatrix}$

j) $\begin{vmatrix} 2 & 1 & 1 \\ 1 & 1 & 1 \\ 2 & 2 & 2 \end{vmatrix}$ k) $\begin{vmatrix} -4 & 1 & 2 \\ 3 & 2 & 1 \\ -1 & -1 & 1 \end{vmatrix}$ l) $\begin{vmatrix} -1 & 3 & 2 \\ 3 & -1 & 1 \\ 6 & -2 & 2 \end{vmatrix}$

9. Let c be a number and multiply each component a_{ij} of a 3×3 matrix A by c, thus obtaining a new matrix which we denote by cA. How does $D(A)$ differ from $D(cA)$?

10. Let x_1, x_2, x_3 be numbers. Show that

$$\begin{vmatrix} 1 & x_1 & x_1^2 \\ 1 & x_2 & x_2^2 \\ 1 & x_3 & x_3^2 \end{vmatrix} = (x_2 - x_1)(x_3 - x_2)(x_3 - x_1).$$

11. Suppose that A^1 is a sum of three columns, say

$$A^1 = B^1 + B^2 + B^3.$$

Using **D1** twice, prove that

$$D(B^1 + B^2 + B^3, A^2, A^3)$$
$$= D(B^1, A^2, A^3) + D(B^2, A^2, A^3) + D(B^3, A^2, A^3).$$

Using summation notation, we can write this in the form

$$D(B^1 + B^2 + B^3, A^2, A^3) = \sum_{j=1}^{3} D(B^j, A^2, A^3),$$

which is shorter. In general, suppose that

$$A^1 = \sum_{j=1}^{n} B^j$$

is a sum of n columns. Using the summation notation, express similarly

$$D(A^1, A^2, A^3)$$

as a sum of (how many?) terms.

12. Let x_j $(j = 1, 2, 3)$ be numbers. Let

$$A^1 = x_1 C^1 + x_2 C^2 + x_3 C^3.$$

Prove that

$$D(A^1, A^2, A^3) = \sum_{j=1}^{3} x_j D(C^j, A^2, A^3).$$

State and prove the analogous statement when

$$A^1 = \sum_{j=1}^{n} x_j C^j.$$

13. State the analogous property to that of Exercise 12 with respect to the second column. Then with respect to the third column.

§6. CRAMER'S RULE

We now come to solving linear equations in three unknowns using determinants. For practical purposes, the simple-minded elimination method of Chapter 2, §2, is the easiest to use when the equations are explicitly given with numerical coefficients, and you don't need any harder theory, like that of determinants. For theoretical purposes, however, and for more advanced computations, the theory of determinant can often be used advantageously, and that is the reason why we now go into it.

We can write a system of linear equations

$$
\begin{aligned}
a_{11}x_1 + a_{12}x_2 + a_{13}x_3 &= b_1 \\
a_{21}x_1 + a_{22}x_2 + a_{23}x_3 &= b_2 \\
a_{31}x_1 + a_{32}x_2 + a_{33}x_3 &= b_3
\end{aligned}
$$

(*)

using the vector notation, in the form

(**) $$x_1 A^1 + x_2 A^2 + x_3 A^3 = B,$$

where A^1, A^2, A^3 are the columns of the matrix of coefficients a_{ij}, and B is the column formed by b_1, b_2, b_3.

Theorem 4. *Let x_1, x_2, x_3 be numbers which are solutions of the system of linear equations above. Let $A = (a_{ij})$ be its matrix of coefficients. If $D(A) \neq 0$, then*

$$
x_1 = \frac{D(B, A^2, A^3)}{D(A^1, A^2, A^3)}, \qquad x_2 = \frac{D(A^1, B, A^3)}{D(A^1, A^2, A^3)},
$$

$$
x_3 = \frac{D(A^1, A^2, B)}{D(A^1, A^2, A^3)}.
$$

Or, written out in terms of the components, say for x_1, we have

$$
x_1 = \frac{\begin{vmatrix} b_1 & a_{12} & a_{13} \\ b_2 & a_{22} & a_{23} \\ b_3 & a_{32} & a_{33} \end{vmatrix}}{\begin{vmatrix} a_{11} & a_{12} & a_{13} \\ a_{21} & a_{22} & a_{23} \\ a_{31} & a_{32} & a_{33} \end{vmatrix}}.
$$

Note. The practicality of our notation now becomes really apparent. It is obviously much longer and more tiring to write out all the components than to write out the expressions with the abbreviation for the columns as in $D(B, A^1, A^2)$, etc.

Proof of Theorem 4. We use the same technique as for the proof of the 2×2 case. We have:

$$D(B, A^2, A^3) = D(x_1 A^1 + x_2 A^2 + x_3 A^3, A^2, A^3)$$
$$= x_1 D(A^1, A^2, A^3) + x_2 D(A^2, A^2, A^3) + x_3 D(A^3, A^2, A^3)$$
$$= x_1 D(A^1, A^2, A^3).$$

The first equality follows from **D1** and **D2**. As an exercise, write out the missing steps (Exercise 1). The second equality follows from **D4**. If $D(A) \neq 0$, then we get the desired expression for x_1. We treat x_2 and x_3 similarly (Exercise 2).

Theorem 4 is known as **Cramer's rule.**

Observe that Theorem 4 does not tell us that a solution exists. It tells us that if a solution exists, then it is given by the formulas as in the theorem. In fact, the following is true.

Theorem 5. *Let A be a 3×3 matrix whose determinant is not 0. Then for any column B there exist numbers x_1, x_2, x_3 such that*

$$x_1 A^1 + x_2 A^2 + x_3 A^3 = B.$$

In other words, the system of linear equations (∗) *has a solution.*

The proof for Theorem 5 is slightly more involved than in the analogous 2×2 case, and we shall omit it.

Example. We solve the following system of linear equations by Cramer's rule.

$$3x + 2y + 4z = 1$$
$$2x - y + z = 0$$
$$x + 2y + 3z = 1$$

We have:

$$x = \frac{\begin{vmatrix} 1 & 2 & 4 \\ 0 & -1 & 1 \\ 1 & 2 & 3 \end{vmatrix}}{\begin{vmatrix} 3 & 2 & 4 \\ 2 & -1 & 1 \\ 1 & 2 & 3 \end{vmatrix}}, \quad y = \frac{\begin{vmatrix} 3 & 1 & 4 \\ 2 & 0 & 1 \\ 1 & 1 & 3 \end{vmatrix}}{\begin{vmatrix} 3 & 2 & 4 \\ 2 & -1 & 1 \\ 1 & 2 & 3 \end{vmatrix}}, \quad z = \frac{\begin{vmatrix} 3 & 2 & 1 \\ 2 & -1 & 0 \\ 1 & 2 & 1 \end{vmatrix}}{\begin{vmatrix} 3 & 2 & 4 \\ 2 & -1 & 1 \\ 1 & 2 & 3 \end{vmatrix}}.$$

Observe how the column

$$B = \begin{pmatrix} 1 \\ 0 \\ 1 \end{pmatrix}$$

shifts from the first column when solving for x, to the second column when solving for y, to the third column when solving for z. The denominator in all three expressions is the same, namely it is the determinant of the matrix of coefficients of the equations.

We know how to compute 3×3 determinants, and we then find

$$x = -\frac{1}{5}, \qquad y = 0, \qquad z = \frac{2}{5}.$$

EXERCISES

1. Fill in the missing steps in the proof of Cramer's rule. Cf. Exercises 11 and 12 of the preceding section.

2. Write out in full the proof of Cramer's rule for x_2 and x_3. It is very similar to the proof for x_1 in the text.

3. Let A^1, A^2, A^3 be columns of a 3×3 matrix A, and assume that there exist numbers x_1, x_2, x_3 not all 0 such that

$$x_1 A^1 + x_2 A^2 + x_3 A^3 = 0.$$

Prove that $D(A) = 0$.

4. Solve the linear equations of Chapter 2, §2 by Cramer's rule.

Index

Index

Answers to Selected Exercises

Answers to Selected Exercises

Chapter 1, §2

1. $(a + b) + (c + d) = a + b + c + d$ by associativity
$$= a + b + d + c = a + d + b + c$$
by commutativity
$$= (a + d) + (b + c) \quad \text{by associativity}$$

3. $(a - b) + (c - d) = a - b + c - d$ by associativity
$$= a + c - b - d \quad \text{by commutativity}$$
$$= (a + c) + (-b - d) \quad \text{by associativity}$$

6. $(a - b) + (c - d) = a - b + c - d$ by associativity
$$= -b + a + c - d = -b + a - d + c$$
$$= -b - d + a + c \quad \text{by commutativity}$$
$$= -(b + d) + (a + c) \quad \text{by associativity and N5}$$

11. It suffices to show that $(a + b + c) + (-a) + (-b) + (-c) = 0$.
But $(a + b + c) + (-a) + (-b) + (-c)$
$$= a + b + c - a - b - c \quad \text{by associativity}$$
$$\left.\begin{array}{l} = a + b - a + c - b - c \\ = a - a + b + c - b - c \\ = a - a + b - b + c - c \end{array}\right\} \text{by commutativity}$$
$$= (a - a) + (b - b) + (c - c) \quad \text{by associativity}$$
$$= 0 + 0 + 0 = 0$$

13. Same type of proof. We must show that $(a - b) + b - a = 0$. We have:
$$a - b + b - a = a - b - a + b$$
$$= a - a - b + b \quad \text{by commutativity}$$
$$= (a - a) + (-b + b) \quad \text{by associativity}$$
$$= 0 + 0 = 0$$

15. $x = -3$ **17.** $x = 5$ **19.** $x = -3$

22. Let $a + b = a + c$. Adding $-a$ to both sides, we get $-a + a + b = -a + a + c$, that is, $(-a + a) + b = (-a + a) + c$ by associativity. Then $0 + b = 0 + c$, i.e., $b = c$.

23. Add $-a$ to both sides of the equation $a + b = a$. We obtain $-a + a + b = -a + a = 0$, so $0 + b = 0$ and $b = 0$.

Chapter 1, §3

1. a) $2^8 3^3 a^7 b^4$ c) $2^{10} 3^3 a^6 b^{10}$ e) $2^4 3^5 a^6 b^8$

2. $(a + b)^3 = (a + b)^2(a + b) = (a^2 + 2ab + b^2)(a + b)$
$$= (a^2 + 2ab + b^2)a + (a^2 + 2ab + b^2)b$$
$$= a^3 + 2aba + b^2a + a^2b + 2ab^2 + b^3$$
$$= a^3 + 3a^2b + 3b^2a + b^3$$
$(a - b)^3 = (a - b)^2(a - b) = (a^2 - 2ab + b^2)(a - b)$
$$= (a^2 - 2ab + b^2)a - (a^2 - 2ab + b^2)b$$
$$= a^3 - 2aba + b^2a - a^2b + 2ab^2 - b^3$$
$$= a^3 - 3a^2b + 3ab^2 - b^3$$

3. $(a + b)^4 = (a + b)^3(a + b) = (a^3 + 3a^2b + 3ab^2 + b^3)(a + b)$
$$= (a^3 + 3a^2b + 3ab^2 + b^3)a + (a^3 + 3a^2b + 3ab^2 + b^3)b$$
$$= a^4 + 4a^3b + 6a^2b^2 + 4ab^3 + b^4$$
$(a - b)^4 = (a - b)^3(a - b) = (a^3 - 3a^2b + 3ab^2 - b^3)(a - b)$
$$= a^4 - 4a^3b + 6a^2b^2 - 4ab^3 + b^4$$

4. $16x^2 - 16x + 4$ **6.** $4x^2 + 20x + 25$ **8.** $x^2 - 1$

12. $x^4 + 2x^2 + 1$ **14.** $x^4 + 4x^2 + 4$ **20.** $2x^3 + 3x^2 - 9x - 10$

22. $6x^3 + 25x^2 + 3x - 4$ **24.** $4x^3 + 3x^2 - 25x + 6$ **30.** 3,200,000

32. a) 145,800,000   b) 48,600,000

Chapter 1, §4

1. a) If $a = 2n$, $b = 2m$, then $a + b = 2n + 2m = 2(n + m)$.
Setting $k = n + m$ gives $a + b = 2k$, whence $a + b$ is even.
b) If $a = 2n + 1$, $b = 2m$, then $a + b = 2n + 1 + 2m = 2(n + m) + 1$.
Setting $k = n + m$ gives $a + b = 2k + 1$, and $a + b$ is odd.

2. Let $a = 2n$, $ab = 2nb$.
Setting $k = nb$ gives $ab = 2k$, whence ab is even.

3. Let $a = 2n$, $a^3 = (2n)^3 = 2^3 n^3$.
Setting $k = 2^2 n^3$ gives $a^3 = 2k$, whence a^3 is even.

4. Let $a = 2n + 1$, $a^3 = (2n + 1)^3 = 8n^3 + 12n^2 + 6n + 1$.
Setting $k = 4n^3 + 6n^2 + 3n$ gives $a^3 = 2k + 1$, whence a^3 is odd.

5. Let $n = 2k$, $(-1)^n = (-1)^{2k} = ((-1)^2)^k = 1^k$, since $(-1)^2 = +1$.
Then $1^k = 1$ for all integers, k.

6. Let $n = 2k + 1$.
Then $(-1)^n = (-1)^{2k+1} = (-1)^{2k}(-1) = 1 \cdot (-1) = -1$.

7. Let $m = 2k + 1$, $n = 2q + 1$.
Then $mn = (2k + 1)(2q + 1) = 4kq + 2q + 2k + 1$.
Setting $t = 2kq + q + k$ gives $mn = 2t + 1$. Hence mn is odd.

8. 4   **10.** 5   **12.** 1   **14.** 2

21. 50 is not divisible by 3. Thus $50 = 3^0 \cdot 50$; 0.

24. a) There exist integers k, q such that $a - b = 5k$, $x - y = 5q$. Then
$(a - b) + (x - y) = 5(k + q)$, whence
$(a + x) - (b + y) = 5(k + q)$, i.e., $a + x \equiv b + y \pmod 5$.

b) There exist integers k, q such that $a = b + 5k$, $x = y + 5q$. Then
$ax = (b + 5k)(y + 5q) = by + 5(ky + bq + 5kq)$.
Setting $t = ky + bq + 5kq$ gives $ax = by + 5t$, i.e., $ax \equiv by \pmod 5$.

26. For any positive integer a, either:

   a) $a = 3k$,   b) $a = 3k + 1$,   or   c) $a = 3k + 2$.

a) Suppose $a = 3k$. Then $a^2 = (3k)^2 = 9k^2 = 3(3k^2)$. a^2 is divisible by 3 and a is divisible by 3.

b) Suppose $a = 3k + 1$. Then $a^2 = (3k + 1)^2 = 9k^2 + 6k + 1 = 3(3k^2 + 2k) + 1$. a^2 is not divisible by 3 and a is not divisible by 3.

c) Suppose $a = 3k + 2$. Then $a^2 = (3k + 2)^2 = 9k^2 + 12k + 4 = 3(3k^2 + 4k + 1) + 1$. a^2 is not divisible by 3 and a is not divisible by 3.

Chapter 1, §5

1. a) $a = \frac{3}{8}$ b) $a = -\frac{35}{3}$ c) $a = -\frac{3}{20}$

2. a) $x = \frac{5}{3}$ b) $x = \frac{5}{2}$ c) $x = -\frac{2}{7}$

3. a) $\frac{2}{5}$ b) $\frac{1}{3}$ c) $\frac{6}{5}$ d) $\frac{10}{3}$

4. Let $b = n/m$ (this is a rational number). Then $ab = m/n \cdot n/m = 1$; $ba = n/m \cdot m/n = 1$.

5. a) 14   b) $\frac{11}{3}$ c) $\frac{9}{32}$ d) $\frac{1}{3}$ e) $\frac{5}{4}$ f) $\frac{1}{3}$ g) $\frac{3}{5}$

6. a) $\frac{50}{63}$ b) $-\frac{20}{31}$

7. a) $5! = 120$; $6! = 720$; $7! = 5{,}040$; $8! = 40{,}320$

b) $\binom{3}{0} = 1$; $\binom{3}{1} = 3$; $\binom{3}{2} = 3$; $\binom{3}{3} = 1$;

$\binom{4}{0} = 1$; $\binom{4}{1} = 4$; $\binom{4}{2} = 6$; $\binom{4}{3} = 4$; $\binom{4}{4} = 1$;

$\binom{5}{0} = 1$; $\binom{5}{1} = 5$; $\binom{5}{2} = 10$; $\binom{5}{3} = 10$; $\binom{5}{4} = 5$; $\binom{5}{5} = 1$

c) $\binom{m}{m - n} = \dfrac{m!}{(m - n)!(m - (m - n))!} = \dfrac{m!}{(m - n)!(m - m + n)!}$
$= \dfrac{m!}{(m - n)!n!} = \binom{m}{n}$

d) $\binom{m}{n} + \binom{m}{n - 1} = \dfrac{m!}{n!(m - n)!} + \dfrac{m!}{(m - n + 1)!(n - 1)!}$
                      [common denominator $n!(m - n + 1)!$]
$= \dfrac{m!(m - n + 1) + m!n}{n!(m - n + 1)!} = \dfrac{m!(m + 1)}{n!(m - n + 1)!}$
$= \dfrac{(m + 1)!}{n!((m + 1) - n)!} = \binom{m + 1}{n}$

8. Suppose that there exists a rational number $a = m/n$, written in lowest form, such that $a^3 = (m/n)^3 = 2$. Then $m^3 = 2n^3$. Thus m^3 is even, and hence m is even (proof similar to the one about m^2 and m). We can write $m = 2p$ for some integer p. Thus $m^3 = 2(4p^3)$. Going back to $m^3 = 2n^3$, this yields $2n^3 = 2(4p^3)$, that is, $n^3 = 4p^3 = 2(2p^3)$. Consequently, n^3 is even and n is even. Thus both m and n are even, which is impossible.

9. If $a^4 = 2$, then $a^4 = (a^2)^2$. But a^2 is also a rational number, and Theorem 4 shows that $(a^2)^2 = 2$ is impossible.

11. b) $a = 1.4141$ **12.** b) $a = 1.7321$ **13.** b) $a = 2.236$

14. b) 1.260   **16.** a) 3 g   b) 3/256 g   c) 3/2,048 g

19. a) $9 \cdot 10^5$ b) $7.29 \cdot 10^5$ e) Between 110 and 120 min

20. b) 22,200   f) Between 4 and 5 mo

21. b) The population triples in 50 yr.

Chapter 1, §6

1. a) $-\frac{15}{19}$ c) $\frac{25}{2}$ e) $-\frac{13}{7}$

2. a) $\dfrac{1}{x + y} - \dfrac{1}{x - y} = \dfrac{(x - y) - (x + y)}{(x + y)(x - y)} = \dfrac{x - y - x - y}{(x^2 - y^2)} = \dfrac{-2y}{x^2 - y^2}$

b) $(x - 1)(1 + x + x^2) = x(1 + x + x^2) - 1(1 + x + x^2)$
$$= x + x^2 + x^3 - 1 - x - x^2$$
$$= (x - x) + (x^2 - x^2) + x^3 - 1$$
Thus $(x - 1)(1 + x + x^2) = x^3 - 1$, or
$$\frac{x^3 - 1}{x - 1} = 1 + x + x^2.$$

c) $(x - 1)(1 + x + x^2 + x^3)$
$$= x(1 + x + x^2 + x^3) - 1(1 + x + x^2 + x^3)$$
$$= x + x^2 + x^3 + x^4 - 1 - x - x^2 - x^3 = x^4 - 1$$
Thus $(x - 1)(1 + x + x^2 + x^3) = x^4 - 1$ or
$$\frac{x^4 - 1}{x - 1} = 1 + x + x^2 + x^3.$$

d) $(x - 1)(x^{n-1} + x^{n-2} + \cdots + x + 1)$
$$= x(x^{n-1} + x^{n-2} + \cdots + x + 1) - (x^{n-1} + x^{n-2} + \cdots + x + 1)$$
$$= x^n + x^{n-1} + \cdots + x - x^{n-1} - x^{n-2} - \cdots - x - 1 = x^n - 1$$
Thus $(x - 1)(x^{n-1} + x^{n-2} + \cdots + x + 1) = x^n - 1$ or
$$\frac{x^n - 1}{x - 1} = x^{n-1} + \cdots + x + 1.$$

3. b) $\dfrac{2x}{x + 5} - \dfrac{3x + 1}{2x + 1} = \dfrac{2x(2x + 1) - (x + 5)(3x + 1)}{(x + 5)(2x + 1)}$
Numerator gives $(4x^2 + 2x) - (3x^2 + x + 15x + 5)$
$$= x^2 - 14x - 5.$$

Denominator gives $2x^2 + x + 10x + 5 = 2x^2 + 11x + 5$.

The quotient is equal to $\dfrac{x^2 - 14x - 5}{2x^2 + 11x + 5}$.

4. a) Use cross multiplication.

$$(x^2 + xy + y^2)(x - y) = (x^2 + xy + y^2) - (x^2 + xy + y^2)y$$
$$= x^3 + x^2y + xy^2 - x^2y - xy^2 - y^3$$
$$= x^3 - y^3$$

b) $(x^3 + x^2y + xy^2 + y^3)(x - y)$
$$= (x^3 + x^2y + xy^2 + y^3)x - (x^3 + x^2y + xy^2 + y^3)y$$
$$= x^4 + x^3y + x^2y^2 + xy^3 - x^3y - x^2y^2 - xy^3 - y^4$$
$$= x^4 - y^4$$

c) $x^2 = \dfrac{(1 - t^2)^2}{(1 + t^2)^2} = \dfrac{1 - 2t^2 + t^4}{(1 + t^2)^2}$, $\qquad y^2 = \dfrac{4t^2}{(1 + t^2)^2}$

Hence $x^2 + y^2 = \dfrac{1 - 2t^2 + t^4 + 4t^2}{(1 + t^2)^2} = \dfrac{1 + 2t^2 + t^4}{(1 + t^2)^2}$

$$= \dfrac{(1 + t^2)^2}{(1 + t^2)^2} = 1.$$

5. b) $(x + 1)(x^4 - x^3 + x^2 - x + 1)$
$$= x(x^4 - x^3 + x^2 - x + 1) + (x^4 - x^3 + x^2 - x + 1)$$
$$= x^5 - x^4 + x^3 - x^2 + x + x^4 - x^3 + x^2 - x + 1$$
$$= x^5 + 1$$

Thus $(x^5 + 1)/(x + 1) = x^4 - x^3 + x^2 - x + 1$.

c) $(x + 1)(x^{n-1} - x^{n-2} + x^{n-3} - \cdots - x + 1)$
$$= x(x^{n-1} - x^{n-2} + \cdots - x + 1) + (x^{n-1} - x^{n-2} + \cdots - x + 1)$$
$$= x^n - x^{n-1} + x^{n-2} + \cdots - x^2 + x + x^{n-1} - x^{n-2}$$
$$+ \cdots + x^2 - x + 1$$
$$= x^n + 1$$

Hence $(x^n + 1)/(x + 1) = x^{n-1} - x^{n-2} + x^{n-3} + \cdots - x + 1$.

6. $25/8$ sec   **7.** a) $9/20$ lb/in^3 b) $9/2$ lb/in^3

8. a) $0°C$ c) $(335/9)°C$ e) $-40°C$

9. a) $32°F$ c) $-40°F$ d) $98.6°F$ e) $104°F$ f) $212°F$

11. a) 9 kg   **12.** $\frac{5}{2}$ hr   **13.** 500 tickets at $\$5.00$ and 800 tickets at $\$2.00$

14. a) 120 g   **15.** a) 18 kg   **18.** a) $\frac{5}{2}$ kg

Chapter 2, §1

1. $x = \frac{5}{3}$ and $y = \frac{1}{3}$ **3.** $x = -2$ and $y = 1$

5. $x = -\frac{1}{2}$ and $y = -\frac{3}{2}$

9. a) Multiply the first equation by d and the second by b. Subtract each side of the second equation from the corresponding side of the first. The terms with y cancel, and you get $adx - bcx = d - 2b$, whence $x = (d - 2b)/(ad - bc)$. Multiply the first equation by c and the second by a. Subtract the first from the second. The terms with x cancel, and you get $ady - bcy = 2a - c$, whence

$$y = \frac{2a - c}{ad - bc}.$$

Chapter 2, §2

1. $x = 1$, $y = \frac{1}{2}$, $z = -\frac{1}{2}$ 3. $x = 51/67$, $y = 29/67$, $z = 25/67$
5. $x = \frac{1}{2}$, $y = -9/16$, $z = 13/16$
9. $x = 205/43$, $y = -130/43$, $z = -94/43$
11. $x = 32/11$, $y = -38/11$, $z = -51/11$

Chapter 3, §2

1. a) a positive and a positive implies by **POS 1** that $aa = a^2$ is positive.
   b) If b is negative, then $-b$ is positive. By **POS 1**, $a(-b)$ is positive. But $a(-b) = -ab$. Hence $-ab$ is positive and ab is negative.
   c) a negative implies $-a$ positive; b negative implies $-b$ positive. By **POS 1**, $(-a)(-b)$ is positive. But $(-a)(-b) = ab$.
2. a) If a^{-1} were negative then $aa^{-1} = 1$ would be negative, which is impossible. Hence a^{-1} is positive.
3. If a^{-1} were positive, then aa^{-1} would be negative, which is impossible because $aa^{-1} = 1$. Hence a^{-1} is negative.
4. $\left(\sqrt{\frac{a}{b}}\right)^2 = \frac{a}{b}$ on one hand; $\left(\frac{\sqrt{a}}{\sqrt{b}}\right)^2 = \frac{(\sqrt{a})^2}{(\sqrt{b})^2} = \frac{a}{b}$ on the other. Since $\dfrac{\sqrt{a}}{\sqrt{b}}$

   is positive, it must be equal to $\sqrt{\frac{a}{b}}$.
5. $\dfrac{1}{1 - \sqrt{2}} = \dfrac{1 + \sqrt{2}}{(1 - \sqrt{2})(1 + \sqrt{2})} = \dfrac{1 + \sqrt{2}}{1 - (\sqrt{2})^2} = \dfrac{1 + \sqrt{2}}{-1}$

   $= -(1 + \sqrt{2})$.
7. $\dfrac{1}{3 + \sqrt{5}} = \dfrac{3 - \sqrt{5}}{3^2 - (\sqrt{5})^2} = \dfrac{3 - \sqrt{5}}{4}$. Hence $c = \frac{3}{4}$ and $d = -\frac{1}{4}$.
10. $(x + y\sqrt{5})(z + w\sqrt{5}) = x(z + w\sqrt{5}) + y\sqrt{5}\,(z + w\sqrt{5})$

    $= xz + xw\sqrt{5} + yz\sqrt{5} + 5yw$
    $= (xz + 5yw) + (xw + yz)\sqrt{5}$

    Let $c = xz + 5yw$, $d = xw + yz$. These are rational numbers since x, y, z, w are rational.
12. a) $\dfrac{x + 1}{2(\sqrt{2x + 3} - 2)}$ c) $\dfrac{-1}{\sqrt{x - h} + \sqrt{x}}$ f) $\dfrac{2}{\sqrt{x + 2h} + \sqrt{x}}$
13. a) $x = 3$ and $x = -1$ c) $x = 7$ and $x = -1$ e) $x = -1$ and $x = -7$
14. b) $x = \frac{1}{3}$ and $x = -1$ d) $x = \frac{2}{3}$ and $x = 0$

15. a) $\dfrac{x-y}{x-2\sqrt{x}\,\sqrt{y}+y}$ c) $\dfrac{1}{x-\sqrt{x+1}\,\sqrt{x-1}}$

   e) $\dfrac{x+y-1}{x+y+3+4\sqrt{x+y}}$

16. a) $\dfrac{x+y+2\sqrt{x}\,\sqrt{y}}{x-y}$ c) $x+\sqrt{x-1}\,\sqrt{x+1}$

17. $\sqrt{x-1}=3+\sqrt{x}$ implies that $x-1=(3+\sqrt{x})^2=9+6\sqrt{x}+x$.
   Thus $\sqrt{x}=-10/6$. This is impossible since $\sqrt{x}$ cannot be negative.

18. $\sqrt{x-1}=3-\sqrt{x}$ implies that $x-1=9-6\sqrt{x}+x$.   Thus,
   $\sqrt{x}=10/6=5/3$, i.e., $x=25/9$.

19. a) No x b) No x c) $x=1$

20. For any number x we know that $|x|=|-x|$, because $|x|=\sqrt{x^2}=\sqrt{(-x)^2}=|-x|$. But $b-a=-(a-b)$, so put $x=a-b$.

Chapter 3, §3

1. a) $2^2a^1b^{-4}$ or $2^23^0a^1b^{-4}$ c) $2^{-1}3^1a^{-2}b^{-2}$ **3.** 8   **4.** No

5. No; for instance, $(\sqrt{2})^5=4\sqrt{2}$. If this were a rational number r, then $\sqrt{2}=r/4$ would be rational, and this is not true.

6. b) 2   c) 27   g) 125   **7.** a) .3   c) .25   **8.** a) $\frac{4}{9}$ b) $\frac{2}{3}$

9. a) $x=2+5^{1/3}$ or $x=2+\sqrt[3]{5}$ c) $x=14/3$ or $x=16/3$

   f) $x=\dfrac{-5\sqrt{8}\pm 1}{3\sqrt{8}}$ or $x=\dfrac{-40\pm\sqrt{8}}{24}$

Chapter 3, §4

1. If $a-b>0$ and $c<0$, then $(a-b)c<0$. But $(a-b)=ac-bc$, so that $ac<bc$.

2. $ac<bc<bd$, using **IN 2** twice.

3. $ac>bc>bd$, using **IN 3** twice.

4. a) x positive implies y positive. Then $1/x$ and $1/y$ are also positive. Multiply each side of the inequality $x<y$ by $1/x$. We get $1=x/x<y/x$. Multiply each side of this last inequality by $1/y$. We get $1/y<1/x$, as desired.

   b) Multiply each side of the inequality $a/b<c/d$ by d, and then by b. You get $ad<bc$. Conversely, multiply each side of the inequality $ad<bc$ by $1/d$ and $1/b$.

5. We must verify that $(b+c)-(a+c)$ is positive. But $(b+c)-(a+c)=b+c-a-c=b-a$, which is positive by assumption. Also, $(b-c)-(a-c)=b-c-a+c=b-a$, which is positive. Hence $(b-c)>(a-c)$.

6. We have $a^2=aa<ab<bb=b^2$, using **IN 2** twice. Next, multiplying each side of the inequality $a^2<b^2$ by a, we get $a^3<ab^2$, and multiply-

ing each side of the inequality $a < b$ by b^2 we get $ab^2 < b^3$, whence $a^3 < b^3$. Continuing in this way yields the general result.

7. Let $r = a^{1/n}$ and $s = b^{1/n}$. If $r \geq s$, then we know from Exercise 6 that $r^n \geq s^n$, or in other words, $a \geq b$. But this contradicts our assumption. Hence $r < s$.

8. a) Use Exercise 4 (b). The inequality $a/b < (a + c)/(b + d)$ is equivalent with $a(b + d) < b(a + c)$ by cross multiplication, and this is equivalent with $ab + ad < ba + bc$, which is equivalent with $ad < bc$, which is true by multiplying each side of the inequality $a/b < c/d$ by b and d.

  As for the inequality on the right-hand side, it is equivalent by cross multiplication with $d(a + c) < c(b + d)$, which is equivalent with $da + dc < cb + cd$, which is equivalent with $da < cb$, which is equivalent by cross multiplication with $a/b < c/d$, which is true by assumption.

  b) The left inequality is equivalent by cross multiplication with $a(b + rd) < b(a + rc)$, which is equivalent with $ab + ard < ba + brc$, which is equivalent with $ard < brc$. Multiplying both sides by r^{-1}, which is positive, this last inequality is equivalent with $ad < bc$, which is true by cross multiplication of the assumed inequality $a/b < c/d$.

  The argument for the right inequality works again by cross multiplication.

  c) By cross multiplication, it suffices to prove that $(a + rc)(b + sd) < (a + sc)(b + rd)$, and this is equivalent with $ab + rcb + asd + rscd < ab + scb + ard + rscd$, which is equivalent with $rcb + asd < scb + ard$, which is equivalent with $r(bc - ad) < s(bc - ad)$. Since $r < s$ and $bc - ad > 0$, this last inequality is true, and our assertion is proved.

9. $3x - 1 > 0$, $3x > 1$, $x > \frac{1}{3}$ **11.** $x > -1$ **13.** $x < -\frac{1}{3}$
19. $-\frac{1}{2} < x < 2$ **21.** $-\frac{1}{2} < x < 0$ **23.** $\frac{1}{3} < x < \frac{1}{2}$
25. $-1 < x < 1$ **27.** $-\sqrt{3} < x < \sqrt{3}$ **29.** $-1 > x > 1$

Chapter 4

1. $\dfrac{-3 \pm \sqrt{17}}{2}$ **2.** $\dfrac{3 \pm \sqrt{17}}{2}$

3. Impossible with real numbers, $(4 \pm \sqrt{-4})/2$ **4.** $x = 5$ and $x = -1$

7. $\dfrac{-3 \pm \sqrt{57}}{6}$

8. Impossible with real numbers, $(-5 \pm \sqrt{-31})/4$

13. $\dfrac{-3 \pm \sqrt{9 + 4\sqrt{2}}}{2}$ **14.** $\dfrac{3 \pm \sqrt{9 + 4\sqrt{5}}}{2}$

Chapter 5, §2

1. a)

b)

c)

d)

e)

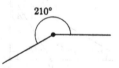

h)

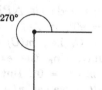

Areas: a) 1 in^2 b) 2 in^2 c) 135/60 in^2

2. a) $\pi r^2 \left(\dfrac{\theta_2 - \theta_1}{360} \right)$ b) $\pi(r_2^2 - r_1^2)$ c) $\pi(r_2^2 - r_1^2)\left(\dfrac{\theta_2 - \theta_1}{360} \right)$

3. a) $4\pi/9$ c) $\pi/3$ **4.** a) 21π c) 32π **5.** a) $16\pi/9$ c) $4\pi/3$

6. a) $7\pi/36$ c) $77\pi/72$

Chapter 5, §3

1. a) $2\sqrt{2}$ c) $4\sqrt{2}$ **2.** a) $\sqrt{5}$ c) $\sqrt{65}$ e) $r\sqrt{34}$

3. a) $\sqrt{3}$ b) $2\sqrt{3}$ e) $r\sqrt{3}$ **4.** a) $\sqrt{50}$ c) $\sqrt{38}$

5. a) $\sqrt{a^2 + b^2 + c^2}$ b) $r\sqrt{a^2 + b^2 + c^2}$

6. $100\sqrt{26}$ ft   **7.** a) $\sqrt{51}$ ft   **8.** a) $10\sqrt{40}$ ft

9. $d(O, P)^2 + d(O, M)^2 = d(P, M)^2$
$d(O, Q)^2 + d(O, M)^2 = d(M, Q)^2$ (I)

If $d(O, P) = d(O, Q)$, that is, $d(O, P)^2 = d(O, Q)^2$, then (I) gives
$d(P, M)^2 = d(M, Q)^2$, that is, $d(P, M) = d(M, Q)$.

10. In the first case, the picture is as follows.

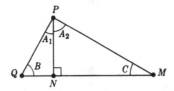

The triangles $\triangle PQN$ and $\triangle PNM$ are right triangles, and hence $m(A_1) + m(B) = 90°$, $m(A_2) + m(C) = 90°$. Adding these, we find $m(A_1) + m(A_2) + m(B) + m(C) = 180°$. But $m(A) = m(A_1) + m(A_2)$, so that we proved what we wanted.

In the second case, the picture is as follows.

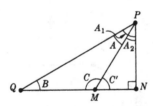

Then $\triangle PQN$ is a right triangle, and so is $\triangle PMN$, with right angle at N. Hence (*) $m(A_1) + m(B) = 90°$, $m(A_2) + m(C') = 90°$, where C' is the supplementary angle to C, i.e., $m(C) + m(C') = 180°$, so that $m(C) = 180° - m(C')$. Subtracting the expressions in (*), we find $m(A_1) - m(A_2) + m(B) - m(C') = 0°$. But $m(A) = m(A_1) - m(A_2)$. Substituting the value for $m(C')$, we get $m(A) + m(B) + m(C) = 180°$.

11. a)

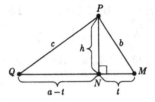

Area of $\triangle PQN = \frac{1}{2}h(a - t)$
Area of $\triangle PMN = \frac{1}{2}ht$
Area of $\triangle PQM$ = area of $\triangle PQN$ + area of $\triangle PMN$
                    $= \frac{1}{2}h(a - t + t) = \frac{1}{2}ha$

b)

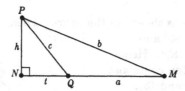

Area of $\triangle PNM = \frac{1}{2}h(a + t)$
Area of $\triangle PNQ = \frac{1}{2}ht$
Area of $\triangle PQM$ = area of $\triangle PNM$ − area of $\triangle PNQ$
$\qquad\qquad = \frac{1}{2}h(a + t − t) = \frac{1}{2}ha$

12. a) By Pythagoras $d(P, M)^2 = d(P, Q)^2 + d(Q, M)^2$. Then $d(P, M)^2 \geq d(P, Q)^2$. Thus $d(P, M) \geq d(P, Q)$, and similarly $d(P, M) \geq d(Q, M)$.

b) $d(P, M)^2 = d(P, Q)^2 + d(Q, M)^2$. The distance is minimum when $d(Q, M) = 0$, i.e., $d(Q, M)^2 = 0$. This occurs exactly when $M = Q$.

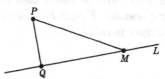

13. a) We can draw a perpendicular, from a point M on K, to L_2 and this line is also perpendicular to L_1. Then $\triangle MQ'P'$ is a right triangle, so $m(C) + m(A) + 90° = 180°$. Also $\triangle MQP$ is a right triangle, so $m(C) + m(B) + 90° = 180°$. Combining these equations gives $m(A) = m(B)$.

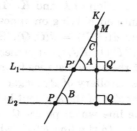

b) See preceding Theorem 1, §3.

c) Let B be the angle as shown on the figure. Then
$m(A) + m(B) = 180°$ and
$m(A') + m(B) = 180°$. Hence
$m(A) = m(A')$. You could also
argue by using (a) and (b).

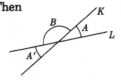

Chapter 6, §1

1. a) All points of the plane are fixed.
   b) The center of the reflection is the only fixed point.
   c) All the points on the reflection line are fixed.
   d) The center of the rotation is the only fixed point.
   e) No fixed point.
   f) The center of the dilation is the only fixed point if $r \neq 1$. If $r = 1$,
2.   the dilation is the identity.
   a) $(-1)\, 360 + 330$ c) $(-1)\, 360 + 180$ g) $(0)\, 360 + 120$
   i) $(-2)\, 360 + 320$

Chapter 6, §2

4. The lines $F(L)$ and $F(K)$ cannot have a point in common, otherwise this point would be of the form $F(P) = F(Q)$ for some point P on L and Q on K. But the distance between P and Q is the same as the distance between $F(P)$ and $F(Q)$, and it would then follow that $d(P, Q) = 0$, so $P = Q$, which is impossible. Hence $F(L)$ and $F(K)$ have no point in common, and are therefore parallel.

5.

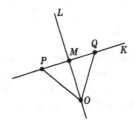

Let M be the point of intersection of L and K. Let O be a point of L, not on K. Let P, Q be points on K lying on opposite sides of M, at the same distance from M. Thus $d(O, P) = d(O, Q)$. Since F is an isometry, we have $d(F(O), F(P)) = d(F(O), F(Q))$. Furthermore, $F(P)$ and $F(Q)$ are distinct, lie on $F(K)$, at the same distance from $F(M)$, again because F is an isometry. From the corollary of Pythagoras, it follows that $F(O)$ lies on the perpendicular bisector of the segment between $F(P)$ and $F(Q)$. The line $F(L)$ is the unique line which passes through $F(O)$ and $F(M)$, and is therefore perpendicular to the line $F(K)$, which is the unique line passing through $F(P)$ and $F(Q)$.

6. Theorems 1 and 2 are valid.

Theorem 4. If P, Q, R, S are four points which do not lie in a plane and are fixed by an isometry f of 3-space, then f is the identity.

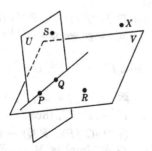

Proof. Observe first that P, Q, R do not lie on a line. Otherwise, this line and S lie in a plane, which would contradict the assumption of the theorem. Let V be the plane passing through P, Q, R. Arguing as above, the points S, P, Q do not lie on a line. Let U be the plane passing through S, P, Q. (We assume that if three points are not on a line, then there exists one and only one plane passing through them.) By Theorem 3, the isometry f leaves U and V fixed. Let X be a point in 3-space. We want to show that $f(X) = X$. If X lies in U or V, we are done. Assume that X does not lie in U or V. If X lies on the line between S and R, then X is fixed by Theorem 2. Assume that X does not lie on the line between S and R. Let W be the plane passing through X, S, R. We assume that the intersection of two planes is a line. Let L be the intersection of W and U. Since L lies in U, it is fixed by f. Since R is not in the plane U, it follows that R is not in L. We assume that given a line and a point not on the line, there exists a unique plane passing through the line and the point. Thus W is the unique plane passing through R and L. Any two points on L are fixed by f, and hence by Theorem 3, the plane W is fixed by f. Since X lies in W, we conclude that X is fixed by f, thereby proving Theorem 4.

Chapter 6, §3

1. $n = 2$

5. a) 4   c) 8   e) 24

Chapter 6, §4

1. a) $F(S)$ is contained in S'.   *Proof.* Let Q be on S, $d(P, Q) = r$. Since F is an isometry, we have $d(F(P), F(Q)) = r$ and $F(Q)$ is on S'. S' is contained in $F(S)$.   *Proof.* Let Q' be on S'. Using F^{-1}, we have similarly $F^{-1}(Q')$ on S, and thus $F(F^{-1}(Q')) = Q'$ is in $F(S)$.

2. There exists a translation T such that $T(P') = P$. This reduces our proof to the case when $P = P'$, which we now assume. Since $d(P, Q) = d(P, Q')$, there exists a rotation R with respect to P such that $R(Q) = Q'$. This concludes the proof.

3. Since $F \circ H = F \circ G$, we get $F^{-1} \circ (F \circ H) = F^{-1} \circ (F \circ G)$. Using the associativity of mappings we get $(F^{-1} \circ F) \circ H = (F^{-1} \circ F) \circ G$, whence $H = G$.

4. a) $F^2 = I$ and $F^3 = I$. But $F^3 = F \circ F^2$, thus $F^3 = I = F \circ I$, i.e. $F = I$.
$F^8 = F^3 \circ F^5$, thus $I = F^3 \circ I$, i.e. $F^3 = I$. $F^5 = F^3 \circ F^2$, thus $I = I \circ F^2$, i.e. $F^2 = I$. $F^3 = F \circ F^2$, thus $I = F \circ I$, i.e. $F = I$.

5. $F(P) = G(P)$ implies $P = (F^{-1} \circ G)(P)$. $F(Q) = G(Q)$ implies $Q = (F^{-1} \circ G)(Q)$. $F(M) = G(M)$ implies $M = (F^{-1} \circ G)(M)$. $F^{-1} \circ G$ is an isometry with 3 fixed points P, Q, M. Hence $F^{-1} \circ G = I$, whence $F \circ F^{-1} \circ G = F$, i.e. $F = G$.

8. $F^{-1} = F^{n-1}$

9. $V = \begin{bmatrix} 1 & 2 & 3 & 4 \\ 2 & 1 & 4 & 3 \end{bmatrix}$, $H \circ V = \begin{bmatrix} 1 & 2 & 3 & 4 \\ 3 & 4 & 1 & 2 \end{bmatrix}$

10. $H \circ G \circ H = G^3$, $n = 3$, $G = \begin{bmatrix} 1 & 2 & 3 & 4 \\ 2 & 3 & 4 & 1 \end{bmatrix}$,

$G^3 = \begin{bmatrix} 1 & 2 & 3 & 4 \\ 4 & 1 & 2 & 3 \end{bmatrix}$, $H \circ G = \begin{bmatrix} 1 & 2 & 3 & 4 \\ 3 & 2 & 1 & 4 \end{bmatrix}$,

$H \circ G^3 = \begin{bmatrix} 1 & 2 & 3 & 4 \\ 1 & 4 & 3 & 2 \end{bmatrix}$, $G^2 \circ H = \begin{bmatrix} 1 & 2 & 3 & 4 \\ 2 & 1 & 4 & 3 \end{bmatrix}$

11.

	I	G	G^2	G^3	H	HG	HG^2	HG^3
I	I	G	G^2	G^3	H	HG	HG^2	HG^3
G	G	G^2	G^3	I	HG^3	H	HG	HG^2
G^2	G^2	G^3	I	G	HG^2	HG^3	H	HG
G^3	G^3	I	G	G^2	HG	HG^2	HG^3	H
H	H	HG	HG^2	HG^3	I	G	G^2	G^3
HG	HG	HG^2	HG^3	H	G^3	I	G	G^2
HG^2	HG^2	HG^3	H	HG	G^2	G^3	I	G
HG^3	HG^3	H	HG	HG^2	G	G^2	G^3	I

13. $I = \begin{bmatrix} 1 & 2 & 3 \\ 1 & 2 & 3 \end{bmatrix}$, $VG = \begin{bmatrix} 1 & 2 & 3 \\ 3 & 2 & 1 \end{bmatrix}$, $VG^2 = \begin{bmatrix} 1 & 2 & 3 \\ 2 & 1 & 3 \end{bmatrix}$

14. $k = 5$, $G^6 = I$

15. b) $G^2 = \begin{bmatrix} 1 & 2 & 3 & 4 & 5 & 6 \\ 3 & 4 & 5 & 6 & 1 & 2 \end{bmatrix}$, $\quad H = \begin{bmatrix} 1 & 2 & 3 & 4 & 5 & 6 \\ 4 & 3 & 2 & 1 & 6 & 5 \end{bmatrix}$,

$HG = \begin{bmatrix} 1 & 2 & 3 & 4 & 5 & 6 \\ 3 & 2 & 1 & 6 & 5 & 4 \end{bmatrix}$

c) $VG = \begin{bmatrix} 1 & 2 & 3 & 4 & 5 & 6 \\ 6 & 5 & 4 & 3 & 2 & 1 \end{bmatrix}$, $\quad VG^4 = \begin{bmatrix} 1 & 2 & 3 & 4 & 5 & 6 \\ 3 & 2 & 1 & 6 & 5 & 4 \end{bmatrix}$

Chapter 6, §5

1. By Theorems 4, 5, and 6 an isometry F can be written as a composite of isometries F_1, or $F_1 \circ F_2$, or $F_1 \circ F_2 \circ F_3$, such that each one of F_1, F_2, F_3 has an inverse (because they are the identity, or a translation, or a reflection, or a rotation). Then F itself has an inverse, given in these cases by F_1^{-1}, $F_2^{-1} \circ F_1^{-1}$, and $F_3^{-1} \circ F_2^{-1} \circ F_1^{-1}$.

Chapter 6, §6

2. S congruent to S' means that there exists an isometry F_1 such that $F_1(S) = S'$. Furthermore, S' congruent to S'' means that there exists an isometry F_2 such that $F_2(S') = S''$. But then $(F_2 \circ F_1)(S) = F_2(F_1(S)) = S''$. Since $F_2 \circ F_1$ is an isometry, it follows that S is congruent to S''.

3. Let the two squares have corners at P, Q, M, N and P', Q', M', N' respectively, as shown. Since $\overline{PQ}$ and $\overline{P'Q'}$ have the same length, there exists an isometry which maps P on P' and Q on Q', say by Theorem 8.

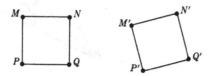

Thus we may assume that $P = P'$ and $Q = Q'$. In this case, either $M = M'$ or $M \neq M'$. If $M = M'$, then N is the point of intersection of the line through M parallel to $\overline{PQ}$, and the line through Q, perpendicular to $\overline{PQ}$. Similarly, N' is this same point of intersection, so that $N = N'$. It follows that in the present case, the squares coincide. If $M \neq M'$, then the reflection through the line L_{PQ} maps M on M', and reduces our problem to the preceding case.

5.

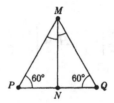

We draw the triangle $\triangle PQM$. Let the line perpendicular to $\overline{PQ}$, passing through M, intersect $\overline{PQ}$ at the point N, as shown. Then $\triangle MNP$ and $\triangle MNQ$ are right triangles. By the known theorem on the sum of the angles of a right triangle, it follows that the angles $\angle PMN$ and $\angle NMQ$ both have 30°. Hence if we reflect the line L_{PM} through the line L_{MN}, we obtain the line L_{MQ}. This reflection maps the line L_{PQ} on itself. Since P is the intersection of L_{PM} and L_{PQ}, its image by the reflection is the intersection of L_{MQ} and L_{PQ}, which is Q. Hence our reflection maps $\overline{MP}$ on $\overline{MQ}$, and these two sides therefore have the same length. The same argument applies to another pair of sides.

7. By an isometry we can map P on P' and Q on Q'. This reduces us to the case when $P = P'$ and $Q = Q'$, which we now assume. The picture is then as follows.

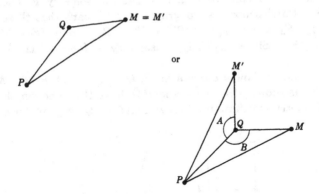

The angles of the two triangles with vertex at Q have the same measure. If M and M' lie on the same side of the line L_{PQ}, then they lie on the same ray with vertex at Q. Since $d(Q, M) = d(Q, M')$, it follows that $M = M'$, and we are done. If M and M' lie on opposite sides of the line L_{PQ}, then we reflect through this line to get them on the same side, and this reduces our problem to the case already taken care of.

9. It is the set of all line segments $\overline{PQ}$, where P lies on one side of the square and Q lies on the other. There are other ways also. For instance, the set of all line segments from one corner to the points on the opposite two sides. These are shown on the next figures.

Chapter 7, §1

1. a) 1 in² b) 4 in²
2. a) 4/3 in² c) 36/3 in² d) $\frac{1}{3}$ in²
3. Multiplied by ab.
4. Area is πab; each square of area 1 is dilated into a square of area ab. Hence the area π of the disc of radius 1 is dilated into πab.

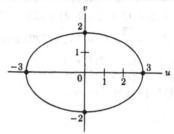

5. a) 28π c) $\pi\sqrt{18}$
6. a) $\pi/\sqrt{12}$ c) $\pi/6$
7. $F_r(x, y, z) = (rx, ry, rz)$.
8. a) Each side dilates by r, thus the volume is multiplied by r^3.
   b) Using the approximation by small cubes, the volume will be multiplied by r^3.
   c) $\frac{4}{3}\pi r^3$
9. $x^2 + y^2 + z^2 = 1$ 10. $V = abc$ 11. c) $abcV$ 12. 56π
13. $\frac{20}{3}\sqrt{6}\,\pi$ 14. $\frac{4}{3}\pi abc$

Chapter 8, §1

3. $x < 0$ and $y > 0$
4. $x < 0$ and $y < 0$

Chapter 8, §2

1. $\sqrt{17}$ 3. $\sqrt{61}$ 5. $\sqrt{106}$

11. $d(P, Q) = 0$, so that $d(P, Q)^2 = 0 = (x_1 - x_2)^2 + (y_1 - y_2)^2$. Thus $(x_1 - x_2)^2 = -(y_1 - y_2)^2$ is impossible with real numbers unless $x_1 - x_2 = 0$ and $y_1 - y_2 = 0$, i.e. $x_1 = x_2$ and $y_1 = y_2$, which means $P = Q$.

12. $d(A, B) = \sqrt{(a_1 - a_2)^2 + (b_1 - b_2)^2}$
$$d(rA, rB) = \sqrt{(ra_1 - ra_2)^2 + (rb_1 - rb_2)^2}$$
$$= \sqrt{r^2[(a_1 - a_2)^2 + (b_1 - b_2)^2]}$$

Thus $d(rA, rB) = r\, d(A, B)$.

13. a) $\sqrt{41}$ c) $\sqrt{50}$

Chapter 8, §3

1. $(x + 3)^2 + (y - 1)^2 = 4$ **3.** $(x + 1)^2 + (y + 2)^2 = \frac{1}{9}$

7. $C = (1, 2)$ and $r = 5$ **9.** $C = (-1, 9)$ and $r = \sqrt{8}$

11. $C = (5, 0)$ and $r = \sqrt{10}$

13. $(x + 1)^2 + y^2 = 6$; $C = (-1, 0)$ and $r = \sqrt{6}$

15. $(x + 2)^2 + (y - 2)^2 = 28$; $C = (-2, 2)$ and $r = \sqrt{28}$

17. $(x - 1)^2 + (y + 5/2)^2 = \dfrac{133}{4}$; $C = (1, -5/2)$ and $r = \dfrac{\sqrt{133}}{2}$

20. a) $(x - 1)^2 + (y + 3)^2 + (z - 2)^2 = 1$
    c) $(x + 1)^2 + (y - 1)^2 + (z - 4)^2 = 9$
    e) $(x + 2)^2 + (y + 1)^2 + (z + 3)^2 = 4$

Chapter 8, §4

2. By cross multiplication for inequalities, it suffices to prove that $(1 - s^2)(1 + t^2) > (1 + s^2)(1 - t^2)$, which is equivalent to $1 - s^2 + t^2 - s^2t^2 > 1 + s^2 - t^2 - s^2t^2$. This is equivalent to $2t^2 > 2s^2$, which is true because $0 \leq s < t$.

3. c) $x = 15/17$ and $y = 8/17$

4. a) $\dfrac{1 - t^2}{1 + t^2}$ approaches -1 b) $\dfrac{1 - t^2}{1 + t^2}$ approaches -1

5. In both cases $\dfrac{2t}{1 + t^2}$ approaches 0.

Chapter 9, §1

1. a) $cA = (-12, 20)$ c) $cA = (-8, -10)$ e) $cA = (4, 5)$

2. $bA = (ba_1, ba_2)$ and $cA = (ca_1, ca_2)$. $bA = cA$ gives $ba_1 = ca_1$ and $ba_2 = ca_2$, whence $b = c$.

3. $d^2(-A, -B) = (-a_1 - (-b_1))^2 + (-a_2 - (-b_2))^2$
$= (b_1 - a_1)^2 + (b_2 - a_2)^2 = d^2(A, B)$

4. a) $cA = (ca_1, ca_2, ca_3)$, stretching all the coordinates by c

b) $-A = (-a_1, -a_2, -a_3)$

Chapter 9, §2

1. $A + B = (4, 6)$ and $A - B = (-2, 2)$

3. $A + B = (2, 3)$ and $A - B = (-4, 1)$

5. $A + B = (-2, 3)$ and $A - B = (0, -1)$

11. We have $T_A(P) = P + A$ and $T_A(Q) = Q + A$. Hence
$d(T_A(P), T_A(Q)) = |T_A(P) - T_A(Q)| = |P + A - Q - A|$
$= |P - Q| = d(P, Q)$.

Under translation by A, we have $d(M, O) = d(M + A, A)$, that is, $|M| = |M + A - A|$, valid for all points of the disc.

14. $R_Q(P) = Q - A = Q - (P - Q) = Q - P + Q = 2Q - P$

15. a) Reflection of M through O is $-M$. The translation of $-M$ by $2Q$ is $2Q - M$.

b) By Exercise 14, we must have $-P + A = 2Q - P$. Hence $Q = \frac{1}{2}A$.

16. a) $F_{r,Q}(P) = r(P - Q) + Q$

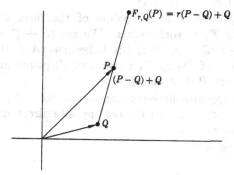

b) Translation by $(1 - r)Q$

17. Reflection of M through Q is $2Q - M$. If $d(M, A) = r$, that is $|M - A| = r$, then $|(2Q - M) - (2Q - A)| = |M - A| = r$. The reflection of $S(r, A)$ is the circle of center $2Q - A$ and of radius r.

18. By $-A$, because $T_{-A} \circ T_A = T_A \circ T_{-A} = I$.

19. a) $F_r^{-1} = F_{1/r}$, because

$$(F_r \circ F_{1/r})(P) = r \circ \left(\frac{1}{r} P\right) = P,$$

$$(F_{1/r} \circ F_r)(P) = \frac{1}{r} \circ (rP) = P$$

20. $(T_A \circ T_B)(P) = T_A(P + B) = (P + B) + A = P + (B + A)$
$$= T_{A+B}(P); \text{ translation by } A + B$$

21. a) $R(P) = -P$, so that $(R \circ R)(P) = -(-P) = P$ and $R^{-1} = R$

    b) $(R \circ T_A \circ R^{-1})(P) = (R \circ T_A)(-P) = R(A - P) = P - A$
$$= T_A(P); \text{ translation by } -A$$

22. a) Image is $(1, -2)$ c) Image is $(-2, 4)$

23. $|R_z(P) - R_z(Q)|^2 = (p_1 - q_1)^2 + (-p_2 - (-q_2))^2$
$$= (p_1 - q_1)^2 + (-p_2 + q_2)^2 = |P - Q|^2$$

25. a) $T_A(P) = P$ means $P + A = P$, in which case $A = 0$, i.e. $T_A = I$.
Then all points are fixed.

    b) $R_o(P) = -P$ means $-P = P$, i.e. $P = O$.

    c) $R_P(M) = M$ means $2P - M = M$, i.e. $P = M$.

    d) $R_z(P) = P$ means $(x, -y) = (x, y)$, i.e. $y = 0$.

    e) $R_v(P) = P$ means $(-x, y) = (x, y)$, i.e. $x = 0$.

33. $r = |A|$; $\left| \dfrac{1}{r} A \right| = \left(\dfrac{1}{r} \right) |A|$ (since $r > 0$) $= \dfrac{|A|}{|A|} = 1$

Chapter 10, §2

1. a) $(2, 2)$ b) $(5/3, 3)$ c) $(7/3, 1)$

5. The segment $\overline{PQ}$ consists of all points of the form $tP + (1 - t)Q$, $0 \leq t \leq 1$. Apply T_A to such points. We get $tP + (1 - t)Q + A = t(P + A) + (1 - t)(Q + A)$, $0 \leq t \leq 1$, because $tA + (1 - t)A = A$. Therefore the image of $\overline{PQ}$ by T_A consists of all points on the segment whose end points are $P + A$ and $Q + A$.

12. $\overrightarrow{PQ}$ and $\overrightarrow{NM}$ have opposite directions means that $(Q - P) = c(M - N)$, for some number $c < 0$. A and B have opposite directions means $A = cB$, for some number $c < 0$.

13. a) $(\frac{5}{3}, 2, \frac{13}{3})$ b) $(\frac{1}{3}, 3, \frac{11}{3})$

Chapter 10, §3

1. a) $(2t + 1, 6t - 1)$ b) $(\frac{4}{3}, 0)$ c) $(0, -4)$

3. a) $(3t - 4, 3t - 2)$ b) $(-2, 0)$ c) $(0, 2)$

7. $(43, -15)$ **8.** $(-\frac{1}{7}, -\frac{5}{7})$ **11.** a) $(\frac{7}{4}, \frac{5}{4})$

12. A and B are parallel if and only if $A = cB$, with some number $c \neq 0$; that is, $(a_1, a_2) = (cb_1, cb_2)$. This means $a_1 = cb_1$ and $a_2 = cb_2$, or in other words, $a_1a_2 = a_2cb_1 = a_1cb_2$, whence $c(a_2b_1 - a_1b_2) = 0$. This implies $a_1b_2 - a_2b_1 = 0$.

13. The first line consists of all points $(p_1 + ta_1, p_2 + ta_2)$, t in $\mathbf{R}$.
The second line consists of all points $(q_1 + sb_1, q_2 + sb_2)$, t in $\mathbf{R}$.

The common point is determined by the values of t, s such that

$$\begin{cases} p_1 + ta_1 = q_1 + sb_1 \\ p_2 + ta_2 = q_2 + sb_2 \end{cases} \text{ or } \begin{cases} ta_1 - sb_1 = q_1 - p_1 \\ ta_2 - sb_2 = q_2 - p_2. \end{cases}$$

Since $a_1b_2 \neq a_2b_1$ [Exercise 12, lines not parallel] we get the values

$$s = \frac{a_2(p_1 - q_1) - a_1(p_2 - q_2)}{a_2b_1 - a_1b_2} \text{ and } t = \frac{b_1(p_2 - q_2) - b_2(p_1 - q_1)}{a_2b_1 - a_1b_2}.$$

14. a) $\left(\dfrac{-1 - 3\sqrt{159}}{5}, \dfrac{-3 + \sqrt{159}}{5}\right)$ and $\left(\dfrac{-1 + 3\sqrt{159}}{5}, \dfrac{-3 - \sqrt{159}}{5}\right)$

b) $\left(\dfrac{5 + \sqrt{103}}{2}, \dfrac{-5 + \sqrt{103}}{2}\right)$ and $\left(\dfrac{5 - \sqrt{103}}{2}, \dfrac{-5 - \sqrt{103}}{2}\right)$

15. a) $\left(\dfrac{-1 - 6\sqrt{6}}{5}, \dfrac{-3 + 2\sqrt{6}}{5}\right)$ and $\left(\dfrac{-1 + 6\sqrt{6}}{5}, \dfrac{-3 - 2\sqrt{6}}{5}\right)$

b) No intersection

16. a), b), c), d)   No intersection

18. The line is described by $(p + at, q + bt)$, t in $\mathbf{R}$. The intersections are given by those values of t such that $(p + at)^2 + (q + bt)^2 = r^2$, that is $(a^2 + b^2)t^2 + 2(ap + bq)t + (p^2 + q^2 - r^2) = 0$. Then

$$t = \frac{-(ap + bq) \pm \sqrt{(ap + bq)^2 + (a^2 + b^2)(r^2 - p^2 - q^2)}}{a^2 + b^2}.$$

The square root is defined since $(ap + bq)^2 \geq 0$, $(a^2 + b^2) \geq 0$, $r^2 \geq p^2 + q^2$. Finally $P + tA$ are the two points, for the two values of t above.

Chapter 10, §4

1. $3x + 4y = 13$ **2.** $2x + 7y = 31$ **3.** $3x - 11y = 10$

Chapter 11, §1

1. a) $\pi/12$ c) $7\pi/12$ **2.** a) $\pi/9$ b) $2\pi/9$
3. a) $315°$ c) $100°$ e) $120°$ g) $300°$

Chapter 11, §2

1. a)

n	1	2	3	4	5	6	7	8	9	10	11	12
$\sin\dfrac{n\pi}{6}$	$\frac{1}{2}$	$\sqrt{3}/2$	1	$\sqrt{3}/2$	$\frac{1}{2}$	0	$-\frac{1}{2}$	$-\sqrt{3}/2$	-1	$-\sqrt{3}/2$	$-\frac{1}{2}$	0
$\cos\dfrac{n\pi}{6}$	$\sqrt{3}/2$	$\frac{1}{2}$	0	$-\frac{1}{2}$	$-\sqrt{3}/2$	-1	$-\sqrt{3}/2$	$-\frac{1}{2}$	0	$\frac{1}{2}$	$\sqrt{3}/2$	1
$\dfrac{n\pi}{6}$	30°	60°	90°	120°	150°	180°	210°	240°	270°	300°	330°	360°

b)

n	1	2	3	4	5	6	7	8
$\sin\dfrac{n\pi}{4}$	$\sqrt{2}/2$	1	$\sqrt{2}/2$	0	$-\sqrt{2}/2$	-1	$-\sqrt{2}/2$	0
$\cos\dfrac{n\pi}{4}$	$\sqrt{2}/2$	0	$-\sqrt{2}/2$	-1	$-\sqrt{2}/2$	0	$\sqrt{2}/2$	1
$\dfrac{n\pi}{4}$	45°	90°	135°	180°	225°	270°	315°	360°

2. a)

n	1	2	3	4	5	6	7	8	9	10	11	12
$\sin\dfrac{-n\pi}{6}$	$-\tfrac{1}{2}$	$-\sqrt{3}/2$	-1	$-\sqrt{3}/2$	$-\tfrac{1}{2}$	0	$\tfrac{1}{2}$	$\sqrt{3}/2$	1	$\sqrt{3}/2$	$\tfrac{1}{2}$	0
$\cos\dfrac{-n\pi}{6}$	$\sqrt{3}/2$	$\tfrac{1}{2}$	0	$-\tfrac{1}{2}$	$-\sqrt{3}/2$	-1	$-\sqrt{3}/2$	$-\tfrac{1}{2}$	0	$\tfrac{1}{2}$	$\sqrt{3}/2$	1
$\dfrac{-n\pi}{6}$	330°	300°	270°	240°	210°	180°	150°	120°	90°	60°	30°	0°

b)

n	1	2	3	4	5	6	7	8
$\sin\dfrac{-n\pi}{4}$	$-\sqrt{2}/2$	-1	$-\sqrt{2}/2$	0	$\sqrt{2}/2$	1	$\sqrt{2}/2$	0
$\cos\dfrac{-n\pi}{4}$	$\sqrt{2}/2$	0	$-\sqrt{2}/2$	-1	$-\sqrt{2}/2$	0	$\sqrt{2}/2$	1
$\dfrac{-n\pi}{4}$	315°	270°	225°	180°	135°	90°	45°	0°

3. Sine is negative. Cosine is positive.

4. a) $400\sqrt{3}$ ft

5. a) 5/3 mi

6. a) 1 mi

7. a) $x = \sqrt{2}$ and $y = \sqrt{2}$

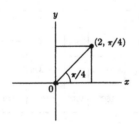

d) $x = -\sqrt{3}$ and $y = -1$

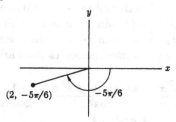

(2, $-5\pi/6$) $-5\pi/6$

8. a) $r = \sqrt{2}$, $\theta = \pi/4$ c) $r = 6$, $\theta = \pi/3$
9. a) $40/3 = 13.3$ b) 10 c) $5/2 = 2.5$ **10.** a) $.8$ b) 1.8
11. a) $.3$ b) 1.2 **12.** a) $\pi/6$ c) $\pi/3$
13. a) $-\pi/6$ c) $-\pi/3$ e) $-\pi/4$

Chapter 11, §3

2.

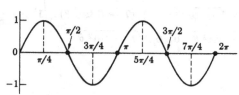

3. b)

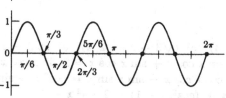

Chapter 11, §4

1.

n	1	2	3	4	5	6	7	8	9	10	11	12
$\tan\dfrac{n\pi}{6}$	$\sqrt{3}/3$	$\sqrt{3}$		$-\sqrt{3}$	$-\sqrt{3}/3$	0	$\sqrt{3}/3$	$\sqrt{3}$		$-\sqrt{3}$	$-\sqrt{3}/3$	0
$\dfrac{n\pi}{6}$	30°	60°	90°	120°	150°	180°	210°	240°	270°	300°	330°	360°

n	1	2	3	4	5	6	7	8
$\tan\dfrac{n\pi}{4}$	1		-1	0	1		-1	0
$\dfrac{n\pi}{4}$	45°	90°	135°	180°	225°	270°	315°	360°

The tangent is not defined at those points where the space is left blank.

2. Note that $1/\cos x$ decreases from arbitrarily large values to 1, in the interval $-\pi/2 \leq x \leq 0$. Furthermore, $-\sin x$ decreases from 1 to 0 in this interval. Since $\cos x$ and $-\sin x$ are both positive in this interval, we conclude that $-\tan x = -\sin x/\cos x$ decreases from arbitrarily large values to 0. Hence $\tan x$ itself increases from arbitrarily large negative values to 0.

5. $1 + \tan^2 x = 1 + \dfrac{\sin^2 x}{\cos^2 x} = \dfrac{\sin^2 x + \cos^2 x}{\cos^2 x} = \dfrac{1}{\cos^2 x} = \sec^2 x$

7. a) $\sqrt{2}/2$ and $-\sqrt{2}/2$ c) $\tfrac{1}{2}$ and $-\tfrac{1}{2}$

8. a) $\sqrt{2}/2$ and $-\sqrt{2}/2$ c) $\sqrt{3}/2$ and $-\sqrt{3}/2$

9. a) 500 ft   c) $\dfrac{500\sqrt{3}}{3}$ ft   **11.** a) $\dfrac{1}{2\sqrt{100 + \tfrac{1}{4}}}$

12. a) $2\sqrt{3}$ c) $7 - 2\sqrt{3}$ f) $16 - 7\sqrt{3}$

14. a) $6 \tan \theta$ b) $6\sqrt{1 + \tan^2 \theta}$ c) $(7 - 6 \tan \theta)\sqrt{1 + \dfrac{1}{\tan^2 \theta}}$

15. a) $(6 \sin \theta)/\sqrt{1 - \sin^2 \theta}$ b) $6/\sqrt{1 - \sin^2 \theta}$

Chapter 11, §5

1. $\dfrac{\sqrt{2}}{4}[1 + \sqrt{3}]$

3. a) $(\sqrt{3} - 1)(\sqrt{2}/4)$ c) $(\sqrt{3} + 1)(\sqrt{2}/4)$ e) $(\sqrt{3} - 1)(\sqrt{2}/4)$

4. a) $\sin (x + x) = \sin x \cos x + \sin x \cos x = 2 \sin x \cos x$

b) $\cos (x + x) = \cos x \cos x - \sin x \sin x = \cos^2 x - \sin^2 x$

c) $\cos 2x = \cos^2 x + (\cos^2 x - 1) = 2 \cos^2 x - 1$,

    i.e. $\cos^2 x = \dfrac{1 + \cos 2x}{2}$

d) $\cos 2x = (1 - \sin^2 x) - \sin^2 x = 1 - 2 \sin^2 x$,

    i.e. $\sin^2 x = \dfrac{1 - \cos 2x}{2}$

5. a) $1.4\sqrt{0.51}$ b) 0.96 **6.** a) 0.02 b) 0.88 c) 0.68

7. a) $\sqrt{0.85}$ b) $\sqrt{0.8}$

10. $\sin 3x = 3 \sin x \cos^2 x - \sin^3 x$

$\sin 4x = 4 \sin x \cos^3 x - 4 \sin^3 x \cos x$

$\sin 5x = \sin x [5 \cos^4 x - 10 \sin^2 x \cos^2 x - \sin^4 x]$

12. $\pi/4$, because $2 \sin \theta \cos \theta = \sin 2\theta$ has a maximum value 1 for $\theta = \pi/4$.

13. a) $\sin (m + n)x = \sin mx \cos nx + \sin nx \cos mx$,

    $\sin (m - n)x = \sin mx \cos nx - \sin nx \cos mx$.

    Adding gives $\sin mx \cos nx = \tfrac{1}{2}[\sin (m + n)x + \sin (m - n)x]$.

Chapter 11, §6

1. $\begin{pmatrix} -1 & 0 \\ 0 & -1 \end{pmatrix}$ 3. $\begin{pmatrix} \dfrac{1}{2} & -\dfrac{\sqrt{3}}{2} \\ \dfrac{\sqrt{3}}{2} & \dfrac{1}{2} \end{pmatrix}$ 5. $\begin{pmatrix} -\dfrac{\sqrt{2}}{2} & -\dfrac{\sqrt{2}}{2} \\ \dfrac{\sqrt{2}}{2} & -\dfrac{\sqrt{2}}{2} \end{pmatrix}$

7. $\begin{pmatrix} 0 & +1 \\ -1 & 0 \end{pmatrix}$

11. 1. $x' = -3, y' = -1$
3. $x' = (3 - \sqrt{3})/2, y' = (3\sqrt{3} + 1)/2$
5. $x' = -2\sqrt{2}, y' = \sqrt{2}$

12. 1. $x' = -5, y' = 2$
3. $x' = (5 + 2\sqrt{3})/2, y' = (5\sqrt{3} - 2)/2$
5. $x' = -3\sqrt{2}/2, y' = 7\sqrt{2}/2$

21. $D_r = \begin{pmatrix} r & 0 \\ 0 & r \end{pmatrix}$ and $F_{a,b} = \begin{pmatrix} a & 0 \\ 0 & b \end{pmatrix}$

22. $G_\varphi = \begin{pmatrix} \cos\varphi & -\sin\varphi \\ \sin\varphi & \cos\varphi \end{pmatrix}$ $G_\psi = \begin{pmatrix} \cos\psi & -\sin\psi \\ \sin\psi & \cos\psi \end{pmatrix}$

$G_\varphi G_\psi = \begin{pmatrix} \cos(\varphi + \psi) & -\sin(\varphi + \psi) \\ \sin(\varphi + \psi) & \cos(\varphi + \psi) \end{pmatrix}$

Chapter 12, §1

1. a)

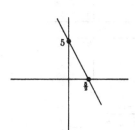

2. a)

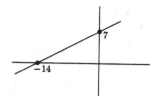

3. a) $(\tfrac{7}{8}, \tfrac{11}{8})$ 4. a) $y = -\tfrac{8}{3}x - \tfrac{5}{3}$ b) $y = -\tfrac{3}{2}x - 4$

5. a) $x = \sqrt{2}$ b) $y = \dfrac{9}{3 + \sqrt{3}}x + \dfrac{27}{3 + \sqrt{3}} + 5$ 8. a) $(\tfrac{3}{5}, \tfrac{2}{5})$

9. $y = 4x - 3$

11. $y = -\frac{1}{2}x + 3 + \dfrac{\sqrt{2}}{2}$

13.

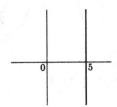

15.

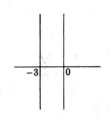

17.

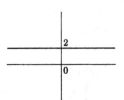

19. $-\frac{1}{4}$ **20.** -8 **21.** $\dfrac{-2}{\sqrt{2} - 2}$

23. $y = \dfrac{2}{\sqrt{2} - \pi}\, x - \dfrac{2\pi}{\sqrt{2} - \pi} + 1$

26. $y = (3 + \sqrt{2})x + 3 + 2\sqrt{2}$

27. a)

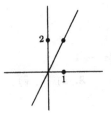

c)

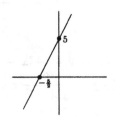

28. If they are parallel, then $a = c$. Any point (x, y) in common would be such that $ax + b = ax + d$, so $b = d$ which is impossible. If they are not parallel, then $a \neq c$. Let (x, y) be a point in common. Then $ax + b = cx + d$, whence $x(a - c) = d - b$, so $x = (d - b)/(a - c)$. Then $y = ax + b = (ad - ab)/(a - c) + b = (ad - bc)/(a - c)$. Substituting in the original equations shows that these values of x, y satisfy the equations.

29. a) $(-4, -7)$ c) $(-\frac{1}{3}, \frac{7}{3})$ **30.** a_2/a_1 if $a_1 \neq 0$. Independent of P.

31. a) $(\frac{4}{5}, \frac{2}{5})$ and $(0, -1)$ c) $(1, 1)$ and $(\frac{1}{5}, -\frac{7}{5})$

Chapter 12, §2

5. $x' = x - 2$, $y' = y + 1$, and $x'^2 + y'^2 = 25$
6. $x' = x$, $y' = y - 1$, and $x'^2 + y'^2 = 9$
8. $y' = y + \frac{25}{8}$, $x' = x + \frac{1}{4}$, and $y' = 2x'^2$
10. $y' = y + 4$, $x' = x - 1$, and $y' = x'^2$

Chapter 12, §3

1. Center $(0, 0)$

Extremities $\begin{cases} (0, 2) & \text{and} & (0, -2) \\ (-4, 0) & \text{and} & (4, 0) \end{cases}$

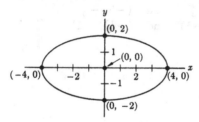

2. Center $(3, 1)$

Extremities $\begin{cases} (1, 1) & \text{and} & (5, 1) \\ (3, 4) & \text{and} & (3, -2) \end{cases}$

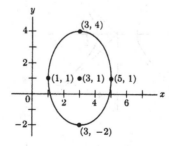

7. Center $(0, 0)$

Extremities $\begin{cases} (0, \frac{1}{3}) & \text{and} & (0, -\frac{1}{3}) \\ (-\frac{1}{2}, 0) & \text{and} & (\frac{1}{2}, 0) \end{cases}$

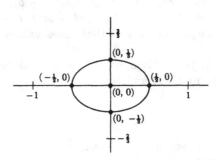

8. Center $(0, 0)$

Extremities $\begin{cases} (0, \frac{1}{4}) & \text{and} & (0, -\frac{1}{4}) \\ (-\frac{1}{5}, 0) & \text{and} & (\frac{1}{5}, 0) \end{cases}$

11. Center $(1, -2)$

Extremities $\begin{cases} (\frac{1}{2}, -2) & \text{and} & (\frac{3}{2}, -2) \\ (1, -\frac{7}{4}) & \text{and} & (1, -\frac{9}{4}) \end{cases}$

12. Center $(-3, -1)$

Extremities $\begin{cases} (-3, -\frac{2}{3}) & \text{and} & (-3, -\frac{4}{3}) \\ (-\frac{13}{4}, -1) & \text{and} & (-\frac{11}{4}, -1) \end{cases}$

Chapter 12, §5

2. $u^2 - v^2 = 2$

3. a) $u^2 - v^2 = 2$

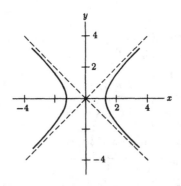

b) $v^2 - u^2 = 2$

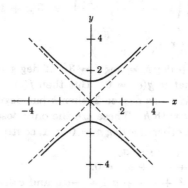

5. a) $u^2 - v^2 = 2$ b) $u^2 - v^2 = 4$
6. a) $v^2 - u^2 = 2$ b) $v^2 - u^2 = 4$
8. For any point (x, y) we have $F_r((x, y)) = (rx, ry)$ and

$$G_\theta(F_r((x, y))) = (rx \cos \theta - ry \sin \theta, \ rx \sin \theta + ry \cos \theta).$$

If you compute the coordinates for $F_r(G_\theta((x, y)))$ in a similar way, you will find the same expression.

Chapter 13, §1

1. $f(\tfrac{3}{4}) = \tfrac{4}{3}$, $f(-\tfrac{2}{3}) = -\tfrac{3}{2}$ **2.** $x \neq \pm\sqrt{2}$, $f(5) = \tfrac{1}{23}$
3. All real numbers, $f(27) = 3$ **4.** a) $f(1) = 1$
5. a) $f(\tfrac{1}{2}) = 1$ b) $f(2) = 4$ **7.** $x \geq 0$, $f(16) = 2$
8. a) Odd   b) Even
9. Let $E(x) = (f(x) + f(-x))/2$ and $O(x) = (f(x) - f(-x))/2$. Then $E(x) + O(x) = f(x)$. Furthermore, $E(-x) = (f(-x) + f(x))/2 = E(x)$, so E is even; $O(-x) = (f(-x) - f(x))/2 = -O(x)$, so O is odd.
10. a) Odd   b) Even   c) Odd
12. a) Even. *Proof:* Let f, g be odd functions. Then

$$(fg)(-x) = f(-x)g(-x) = -f(x)(-g(x)) = f(x)g(x) = (fg)(x).$$

Chapter 13, §2

1. a) 2   b) 5   d) 7   e) 6
   h) $f(x)g(x) = (a_n x^n + a_{n-1} x^{n-1} + \cdots + a_0)(b_m x^m + b_{m-1} x^{m-1} + \cdots + b_0)$ with $a_n \neq 0$ and $b_m \neq 0$. Thus $\deg f = n$ and $\deg g = m$. Then $f(x)g(x) = a_n b_m x^{m+n} + $ terms of lower degree. Hence $\deg (fg) = m + n = \deg f + \deg g$, because the leading coefficient $a_n b_m$ is $\neq 0$.

2. a) $\left(x - \dfrac{-3 - \sqrt{17}}{2}\right)\left(x - \dfrac{-3 + \sqrt{17}}{2}\right)$ **e)** $3(x + 1)(x - \tfrac{1}{3})$

   g) $3\left(x + \dfrac{3 + \sqrt{57}}{6}\right)\left(x + \dfrac{3 - \sqrt{57}}{6}\right)$

3. We must have $\deg g + \deg h = \deg f = 3$. If $\deg g = 1$, then g has a root (why?), say c, that is $g(c) = 0$. But then $f(c) = g(c)h(c) = 0$, so that c is a root of f. If g has degree 2, then h has degree 1, and we can argue in the same way with h. These are the only possibilities, because g cannot have degree 3, otherwise $\deg h = 0 < 1$, contrary to assumption.

5. $f(x) = x^n + a_{n-1}x^{n-1} + \cdots + a_0;$
    $f(c) = c^n + a_{n-1}c^{n-1} + \cdots + a_0 = 0.$
    Thus $c[c^{n-1} + a_{n-1}c^{n-2} + \cdots + a_1] = -a_0$, and c divides a_0.

6. a) 1   **b)** 1 and -1 **c)** 1
    d) 1 if n is odd. 1 and -1 if n is even.   **f)** No root

7. a) $q(x) = 4x^2 + 8x + 15$ and $r(x) = 32$
    b) $q(x) = 4x$ and $r(x) = 3x + 2$
    c) $q(x) = 4x$ and $r(x) = -5x + 2$

8. a) $q(x) = 6x - 1$ and $r(x) = x^2 + 4x + 4$
    b) $q(x) = 6x^2 - x + 31$ and $r(x) = 160 - 7x$
    c) $q(x) = 6x^3 - 13x^2 + 25x - 52$ and $r(x) = 109$
    d) $q(x) = 2x^3 - x^2 + \tfrac{2}{3}x - \tfrac{8}{9}$ and $r(x) = \tfrac{53}{9}$

9. Rational functions.

   a) $\dfrac{3x^2 - 2x + 23}{x^2 - 2x - 15}$ **b)** $\dfrac{11x^2 - 3x - 10}{6x^2 + 10x + 4}$ **c)** $\dfrac{4x^3 + 16x^2 + x + 9}{x^2 + 16x + 5}$

Chapter 13, §3
9.

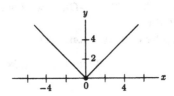

10.

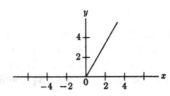

12.

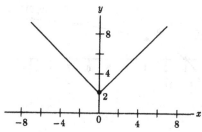

25.

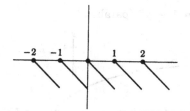

28.

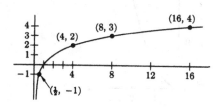

Chapter 13, §4

3. b^x is steeper.

4. a) $5/a^3$

5. $P(t) = 10^5(3^{t/50})$, and t is in years.

   a) $9 \cdot 10^5$ b) $27 \cdot 10^5$ c) $81 \cdot 10^5$

Chapter 13, §5

1. a)

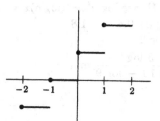

b)

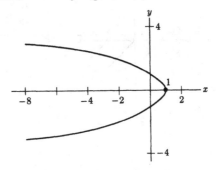

2. a) 6   b) −3   c) 2

3. $e^{(\log a)x} = (e^{\log a})^x = a^x$. Hence $\log a^x = (\log a)x = x \log a$.

   a) .36   b) 1.5   c) 0.1   d) 0.4   e) 1.8

4. $c = 5e^{-4}$ **5.** $t = \frac{1}{5} \log 2$

6. $12 \dfrac{\log 10}{\log 2}$ min   **8.** $\dfrac{3 \log 2}{\log 10 - \log 9}$ min

10. 20.7   **11.** 11.05 hr

Chapter 14, §1

2. a) If $t = y$, then $x = 1 - y^2$ (parabola).

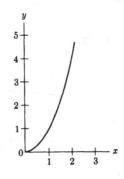

b) If $x = t^2$, then $y = x^2$ with $y \geq 0$ and $x \geq 0$, half a parabola.

3. a) $\frac{5}{4}\sqrt{2}$ sec   d) $\sqrt{\frac{5}{4}}$

4. a) 1

b) The points $(x(t), y(t))$ satisfy the equation $y = 6 - x^2$, which is a parabola.

c) $\dfrac{\sqrt{6}}{2}$

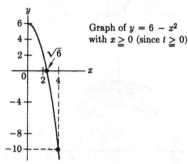

Graph of $y = 6 - x^2$
with $x \geq 0$ (since $t \geq 0$)

6. Image of a line $y = c$ is a circle centered at the origin of radius 2^c. Image of a line $x = c$ is a ray of slope tan c. The ray is open, i.e. does not contain the origin, because for all numbers y, the number 2^y is > 0!!

a)

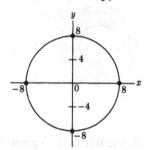

c)

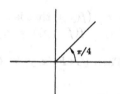

7. a)

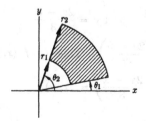

c)

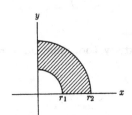

d)

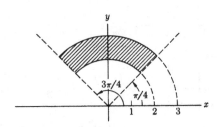

e)

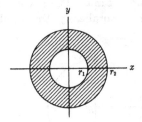

Chapter 14, §2

1. $h \circ f = h \circ g$. Multiplying on the left by h^{-1} gives $h^{-1} \circ h \circ f = h^{-1} \circ h \circ g$, so $I_T \circ f = I_T \circ g$, i.e. $f = g$.

2. a) $f(x) = f(y)$; $f^{-1}(f(x)) = f^{-1}(f(y))$, that is $x = y$
   b) Let $x = f^{-1}(z)$. Then $f(x) = f(f^{-1}(z)) = z$.

3. Let $G: (x, y) \mapsto \left(\dfrac{x}{a}, \dfrac{y}{b}\right)$. Then $(G \circ F)(x, y) = G(ax, by) = (x, y)$

$(F \circ G)(x, y) = F\left(\dfrac{x}{a}, \dfrac{y}{b}\right) = (x, y)$.

4. $\left.\begin{array}{l} u = 2x - y \\ v = y + x \end{array}\right\}$ i.e. $x = \dfrac{u + v}{3}$ and $y = \dfrac{2v - u}{3}$

Let $g: (u, v) \mapsto \left(\dfrac{u + v}{3}, \dfrac{2v - u}{3}\right)$. Then $(g \circ f)(x, y) = (x, y)$

$(f \circ g)(x, y) = f\left(\dfrac{x + y}{3}, \dfrac{2y - x}{3}\right) = (x, y)$.

6. a) $f^5 = f^3 \circ f^2 = I = I \circ f^2$, thus $f^2 = I$
$f^3 = f \circ f^2 = I = f \circ I$, thus $f = I$

7. a) $f^6 \circ g^{-3}$ c) $f \circ g^0 = f$

8. a) $g^{-1} \circ f^{-1}$ is the inverse; $(g^{-1} \circ f^{-1}) \circ (f \circ g) = g^{-1} \circ g = I$
$$(f \circ g) \circ (g^{-1} \circ f^{-1}) = f \circ f^{-1} = I$$

b) $f_m^{-1} \circ f_{m-1}^{-1} \circ \cdots \circ f_1^{-1}$ is the inverse; $(f_1 \circ f_2 \cdots \circ f_m)(f_m^{-1} \circ \cdots \circ f_1^{-1}) = I$

$$(f_m^{-1} \circ \cdots \circ f_1^{-1})(f_1 \circ f_2 \cdots \circ f_m) = I$$

Chapter 14, §3

1. a) $\begin{bmatrix} 1 & 2 & 3 \\ 2 & 1 & 3 \end{bmatrix}\begin{bmatrix} 1 & 2 & 3 \\ 1 & 3 & 2 \end{bmatrix}$; sign $+1$

c) It is already a transposition; sign -1

e) $\begin{bmatrix} 1 & 2 & 3 & 4 \\ 2 & 1 & 3 & 4 \end{bmatrix}\begin{bmatrix} 1 & 2 & 3 & 4 \\ 1 & 2 & 4 & 3 \end{bmatrix}$; sign $+1$

h) $\begin{bmatrix} 1 & 2 & 3 & 4 \\ 2 & 1 & 3 & 4 \end{bmatrix}\begin{bmatrix} 1 & 2 & 3 & 4 \\ 3 & 2 & 1 & 4 \end{bmatrix}\begin{bmatrix} 1 & 2 & 3 & 4 \\ 1 & 2 & 4 & 3 \end{bmatrix}$; sign -1

2. a) $\begin{bmatrix} 1 & 2 & 3 & 4 & 5 \\ 1 & 2 & 5 & 4 & 3 \end{bmatrix}\begin{bmatrix} 1 & 2 & 3 & 4 & 5 \\ 4 & 2 & 3 & 1 & 5 \end{bmatrix}\begin{bmatrix} 1 & 2 & 3 & 4 & 5 \\ 2 & 1 & 3 & 4 & 5 \end{bmatrix}$; sign -1

c) $\begin{bmatrix} 1 & 2 & 3 & 4 & 5 \\ 4 & 2 & 3 & 1 & 5 \end{bmatrix}\begin{bmatrix} 1 & 2 & 3 & 4 & 5 \\ 2 & 1 & 3 & 4 & 5 \end{bmatrix}\begin{bmatrix} 1 & 2 & 3 & 4 & 5 \\ 1 & 5 & 3 & 4 & 2 \end{bmatrix}\begin{bmatrix} 1 & 2 & 3 & 4 & 5 \\ 3 & 2 & 1 & 4 & 5 \end{bmatrix}$; sign $+1$

3. a) $\begin{bmatrix} 1 & 2 & 3 & 4 & 5 & 6 \\ 1 & 2 & 3 & 5 & 4 & 6 \end{bmatrix}\begin{bmatrix} 1 & 2 & 3 & 4 & 5 & 6 \\ 1 & 2 & 3 & 6 & 5 & 4 \end{bmatrix}\begin{bmatrix} 1 & 2 & 3 & 4 & 5 & 6 \\ 3 & 2 & 1 & 4 & 5 & 6 \end{bmatrix}$; sign -1

c) $\begin{bmatrix} 1 & 2 & 3 & 4 & 5 & 6 \\ 1 & 6 & 3 & 4 & 5 & 2 \end{bmatrix}\begin{bmatrix} 1 & 2 & 3 & 4 & 5 & 6 \\ 4 & 2 & 3 & 1 & 5 & 6 \end{bmatrix}\begin{bmatrix} 1 & 2 & 3 & 4 & 5 & 6 \\ 1 & 5 & 3 & 4 & 2 & 6 \end{bmatrix}\begin{bmatrix} 1 & 2 & 3 & 4 & 5 & 6 \\ 3 & 2 & 1 & 4 & 5 & 6 \end{bmatrix}$; sign $+1$

4. a) $\begin{bmatrix} 1 & 2 & 3 \\ 3 & 1 & 2 \end{bmatrix}$ c) $\begin{bmatrix} 1 & 2 & 3 \\ 3 & 2 & 1 \end{bmatrix}$ e) $\begin{bmatrix} 1 & 2 & 3 & 4 \\ 2 & 1 & 4 & 3 \end{bmatrix}$

5. a) $\begin{bmatrix} 1 & 2 & 3 & 4 & 5 \\ 4 & 1 & 5 & 2 & 3 \end{bmatrix}$ c) $\begin{bmatrix} 1 & 2 & 3 & 4 & 5 \\ 4 & 3 & 1 & 5 & 2 \end{bmatrix}$ e) $\begin{bmatrix} 1 & 2 & 3 & 4 & 5 \\ 2 & 1 & 4 & 5 & 3 \end{bmatrix}$

6. a) $\begin{bmatrix} 1 & 2 & 3 & 4 & 5 & 6 \\ 3 & 2 & 1 & 5 & 6 & 4 \end{bmatrix}$ c) $\begin{bmatrix} 1 & 2 & 3 & 4 & 5 & 6 \\ 4 & 6 & 1 & 3 & 2 & 5 \end{bmatrix}$ e) $\begin{bmatrix} 1 & 2 & 3 & 4 & 5 & 6 \\ 3 & 4 & 2 & 5 & 1 & 6 \end{bmatrix}$

7. a) [1 2 3] c) [1 3] [2] e) [1 2] [3 4] g) [1 4 3] [2]

8. a) [1 2 4] [3 5] c) [1 3 2 5 4] d) [1 4] [2 5] [3]

9. a) [1 3] [2] [4 6 5] b) [1 2 6 4 5 3] e) [1 3 4] [2 5 6]

10. Let S_e be the set of all *distinct* even permutations.
Let $S_e = \{\sigma_1, \sigma_2, \ldots, \sigma_m\}$ and let $S' = \{\tau\sigma_1, \tau\sigma_2, \ldots, \tau\sigma_m\}$, with τ being a transposition. Each σ_i is even. Hence $\tau\sigma_i$ is odd. We now prove that all the elements of the set S' are distinct. If $\tau\sigma_k = \tau\sigma_j$, then multiplying on the left by τ shows that $\sigma_k = \sigma_j$, so that $k = j$ and $\sigma_k = \sigma_j$. Finally, we must prove that S' contains all odd permutations. Let σ be an odd permutation. Then $\tau\sigma$ is even and $\tau\sigma = \sigma_k$ for some k. Hence $\sigma = \tau^{-1}\sigma_k = \tau\sigma_k$, and σ is in S'.

11. Property is true for $n = 1$. Assume it for n. Suppose that $\tau_i\sigma = \tau_j\sigma'$ for some indices i, j and some σ, σ' in S_n. Since $\tau_i\sigma(n + 1) = i$ and $\tau_j\sigma'(n + 1) = j$, it follows that $\tau_i = \tau_j$. Multiplying on the left by $\tau_i \, (=\tau_j)$, we conclude that $\sigma = \sigma'$. If $\sigma' = \tau_i\sigma$, then again looking at the effect on the number $n + 1$, we conclude that this cannot be so. Hence the permutations $\sigma, \tau_1\sigma, \ldots, \tau_n\sigma$ with σ in S_n are all distinct.

Furthermore, if γ is a permutation of J_{n+1}, then either γ leaves $n + 1$ fixed, in which case γ is already in S_n, or $\gamma(n + 1) = i$ for some i with $1 \leq i \leq n$. In this case $\tau_i\gamma$ leaves $n + 1$ fixed, so $\tau_i\gamma = \sigma$ is an element of S_n, and $\gamma = \tau_i\sigma$. Thus we have found all the permutations of J_{n+1}. We assume by induction that there are $n!$ permutations in S_n. To each σ in S_n we have associated the $n + 1$ permutations $\sigma, \tau_1\sigma, \ldots, \tau_n\sigma$ in S_{n+1}. Hence the total number of permutations in S_{n+1} is

$$(n + 1)n! = (n + 1)!.$$

Chapter 15, §1

1. a) $(-1 - 3i)/10$ c) $3 + i$ e) $6\pi + i(7 + \pi^2)$
2. a) $(1 - i)/2$ c) $(3 + 4i)/5$ e) $1 - i$ **3.** 1
4. $z = x + iy$ and $w = u + iv$; thus $\bar{z} = x - iy$ and $\bar{w} = u - iv$

$$zw = (xu - yv) + i(yu + xv) \text{ and } \overline{zw}(xu - yv) - i(yu + xv)$$
$$\bar{z}\bar{w} = (x - iy)(u - iv) = (xu - yv) - i(yu + xv)$$
$$\bar{z} + \bar{w} = (x - iy) + (u - iv) = (x + u) - i(y + v)$$
$$\overline{z + w} = \overline{[(x + u) + i(y + v)]} = (x + u) - i(y + v)$$
$$\bar{\bar{z}} = \overline{(x - iy)} = x + iy = z$$

5. $\mathrm{Im}(z) = y$; we always have $y \leq |y|$, thus $\mathrm{Im}(z) \leq |\mathrm{Im}(z)|$. Next, $y^2 \leq x^2 + y^2$ because $x^2 \geq 0$. Hence $|y| = \sqrt{y^2} \leq \sqrt{x^2 + y^2} = |z|$. These two inequalities together constitute the desired inequalities, namely $\mathrm{Im}(z) \leq |\mathrm{Im}(z)| \leq |z|$.

Chapter 15, §2

1. a) $\sqrt{2}\, e^{i\pi/4}$ c) $3e^{i\pi}$ e) $2e^{-i\pi/3}$
2. a) -1 c) $\dfrac{3\sqrt{2} + i3\sqrt{2}}{2}$ e) $\dfrac{1 + i\sqrt{3}}{2}$

3. Suppose α, β are complex numbers such that $\alpha^2 = \beta^2 = z$. Then $\alpha^2 - \beta^2 = 0$ so that $(\alpha + \beta)(\alpha - \beta) = 0$. Hence $\alpha = \beta$ or $\alpha = -\beta$. So there are at most two such complex numbers. But if $z = re^{i\theta}$, then $\alpha = r^{1/2}e^{i\theta/2}$ and $-\alpha$ have squares equal to z.

4. Let $z = re^{i\theta}$. Let $w_k = r^{1/n}e^{i\theta/n + 2k\pi i/n}$ where $k = 1, \ldots, n$. Then $w_1, \ldots, w_n$ have n-th powers equal to z.

5. $w_n = e^{2k\pi i/n}$ with k from 0 to $(n - 1)$

6. $\left. \begin{aligned} e^{i\theta} &= \cos\theta + i\sin\theta \\ e^{-i\theta} &= \cos\theta - i\sin\theta \end{aligned} \right\}$. Adding the two equations yields

$\cos\theta = \dfrac{e^{i\theta} + e^{-i\theta}}{2}$. Subtracting them yields $\sin\theta = \dfrac{1}{2i}[e^{i\theta} - e^{-i\theta}]$.

Chapter 16, §1

1. Let $A(n) = \displaystyle\sum_{k=0}^{n-1} (2k + 1)$. We want to show $A(n) = n^2$. For $A(1)$ we have $1 = 1$. By induction $A(n+1) = A(n) + 2n + 1 = n^2 + 2n + 1 = (n + 1)^2$.

2. b) Let $A(n) = \displaystyle\sum_{k=1}^{n} k^3$. We want to prove that $A(n) = \left[\dfrac{n(n + 1)}{2}\right]^2$.

For $A(1)$ we have $1 = \left[\dfrac{1(2)}{2}\right]^2 = 1$. By induction, $A(n + 1) = A(n) + (n + 1)^3 = \frac{1}{4}(n + 1)^2[n^2 + 4n + 4]$. Thus $A(n + 1) = \frac{1}{4}(n + 1)^2(n + 2)^2 = \left[\dfrac{(n + 1)(n + 2)}{2}\right]^2$.

5. Let $A(n)$ be the product $\displaystyle\prod_{k=0}^{n} (1 + x^{2^k})$. We want to show

$$A(n) = \frac{1 - x^{2^{(n+1)}}}{1 - x}. \text{ We have } A(0) = 1 + x = \frac{1 - x^2}{1 - x}.$$

By induction $A(n + 1) = A(n)(1 + x^{2^{(n+1)}})$

$$= \frac{(1 - x^{2^{(n+1)}})(1 + x^{2^{(n+1)}})}{1 - x}$$

$$= \frac{1 - x^{2^{(n+1)}+1}}{1 - x} = \frac{1 - x^{2^{(n+2)}}}{1 - x}.$$

6. Let $A(n)$ be the property: $f(x^n) = nf(x)$. $A(1)$ is true since $f(x) = f(x)$. By induction let us prove $A(n + 1)$. We have:

$$f(x^{n+1}) = f(x^n x) = f(x^n)f(x) = nf(x) + f(x) = (n + 1)f(x).$$

8. a) $\dfrac{n(2n+1)(4n+1)}{3}$

   c) $4n^2$ e) $\dfrac{[m(n-1)]^2}{4}$

9. a) $\dbinom{n}{k} = \dfrac{n!}{k!(n-k)!}$

$$= \dfrac{n!}{(n-k)!k!} = \dbinom{n}{n-k}$$

   b) $\dbinom{n}{k-1} + \dbinom{n}{k} = \dfrac{n!}{(k-1)!(n-k+1)!} + \dfrac{n!}{k!(n--k)!}$

$$= \dfrac{n!}{(k-1)!(n-k)!}\left[\dfrac{1}{n-k+1} + \dfrac{1}{k}\right]$$

$$= \dfrac{n!}{(k-1)!(n-k)!} \cdot \dfrac{n+1}{k(n-k+1)}$$

$$= \dfrac{(n+1)!}{k!(n-k+1)!}$$

$$= \dbinom{n+1}{k}$$

   c) Let $A(n)$ be the property: $(x+y)^n = \sum\limits_{k=0}^{n} \dbinom{n}{k} x^k y^{n-k}$. $A(1)$ is true,
because $(x+y) = \dbinom{1}{0}y + \dbinom{1}{1}x = x+y$. $A(n+1)$ is proved
using induction:

$$(x+y)^{n+1} = (x+y)^n(x+y) = \left[\sum_{k=0}^{n}\dbinom{n}{k}x^k y^{n-k}\right](x+y)$$

$$= \sum_{k=0}^{n}\dbinom{n}{k}x^{k+1}y^{n-k} + \sum_{k=0}^{n}\dbinom{n}{k}x^k y^{n-k+1}$$

$$= \sum_{j=1}^{n+1}\dbinom{n}{j-1}x^j y^{n+1-j} + \sum_{j=0}^{n}\dbinom{n}{j}x^j y^{n+1-j}$$

$$= \sum_{j=0}^{n+1}\dbinom{n+1}{j}x^j y^{n+1-j}$$

10. a) 3   c) 1   e) 4

11. a) $x^3 + 3x^2y + 3xy^2 + y^3$ b) $x^4 + 4x^3y + 6x^2y^2 + 4xy^3 + y^4$
   c) $x^5 + 5x^4y + 10x^3y^2 + 10x^2y^3 + 5xy^4 + y^5$

12. The inductive argument does not apply going from $n = 1$ to $n = 2$.
There is no "middle" ball!!

13. $A(1)$ is obviously true. By induction, there are m^n ways of mapping n
elements into F, and for each of these ways there are m ways of mapping
the remaining element. Thus there are $m^n \cdot m$ ways or m^{n+1} of mapping
E into F.

Chapter 16, §2

1. Divide the height into n segments of length $\dfrac{h}{n}$.

   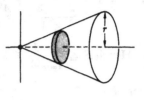

   The radius of a small cylinder is $\dfrac{r(kh)/n}{h} = \dfrac{rk}{n}$.

   The volume of each cylinder is $\pi\left(\dfrac{rk}{n}\right)^2 \cdot \dfrac{h}{n}$.

   Using Exercise 2(a) of §1, the sum of these volumes is

   $$\sum_{k=1}^{n} \pi\left(\frac{rk}{n}\right)^2 \cdot \frac{h}{n} = \frac{\pi h r^2}{n^3}\sum_{k=1}^{n} k^2 = \frac{\pi h r^2}{6}\left[\left(1 + \frac{1}{n}\right)\left(2 + \frac{1}{n}\right)\right].$$

   As n becomes arbitrarily large the sum approaches $\dfrac{\pi h r^2}{3}$.

2. a) 24π c) $3\pi c^3$ 3. a) $\pi/2$ c) $9\pi/2$ d) $\pi h^2/2$
4. a) $2\pi/3$ d) $2\pi r^3/3$ 5. It is half a ball.   6. a) $\frac{1}{3}$ c) 9
7. a) $\frac{1}{4}$ c) 81/4
8. S is mapped into T with $F_{1,3}$: $(x, y) \mapsto (x, 3y)$. The area is 1.
9. a) $c/3$

Chapter 16, §3

1. a) 3/2   c) 5/4   e) 4
2. $(1 - c + c^2 - c^3 + \cdots + (-1)^n c^n)(1 + c) = 1 + (-1)^n c^{n+1}$. Hence

   $$\sum_{k=0}^{n} (-1)^k c^k = \frac{1}{1 + c} + \frac{(-1)^n c^{n+1}}{1 + c}.$$

   But c^{n+1} approaches zero as n becomes large. Hence

   $$\lim_{n \to \infty} \sum_{k=0}^{n} (-1)^k c^k = \frac{1}{1 + c}.$$

   a) 3/4   c) 5/6   e) 4/7
4. $\displaystyle\sum_{k=0}^{n} r e^{ik\theta} = \frac{1 - r^{n+1}e^{i\theta(n+1)}}{1 - re^{i\theta}}$; $\displaystyle\sum_{k=0}^{\infty} r e^{ik\theta} = \frac{1}{1 - re^{i\theta}}$
5. 0
7. $1 + \frac{1}{2} + \frac{1}{3} + \frac{1}{4} + \frac{1}{5} + \cdots \geq 2 \cdot \frac{1}{2} + 2 \cdot \frac{1}{4} + 4 \cdot \frac{1}{8} + 8 \cdot \frac{1}{16} + \cdots$
   $$\geq 1 + \frac{1}{2} + \frac{1}{2} + \frac{1}{2} + \cdots$$

   By additional groups of terms we add at least $\frac{1}{2}$ each time. The sum can be made larger than any given value A by adding $2A$ groups of terms.

8. $S = 1 + \underbrace{\dfrac{1}{2^2} + \dfrac{1}{3^2}}_{\leq \frac{1}{2}} + \underbrace{\dfrac{1}{4^2} + \dfrac{1}{5^2} + \dfrac{1}{6^2} + \dfrac{1}{7^2}}_{\leq \frac{1}{4}} + \underbrace{\dfrac{1}{8^2} + \cdots + \dfrac{1}{15^2}}_{\leq \frac{1}{8}} + \cdots$

Hence $S \leq \displaystyle\sum_{k=0}^{\infty} \dfrac{1}{2^k} \leq 2$.

Chapter 17, §1

1. a) 26   c) -5 **2.** 1   **3.** b) -1 d) 0

Chapter 17, §2

1. $D(B, C' + C'') = \begin{vmatrix} b_1 & c_1' + c_1'' \\ b_2 & c_2' + c_2'' \end{vmatrix} = b_1(c_2' + c_2'') - b_2(c_1' + c_2'')$
$ = (b_1 c_2' - b_2 c_1') + (b_1 c_2'' - b_2 c_1'')$
$ = D(B, C') + D(B, C'')$

3. $D(B, C + xB) = D(B, C) + D(B, xB)$ using **D1**
$ = D(B, C) + xD(B, B)$ using **D2**
$ = D(B, C) + 0 $ using **D4**

4. $D(C, A) = D(xA + yB, A) = D(xA, A) + D(yB, A)$
$ = xD(A, A) + yD(B, A) = yD(B, A)$

Hence $y = \dfrac{D(C, A)}{D(B, A)}$.

5. 1. $x = \frac{5}{3}$ and $y = \frac{1}{3}$ 3. $x = -2$ and $y = 1$ 5. $x = -\frac{1}{2}$ and $y = -\frac{3}{2}$

Chapter 17, §4

1. a) $D(A) = a_{31}\begin{vmatrix} a_{12} & a_{13} \\ a_{22} & a_{23} \end{vmatrix} - a_{32}\begin{vmatrix} a_{11} & a_{13} \\ a_{21} & a_{23} \end{vmatrix} + a_{33}\begin{vmatrix} a_{11} & a_{12} \\ a_{21} & a_{22} \end{vmatrix}$
$ = a_{31}a_{12}a_{23} - a_{31}a_{13}a_{22} - a_{32}a_{11}a_{23} + a_{32}a_{21}a_{13}$
$ + a_{33}a_{11}a_{22} - a_{33}a_{12}a_{21}$

c) $D(A) = a_{13}\begin{vmatrix} a_{21} & a_{22} \\ a_{31} & a_{32} \end{vmatrix} - a_{23}\begin{vmatrix} a_{11} & a_{12} \\ a_{31} & a_{32} \end{vmatrix} + a_{33}\begin{vmatrix} a_{11} & a_{12} \\ a_{21} & a_{22} \end{vmatrix}$
$ = a_{13}a_{21}a_{32} - a_{13}a_{31}a_{22} - a_{23}a_{11}a_{32} + a_{23}a_{31}a_{12}$
$ + a_{33}a_{11}a_{22} - a_{33}a_{21}a_{12}$

2. a) -20 c) 4   e) -76 **3.** a) 140   **4.** abc   **5.** a) -24 c) 90
6. a) $a_{11}a_{22}a_{33}$ b) $a_{11}a_{22}a_{33}$

Chapter 17, §5

1. $D(A^1, B + C, A^3) = -(b_1 + c_1)\begin{vmatrix} a_{21} & a_{23} \\ a_{31} & a_{33} \end{vmatrix} + (b_2 + c_2)\begin{vmatrix} a_{11} & a_{13} \\ a_{31} & a_{33} \end{vmatrix}$
$ - (b_3 + c_3)\begin{vmatrix} a_{11} & a_{13} \\ a_{21} & a_{23} \end{vmatrix}$

$$D(A^1, B, A^3) = -b_1 \begin{vmatrix} a_{21} & a_{23} \\ a_{31} & a_{33} \end{vmatrix} + b_2 \begin{vmatrix} a_{11} & a_{13} \\ a_{31} & a_{33} \end{vmatrix} - b_3 \begin{vmatrix} a_{11} & a_{13} \\ a_{21} & a_{23} \end{vmatrix}$$

$$D(A^1, C, A^3) = -c_1 \begin{vmatrix} a_{21} & a_{23} \\ a_{31} & a_{33} \end{vmatrix} + c_2 \begin{vmatrix} a_{11} & a_{13} \\ a_{31} & a_{33} \end{vmatrix} + c_3 \begin{vmatrix} a_{11} & a_{13} \\ a_{21} & a_{23} \end{vmatrix}$$

You could also get a proof by expanding according to any row or column.

4. a) $D(A^1 + xA^3, A^2, A^3) = D(A^1, A^2, A^3) + D(xA^3, A^2, A^3)$
$$= D(A^1, A^2, A^3) + xD(A^3, A^2, A^3)$$
$$= D(A^1, A^2, A^3) + 0$$

5. $D(A^1 + A^2, A^1 + A^2, A^3) = D(A^1, A^1, A^3) + D(A^1, A^2, A^3)$
$$+ D(A^2, A^1, A^3) + D(A^2, A^3, A^3)$$
$$= 0 + D(A^1, A^2, A^3)$$
$$+ D(A^2, A^1, A^3) + 0$$

6. $D(A^1, A^2, A^3) = -D(A^1, A^3, A^2) = D(A^3, A^1, A^2)$
$$= -D(A^3, A^2, A^1)$$

Since $D({}^t A) = D(A)$, the same holds for rows.

7. If you add a multiple of one row to another, you do not change the value of the determinant. If you interchange two adjacent rows, then the determinant changes by a sign.

8. a) 0 **b)** 24 **c)** -12 **d)** 0 **9.** $D(cA) = c^3 D(A)$

10. $\begin{vmatrix} 1 & x_1 & x_1^2 \\ 1 & x_2 & x_2^2 \\ 1 & x_3 & x_3^2 \end{vmatrix} = \begin{vmatrix} 1 & x_1 & x_1^2 \\ 0 & x_2 - x_1 & x_2^2 - x_1^2 \\ 1 & x_3 & x_3^2 \end{vmatrix} = \begin{vmatrix} 1 & x_1 & x_1^2 \\ 0 & x_2 - x_1 & x_2^2 - x_1^2 \\ 0 & x_3 - x_1 & x_3^2 - x_1^2 \end{vmatrix}$
$$= (x_2 - x_1)(x_3 - x_1)(x_3 + x_1 - x_2 - x_1)$$
$$= (x_2 - x_1)(x_3 - x_1)(x_3 - x_2)$$

11. $D(B^1 + B^2 + B^3, A^2, A^3) = D(B^1, A^2, A^3) + D(B^2 + B^3, A^2, A^3)$
$$= D(B^1, A^2, A^3) + D(B^2, A^2, A^3)$$
$$+ D(B^3, A^2, A^3)$$

$$D\left(\sum_{j=1}^{n} B_j, A^2, A^3\right) = \sum_{j=1}^{n} D(B_j, A^2, A^3)$$

Chapter 17, §6

1. $D(x_1 A^1 + x_2 A^2 + x_3 A^3, A^2, A^3)$
$$= x_1 D(A^1, A^2, A^3) + x_2 D(A^2, A^2, A^3) + x_3 D(A^3, A^2, A^3)$$
$$= x_1 D(A^1, A^2, A^3)$$

2. a) $D(A^1, B, A^3) = x_1 D(A^1, A^1, A^3) + x_2 D(A^1, A^2, A^3)$
$$+ x_3 D(A^1, A^3, A^3)$$
$$= 0 + x_2 D(A^1, A^2, A^3)$$

Hence $x_2 = \dfrac{D(A^1, B, A^3)}{D(A^1, A^2, A^3)}.$

3. $D(x_1A^1 + x_2A^2 + x_3A^3, A^2, A^3)$

$\qquad = D(0, A^2, A^3) = 0$

$\qquad = x_1D(A^1, A^2, A^3) + x_2D(A^2, A^2, A^3) + x_3D(A^3, A^2, A^3)$

$\qquad = x_1D(A^1, A^2, A^3) + 0 + 0$

Since at least one of the number x_1, x_2, x_3 is not zero, we may assume that $x_1 \neq 0$. Then $x_1D(A^1, A^2, A^3) = 0$ implies $D(A^1, A^2, A^3) = 0$.